Neuroplasticity:
The Art and Science of Brain Transformation

Selen Gür Özmen and Aygül Köseoğlu Kurt (eds.)

Neuroplasticity: The Art and Science of Brain Transformation

PETER LANG
Berlin · Bruxelles · Chennai · Lausanne · New York · Oxford

Bibliographic information published by the Deutsche Nationalbibliothek
The German National Library lists this publication in the German National Bibliography; detailed bibliographic data is available on the Internet at http://dnb.d-nb.de.

Library of Congress Cataloging-in-Publication Data
A CIP catalog record for this book has been applied for at the Library of Congress.

Cover image: Yasin Akşit, "The Art and Science"
Cover design by Peter Lang Group AG

ISBN 978-3-631-93017-5 (Print)
ISBN 978-3-631-94869-9 (ePDF)
ISBN 978-3-631-94870-5 (ePUB)
DOI 10.3726/b23540

Published by Peter Lang GmbH, Berlin (Germany)

info@peterlang.com

This publication has been peer reviewed.

www.peterlang.com

Contact for General Product Safety Regulation (GPSR): gpsr@peterlang.com

Figures

Disclosures

Conflict of Interest: The Authors Declared No Conflict of Interest.
Use of Ai for Writing Assistance: The Translation and Text Editing Were Assisted by Artificial Intelligence.
Financial Disclosure: The Authors Declared That This Study Received No Financial Support.
Peer-Review: Externally Peer-Reviewed.

Fundamentals of Neuroplasticity: Concepts and Definitions

Selen Gür Özmen and Asena Himmetoğlu Padgett

1. Introduction

Welcome to *Neuroplasticity: The Art and Science of Brain Transformation.* In this book, we explore how the brain changes itself—how it learns, adapts, heals, and evolves. Neuroplasticity is not just a biological phenomenon; it is the dynamic interplay of structure, function, and experience that makes us who we are.

Imagine deciding to learn the guitar. At first, the strings are uncooperative, your fingers feel clumsy, and the notes seem random. But day after day, your brain silently adapts. Synaptic connections reorganize, motor skills improve, and sounds begin to flow more naturally. What once felt foreign becomes second nature. This transformation isn't magic—it's neuroplasticity at work.

The human brain is astonishingly complex, composed of about 86 billion neurons and trillions of synapses (Herculano-Houzel, 2009). This complexity allows it to think, feel, move—and, most importantly, to change. Neuroplasticity enables the brain to rewire itself in response to experience, environment, and even injury (Fuchs & Flügge, 2014). It is the foundation of learning, memory, recovery, and growth.

In this chapter, we will lay the groundwork for understanding neuroplasticity by examining its historical development, core mechanisms, types, and real-world applications. By grasping these fundamentals, we open the door to a deeper understanding of the brain's resilience and potential.

1.1. Historical Perspective and Core Concepts

The concept of brain plasticity has a rich history dating back to the late nineteenth century. William James first used "plasticity" in Principles of

Psychology (1890) (James, 1890). He meant it to describe habit-related brain changes and suggested that the nervous system adapts to experiences. It strengthens pathways through repeated actions. James's concept of "plasticity" has laid the groundwork for modern behavioral therapies and educational methods. Strategies for developing and altering learning habits are mainly rooted in James's early insights. The historical evolution of neuroplasticity concepts, from William James to Santiago Ramón y Cajal, has been extensively documented (Berlucchi & Buchtel, 2009).

1.2. Pioneers of Neuroplasticity: From Cajal to Hebb

Beyond William James, several key figures shaped the foundational understanding of brain plasticity.

In the late nineteenth century, Eugenio Tanzi suggested that changes at the synapse could underlie learning. His student, Ernesto Lugaro, further proposed that these changes reflected the essence of neural plasticity.

A major turning point came with Santiago Ramón y Cajal, often considered the father of modern neuroscience. Cajal's detailed observations of cortical neurons led him to conclude that the adult brain could undergo structural changes—an idea revolutionary for its time (DeFelipe, 2006). His work laid the groundwork for modern neurorehabilitation and the study of experience-dependent brain changes.

Though less discussed in this context, Sigmund Freud also speculated on the plastic nature of the nervous system in his unpublished manuscript *Project for a Scientific Psychology* (1895). Some scholars view this as a precursor to modern ideas of synaptic modulation (Centonze et al., 2004).

In the mid-twentieth century, Jerzy Konorski and Donald Hebb revitalized interest in synaptic plasticity. Hebb's principle—neurons that fire together wire together—became a cornerstone of modern neuroscience (Hebb, 1949). His theory not only influenced learning and memory research but also inspired the architecture of artificial neural networks.

1.3. Key Neuroplasticity Research Breakthroughs

Several studies reshaped the view on the brain's adaptability. Here are key milestones:

1.3.1. 1960s: Ocular Dominance and Critical Periods

David Hubel and Torsten Wiesel's work in the 1960s was pivotal. They used kittens to show critical periods in visual development. Their research uncovered ocular dominance columns in the visual cortex. These are bands of neurons that respond mainly to one eye. By closing a kitten's eye during a crucial developmental phase, they noted significant changes in the brain's structure. This work proved that early visual experiences could reshape the visual system. They defined critical periods as the brain is most open to environmental influences. Their findings underpin modern treatments for amblyopia (lazy eye) and strabismus, emphasizing early detection and intervention during these critical periods to maximize visual recovery.

1.3.2. The 1970s: Brain Adaptation and Sensory Change

In the 1970s and 1980s, Michael Merzenich showed the brain's ability to adapt to sensory changes (Merzenich et al., 1980). He and his team studied adult monkeys to see how their somatosensory cortex reacted to sensory changes. In a critical experiment, they removed a finger from a monkey and watched how areas for nearby fingers took over the brain areas for that finger. This showed that the adult brain could change significantly, contrary to the belief that it was mostly fixed. Merzenich also found that intense training could lead to brain changes, even without sensory loss. His research has directly influenced the development of brain-training programs such as Fast ForWord, which improve language and learning abilities in children with developmental delays (Merzenich et al., 1996)

1.3.3. 1990s: Adult Neurogenesis

In the late 1990s, Elizabeth Gould's work challenged a crucial belief in neuroscience. It was thought that adults could not grow new neurons. However, her team found new neurons in the hippocampus of adult primates, like marmoset monkeys. This was groundbreaking. It contradicted the idea that adult brains were static. They showed thousands of new neurons form daily in the hippocampus, vital for learning and memory. Her findings have inspired stress management therapies, demonstrating how chronic stress can suppress neurogenesis and how mindfulness and exercise can counteract these effects, promoting better mental health (Gould E, 1999) (Figure 1).

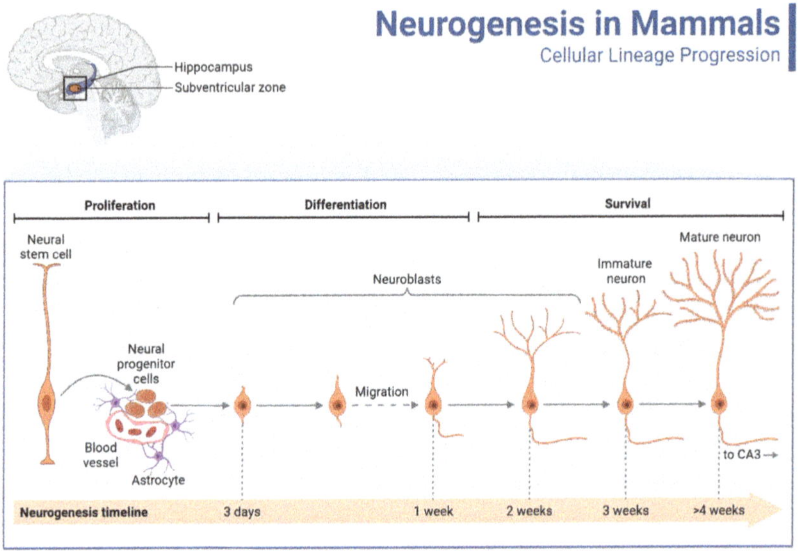

Figure 1: Stages of neurogenesis including proliferation, differentiation, and survival of new neurons in the hippocampus (created with BioRender.com).

2. Types of Neuroplasticity

Neuroplasticity comes in different forms, each aiding the brain's adaptability:

2.1. Structural Plasticity

Structural plasticity refers to the physical changes in neuronal connections and architecture that underlie the brain's adaptability.

- Synaptogenesis: The formation of new synapses, allowing neurons to establish fresh communication pathways. This process is particularly active during learning and memory formation.
- Synaptic Pruning: Eliminating weaker or unused synaptic connections, optimizing neural networks. This mechanism is critical during development and in response to environmental demands. Microglia play a crucial role in synaptic pruning, a process necessary for normal brain development and the optimization of neural circuits (Paolicelli et al., 2011).
- Neurogenesis: The generation of new neurons, predominantly in the hippocampus (critical for learning and memory) and the olfactory bulb

(important for smell). Emerging evidence suggests its role in mood regulation and recovery from stress. The hippocampus, a key site for adult neurogenesis, plays a central role in learning and memory (Kempermann et al., 2015). Studies in rodents have identified mechanisms, such as synaptic remodeling and neurogenesis, which support hippocampal-dependent cognitive functions (Gil-Mohapel, 2016).

- Remodeling: Changes in neuronal structures, such as the growth of dendritic spines and axonal branching, enhance neural circuit efficiency.

2.2. Functional Plasticity

Functional plasticity involves changes in the strength of synaptic connections, affecting how neurons communicate:

This includes mechanisms like long-term potentiation (LTP) and long-term depression (LTD), which are fundamental for memory and learning processes (Bliss & Cooke, 2011) (Figures 2 and 3).

2.3. Network Plasticity

Network plasticity encompasses the reorganization of neural circuits, involving changes in how neurons interact and activate in coordinated networks:

This type of plasticity is crucial for large-scale brain functions, such as decision-making and sensory integration (Bassett et al., 2011).

2.4. Developmental Plasticity

Developmental plasticity occurs during critical periods of development, enabling the brain to adapt to environmental stimuli:

Examples include language acquisition in early childhood and the refinement of sensory and motor systems (Kolb & Gibb, 2011; Werker & Hensch, 2015).

2.5. Injury-Induced Plasticity

This type of plasticity is activated following injuries such as strokes or traumatic brain injuries:

It involves the rerouting of neural circuits to compensate for lost functions. Mechanisms like axonal sprouting and cortical remapping play pivotal roles in recovery (Cramer et al., 2011; Nudo, 2013).

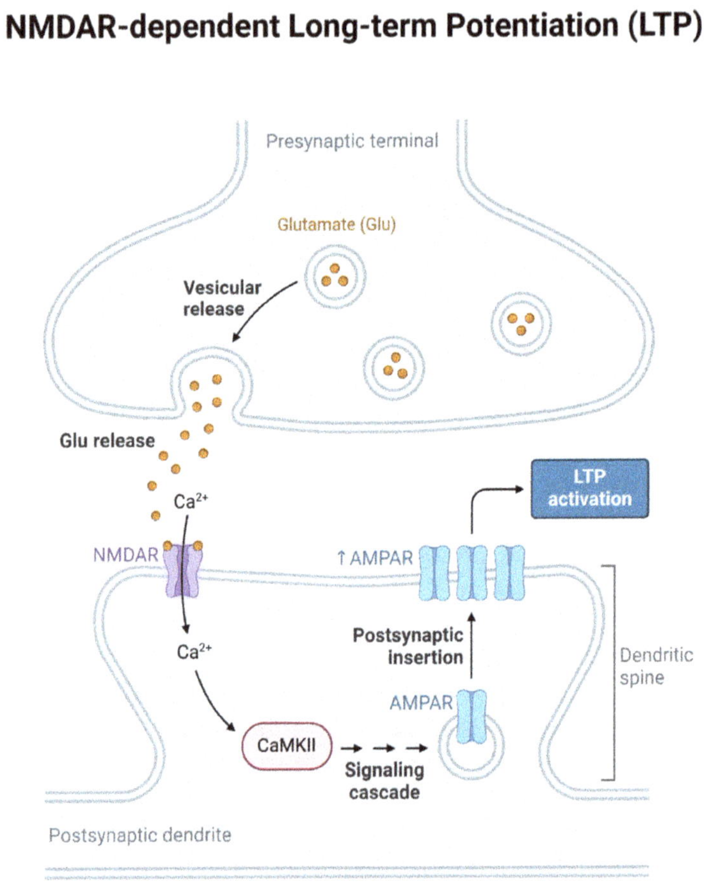

Figure 2: Mechanism of LTP involving NMDAR activation, calcium influx, and AMPAR insertion (created with BioRender.com).

3. Mechanisms of Neuroplasticity

Neuroplasticity works through various cellular and molecular methods:

3.1. Synaptic Plasticity

- Structural changes at the synaptic level, driven by experience, are key mechanisms underlying learning and memory (Holtmaat & Svoboda, 2009).

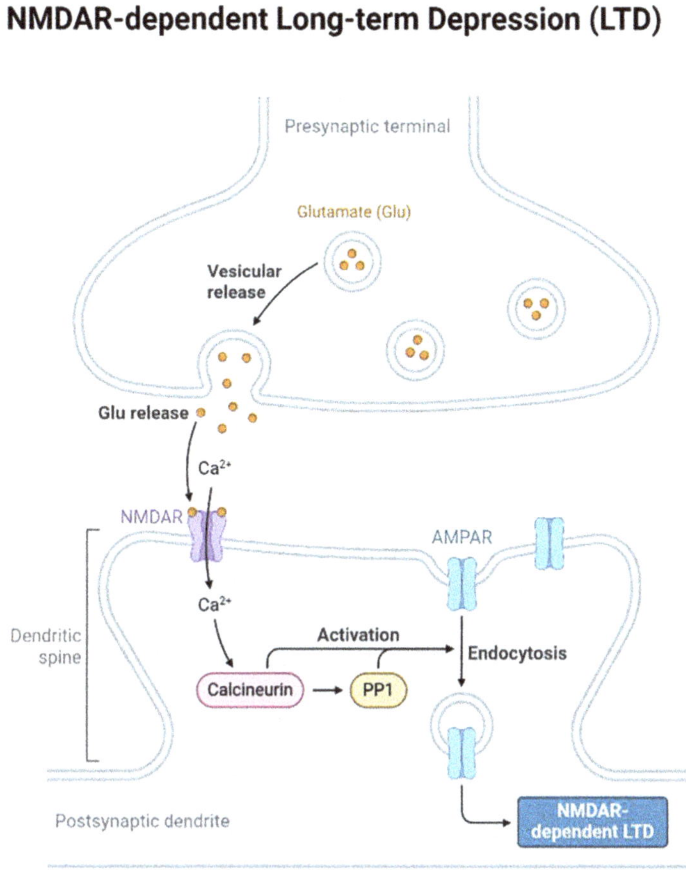

Figure 3: LTD mechanism involving Ca^{2+} influx and AMPAR internalization via calcineurin and PP1 (created with BioRender.com).

- Long-term potentiation (LTP): Strengthens synaptic connections, enhancing signal transmission between neurons (Malenka & Bear, 2004). For example, a student practicing math problems develops stronger neural pathways in regions responsible for numerical reasoning.
- Long-term depression (LTD): Weakens synaptic connections, fine-tuning neural circuits (Collingridge et al., 2010). For example, a previously learned but unused skill, such as a foreign language, may weaken over time due to lack of practice.

- Spike-timing-dependent plasticity (STDP): Adjusts synaptic strength based on the timing of neuronal firing, balancing excitation, and inhibition (Dan & Poo, 2006).

3.2. Biochemical Changes

Biochemical processes enable neurons to respond dynamically to stimuli:

- Activation of neurotrophic factors: Molecules like brain-derived neurotrophic factor (BDNF) and nerve growth factor (NGF) promote neuronal growth and survival (Lu et al., 2014). For instance, physical activity like running increases BDNF levels, promoting memory and cognitive function. Emerging methods, such as focused ultrasound-enhanced intranasal delivery, have shown promise in efficiently delivering neurotrophic factors like BDNF to the brain, potentially enhancing neuroplasticity and recovery from neurological conditions (Chen et al., 2016).
- Gene expression and protein synthesis: Cellular changes supporting long-term adaptations (Sweatt, 2013). Learning new concepts triggers gene expression changes in neurons, strengthening neural circuits involved in learning.
- Neurotransmitter modulation: Alterations in neurotransmitter systems such as glutamate and dopamine adjust neuronal responses (Citri & Malenka, 2008).

3.3. Structural Modifications

Structural changes enable neurons to reconfigure their physical properties:

- Cytoskeletal remodeling: Adjustments in the neuron's internal structure support plasticity (Bosch & Hayashi, 2012). For instance, during spatial learning, the cytoskeleton of neurons in the hippocampus reorganizes, enhancing memory formation.
- Formation and elimination of synapses: Dynamic changes in synaptic connections are fundamental to learning and adaptation (Carmichael & Chesselet, 2002).
- Axonal sprouting and dendritic growth: Extensions of axons and dendrites improve connectivity and network efficiency (Fields, 2015). For example, stroke patients regain lost motor skills through physical therapy, which promotes axonal sprouting in the motor cortex.

3.4. Neuroimaging and Structural Plasticity

Advancements in neuroimaging have provided valuable insights into the dynamic nature of brain plasticity. Studies by Zatorre et al. (2012) reveal that both gray and white matter can undergo structural changes during learning, highlighting the adaptability of neural networks. These findings underscore the importance of structural plasticity in cognitive functions and learning processes.

Neuroimaging techniques such as MRI and diffusion tensor imaging (DTI) have allowed researchers to observe these structural changes in vivo. Gray matter changes are often associated with synaptic remodeling, while white matter plasticity reflects alterations in myelination and axonal connectivity. These adaptations enable the brain to optimize information processing and enhance learning efficiency.

Understanding structural plasticity through neuroimaging not only deepens our knowledge of learning mechanisms but also opens new avenues for interventions in neurodegenerative diseases and cognitive disorders. By identifying patterns of structural change, clinicians can better target therapies that promote brain health and recovery.

4. Factors Influencing Neuroplasticity

Multiple factors can enhance or inhibit neuroplasticity:

4.1. Age

Neuroplasticity is most pronounced during early development, such as in the first few years of life, when the brain forms critical sensory and cognitive circuits. Due to their heightened plasticity during this period, infants can easily acquire multiple languages.

However, adults also retain significant plastic potential. A study by Draganski et al. (2004) demonstrated that adults who learned to juggle showed measurable increases in gray matter in the occipital lobe, proving that the adult brain can still adapt to new challenges (Draganski et al., 2004).

4.2. Critical Periods in Development

Certain stages of brain development are characterized by heightened sensitivity to environmental stimuli, known as critical periods. During these windows, the brain is particularly receptive to forming new neural connections and modifying existing circuits. Hubel and Wiesel's pioneering work (1970) on visual development demonstrated that depriving one eye of visual input during a critical period leads to long-lasting changes in the visual cortex, a phenomenon known as ocular dominance plasticity.

These findings highlight the importance of early experiences in shaping brain structure and function. For example, exposure to language during critical periods is crucial to the development of linguistic skills. Similarly, interventions for conditions like amblyopia are more effective when implemented during these sensitive periods.

Beyond sensory development, critical periods also influence emotional and social behaviors. Research suggests that disruptions in early life experiences, such as neglect or trauma, can have lasting effects on brain function and mental health. Understanding these periods can guide educational and therapeutic strategies to optimize cognitive and emotional outcomes.

4.3. Environment

Enriched environments with stimulating activities, diverse experiences, and social interactions promote neuroplasticity. For instance, animals raised in enriched cages with toys and opportunities for exploration develop more robust synaptic connections and larger hippocampi (van Praag H et al., 2000).

Conversely, environments with limited sensory input or chronic deprivation can lead to synaptic atrophy, negatively affecting cognitive and emotional resilience.

4.4. Brain Size and Neuron Density

While brain size often correlates with cognitive abilities across species, neuron density plays a more critical role in determining brain function. The human brain, although smaller than that of elephants and some whales, contains a higher number of neurons in the cerebral cortex, which underlies our capacity for advanced reasoning and problem solving.

Herculano-Houzel et al. (2014) highlighted that the elephant brain, despite its large size, has fewer neurons compared to humans, emphasizing that neuron density, rather than sheer brain volume, is a better predictor of cognitive capabilities.

This distinction between brain size and neuron density has implications for understanding human intelligence and brain efficiency. Studies on neuron density help explain why some species with smaller brains exhibit superior problem-solving abilities and behavioral flexibility. In particular, the high neuron density in the human brain's prefrontal cortex is linked to advanced executive functions, including planning, decision-making, and social behavior.

Furthermore, understanding neuron density across species offers insights into evolutionary adaptations. For example, humans evolved to maximize cognitive output with an energy-efficient brain, balancing metabolic demands and intellectual capacity. These findings also have implications for neuroscience research, particularly in studying neurodegenerative diseases, where neuron loss directly impacts cognitive performance.

4.5. Experience and Learning

Learning new skills or engaging in novel experiences promotes neural adaptability. For example, London taxi drivers, who memorize the intricate city layout known as "The Knowledge," exhibit increased hippocampal volume compared to non-drivers (Maguire et al., 2000). This showcases how learning directly sculpts brain structure.

Similarly, musicians who practice intensely demonstrate enhanced connectivity between auditory and motor areas, reflecting the brain's ability to rewire itself for expertise (Zatorre et al., 2012). Studies on foraging and navigation in mice demonstrate that animals can develop efficient behavioral strategies in response to complex natural stimuli, showcasing the brain's capacity to adapt to environmental challenges (Gire et al., 2013).

4.6. Physical Activity

Exercise, especially aerobic activity, is a powerful promoter of neuroplasticity. Activities like running or swimming increase BDNF levels, a key protein in synaptic growth and neurogenesis (Cotman & Berchtold, 2002). Physical activity enhances brain health and positively impacts adult learning performance, especially in distance education settings (Gijselaers et al., 2015).

A landmark study by Erickson et al. (2011) found that one year of regular aerobic exercise not only improved memory but also reversed age-related hippocampal shrinkage in older adults, emphasizing the role of physical activity in maintaining brain health.

4.7. Sleep

Sleep supports the maintenance of critical brain states that foster neuroplasticity and cognitive function (Iyer, 2018). Sleep, particularly REM sleep, is crucial in memory consolidation and synaptic optimization. During sleep, the brain prunes unnecessary synapses while strengthening those critical for learning and adaptation (Tononi & Cirelli, 2014).

A notable study by Rasch and Born (2013) demonstrated that individuals who napped after a learning task performed better in memory tests than those who stayed awake, underlining the role of sleep in enhancing neuroplasticity. Studies have shown that sleep quality directly affects psychological well-being and life satisfaction, especially in college students (Jamal, 2022).

4.8. Stress and Depression

Short-term stress, such as preparing for an exam or solving a challenging problem, can enhance plasticity by temporarily boosting brain function by increasing the release of adrenaline and cortisol.

However, chronic stress is detrimental, reducing hippocampal neurogenesis and increasing synaptic loss. Prolonged exposure to stress hormones has been linked to structural changes in the prefrontal cortex and amygdala, contributing to anxiety and cognitive decline (McEwen & Morrison, 2013). Additionally, gene-environment interactions, particularly in neurodevelopmental disorders such as schizophrenia, further illustrate how chronic stress and environmental factors can influence brain structure and function (Burrows & Hannan, 2013). Chronic stress can induce epigenetic changes that contribute to neurodegenerative diseases, including Alzheimer's disease (Anderson, 2015).

Chronic stress leads to synaptic loss in key brain regions such as the hippocampus and prefrontal cortex, which are critical for mood regulation and cognitive function. These structural changes contribute to the onset of depression. Recent studies show that rapid-acting antidepressants, such as ketamine, can restore synaptic density by promoting synaptogenesis and reversing stress-induced deficits (Duman et al., 2016).

4.9. Diet and Nutrition

The link between nutrition and brain function has been well-documented, with certain nutrients shown to enhance neuroplasticity and cognitive function (Gómez-Pinilla, 2008).

A balanced diet rich in omega-3 fatty acids, antioxidants, and other nutrients supports neuroplasticity. For instance:

- Omega-3 fatty acids promote the synthesis of BDNF, which enhances synaptic plasticity.
- Turmeric and black pepper: Studies like Fotuhi et al. (2016) show that combining turmeric with black pepper improves neuroinflammation and cognitive performance.
- Caloric restriction has also been shown to stimulate neurogenesis and enhance cognitive function, likely through metabolic pathways involving sirtuins and ketone bodies.

4.10. Substance Use and Neuroplasticity

Substance abuse, particularly methamphetamine use, has significant neurotoxic effects on the brain's plasticity. Chronic use disrupts motor cortical plasticity, impairing the brain's ability to recover from injury and adapt to new tasks (Du et al., 2017). This can result in long-term cognitive deficits, including memory impairments and reduced motor coordination.

In addition to motor impairments, substance abuse affects the brain's reward system, particularly the dopaminergic pathways. Overstimulation of dopamine release leads to a dysregulated reward system, diminishing the brain's response to natural positive reinforcement.

Addressing substance abuse is crucial for promoting neuroplastic recovery. Rehabilitation strategies, including cognitive-behavioral therapy (CBT) and physical exercise, have shown promise in restoring healthy neural connections and improving cognitive resilience.

5. Implications and Applications

Understanding neuroplasticity has profound implications across various fields:

5.1. Education

Teachers can apply neuroplasticity principles by incorporating active learning, spaced repetition, and multisensory approaches into classrooms. For instance, gamified learning platforms enhance student engagement and retention by harnessing the brain's adaptability. Promoting a growth mindset in educational settings can foster better learning outcomes by applying neuroplastic principles, particularly in subjects perceived as challenging, such as physics (Goldhorn et al., 2023).

5.2. Clinical Neuroscience

Neuroplasticity-based therapies, such as transcranial magnetic stimulation (TMS), are being developed to treat depression, anxiety, and other psychiatric disorders. These approaches modulate neural circuits to restore functional balance. Neuroimaging methods, such as MRI and PET scans, have become invaluable in understanding brain plasticity, but they also pose ethical and practical challenges that need to be addressed (Coutinho et al., 2017). Innovative treatments such as platelet-rich fibrin and silver nanoparticle dressing show how modern therapies can influence wound healing, utilizing neuroplastic principles for faster recovery (Lin et al., 2023). Emerging research highlights the involvement of glial cells in Alzheimer's disease, suggesting that these cells play a more active role in neurodegeneration than previously thought (Smit, 2020).

5.3. Motor Rehabilitation and Neuroplasticity

Recent advancements in neuroplasticity research have highlighted the potential of combining motor imagination with peripheral electrical stimulation to induce long-term depression-like plasticity. Jochumsen et al. (2015) demonstrated that such pairings can lead to lasting changes in motor pathways, offering promising applications for motor rehabilitation, especially in stroke patients and individuals with spinal cord injuries.

Integrating mental practice with physical therapy enhances neuroplastic mechanisms, making rehabilitation more effective. These approaches leverage the brain's ability to adapt through accurate and imagined movements, a concept reshaping traditional physical therapy protocols.

5.4. Cognitive Rehabilitation and Neuroplasticity

Brain-training programs like Lumosity and BrainHQ aim to harness neuroplasticity to improve memory, attention, and problem-solving skills. Although results are mixed, they highlight the potential for targeted cognitive interventions. Harnessing the brain's neuroplastic potential can revolutionize cognitive rehabilitation, enabling personalized interventions that target specific neural circuits to improve mental performance (Mishra & Gazzaley, 2014).

5.5. Neuroplasticity and Emotional Regulation

The brain's capacity to adapt and reorganize itself extends beyond cognitive and motor functions, including emotional regulation. Through neuroplastic mechanisms, the brain adapts how it processes emotions such as fear and anxiety—evolutionarily beneficial responses that may become maladaptive under chronic stress. Lee (2016) highlights that fear and anxiety can reshape neural circuits, particularly in the amygdala and prefrontal cortex, influencing how individuals perceive threats and respond to stress. Longitudinal studies in rodents suggest that fear and anxiety can lead to long-term changes in neural circuits, emphasizing the importance of holistic approaches in studying these emotions (Scheidwasser et al., 2021).

Chronic exposure to stress or trauma can lead to lasting changes in these brain regions, resulting in heightened emotional responses and difficulty regulating emotions. However, the brain's plasticity also provides opportunities for intervention. Therapeutic approaches, such as cognitive-behavioral therapy (CBT) and mindfulness practices, can help rewire maladaptive circuits, promoting healthier emotional responses and psychological well-being.

Understanding the neuroplastic mechanisms underlying emotional regulation can guide the development of targeted treatments for anxiety disorders, post-traumatic stress disorder (PTSD), and depression. These insights are critical for clinicians employing neuroplasticity to improve mental health outcomes.

5.6. Artificial Intelligence

Neuroplasticity has inspired neural network algorithms, enabling artificial intelligence (AI) systems to mimic human learning and adaptability. Technologies like DeepMind's AlphaGo rely on principles derived from the brain's plastic processes (Hassabis, D., 2017).

5.7. Digital Technologies and Neuroplasticity

The digital age has brought profound changes to human cognition, with digital technologies, particularly the internet, influencing how we process information, form memories, and interact socially. Loh and Kanai (2016) argue that the pervasive use of the internet has reshaped human cognition, with significant effects on attention, memory, and social behaviors. This reshaping occurs as the brain adapts to new ways of accessing and consuming information, creating opportunities and challenges for cognitive functioning.

One of the most notable impacts of digital technologies is on attention spans. The constant availability of information and the multitasking demands of digital environments can fragment attention, reducing the brain's capacity for deep focus. At the same time, these technologies offer new ways to enhance learning and memory through interactive and engaging formats.

Digital platforms have also transformed social interactions. While these technologies can foster connections and facilitate communication, they can also alter how individuals perceive social cues and manage relationships. Neuroplastic changes in response to digital interactions highlight the brain's adaptability and raise concerns about potential adverse effects, such as reduced empathy and increased social isolation.

Digital technologies reshape the brain in measurable ways. Recognizing these effects is key to developing strategies for responsible and effective use. Encouraging mindful engagement with digital devices and balancing screen time with offline activities can help mitigate potential cognitive drawbacks while capitalizing on these tools' benefits for learning and social interaction.

6. Conclusion: the Endless Potential of Neuroplasticity

Neuroplasticity lies at the heart of the brain's extraordinary ability to learn, adapt, and recover. It serves not only as a biological foundation for cognitive and emotional functioning but also as a conceptual framework through which we understand how experiences sculpt identity, how rehabilitation unfolds after injury, and how humans continue to evolve across the lifespan. Dissecting its mechanisms provides a window into the brain's dynamic architecture and offers a blueprint for leveraging this malleability to expand human potential.

The implications of neuroplasticity transcend the laboratory bench. In education, it challenges traditional pedagogies by calling for adaptive, learner-centered strategies. In healthcare, it drives the development of cutting-edge interventions for neurodegenerative diseases, psychiatric conditions, and acquired brain injuries. In the realm of technology, neuroplasticity inspires the design of artificial intelligence systems that emulate the brain's learning processes, contributing to the convergence of neuroscience and machine learning.

Yet, the vast promise of neuroplasticity also demands ethical and philosophical scrutiny. What are the boundaries between therapeutic enhancement and cognitive augmentation? How do we ensure equitable access to neuroplasticity-based interventions in a global context rife with socioeconomic disparities? To what extent should we intervene in the brain's natural processes to optimize performance, and who decides the parameters of such interventions?

These questions highlight that neuroplasticity is not merely a scientific phenomenon—it is a societal force. As research advances, the dialogue must expand to include educators, clinicians, ethicists, technologists, and the broader public. Our collective understanding of the brain's adaptability will increasingly shape how we define health, learning, identity, and human agency.

In the chapters that follow, we will explore the molecular and cellular underpinnings of neuroplasticity, unpack its role in behavior and cognition, and examine its application in real-world settings. As we begin this journey, we invite you to consider not only what the brain can become—but also what kind of future we are collectively shaping through our growing ability to rewire it.

This is the transformative promise of neuroplasticity: not just to change the brain, but to reimagine the human condition.

7. Takeaways

1. The Ever-Adaptive Brain
   The brain's architecture is never static—neuroplasticity ensures it remains responsive, flexible, and attuned to life's ever-changing demands.
2. Historical Roots of a Revolution
   From James and Cajal to Hebb, the evolution of neuroplasticity reflects a scientific revolution in understanding how experience molds biology.
3. Multidimensional Nature
   Neuroplasticity manifests across multiple dimensions—structural, functional, developmental, and adaptive—each contributing uniquely to brain flexibility.

4. Underlying Mechanisms
   Synaptic remodeling, neurochemical shifts, and structural transformations collectively empower the brain's ability to learn, heal, and evolve.
5. Breakthrough Discoveries
   Pioneering research—from visual cortex plasticity to adult neurogenesis—redefined the once rigid view of the adult brain.
6. Dynamic Influences
   Age, environment, emotion, nutrition, and behavior all shape how—and how much—the brain can adapt throughout life.
7. Plasticity in Adulthood
   Contrary to old beliefs, the adult brain remains profoundly capable of change, growth, and renewal well into later life.
8. Real-World Applications
   Neuroplasticity serves as a cornerstone for innovation in education, rehabilitation, mental health care, and artificial intelligence.
9. Transformational Power
   Through neuroplasticity, the brain reorganizes itself to support learning, memory, motor skill development, and functional recovery.
10. A Vision for the Future
    The science of neuroplasticity is poised to transform how we think, heal, and connect—ushering in a future defined by cognitive resilience and human potential.

Abbreviations

AI: Artificial intelligence
BDNF: Brain-derived neurotrophic factor
CBT: Cognitive-behavioral therapy
LTD: Long-term depression
LTP: Long-term potentiation
NGF: Nerve growth factor
PTSD: Post-traumatic stress disorder
STDP: Spike-timing-dependent plasticity
TMS: Transcranial magnetic stimulation

Bibliography

1. Anderson, K. (2015). Epigenetics of Neurodegeneration: Quantification of Histone Deacetylase Isoforms and Post-translational Modifications of Histones in Alzheimer's Disease. https://doi.org/10.13016/M2GD9J

2. Berlucchi, G., & Buchtel, H. A. (2009). Neuronal plasticity: Historical roots and evolution of meaning. Experimental Brain Research, 192(3), 307–319. https://doi.org/10.1007/s00221-008-1611-6

3. Bliss, T. V., & Cooke, S. F. (2011). Long-term potentiation and long-term depression: a clinical perspective. *Clinics*, *66*, 3–17. https://doi.org/10.1590/S1807-59322011001300002

4. Bosch, M., & Hayashi, Y. (2012). Structural plasticity of dendritic spines. *Current opinion in neurobiology*, *22*(3), 383–388. https://doi.org/10.1016/j.conb.2011.09.002

5. Burrows, E., & Hannan, A. (2013). Decanalization mediating gene-environment interactions in schizophrenia and other psychiatric disorders with neurodevelopmental etiology. Frontiers in Behavioral Neuroscience, (), n/a.Current Opinion in Neurobiology, 22(3), 383–388. https://doi.org/10.3389/fnbeh.2013.00157

6. Carmichael, S. T., & Chesselet, M. F. (2002). Synchronous neuronal activity signals axonal sprouting after cortical lesions in the adult. Journal of Neuroscience, 22(14), 6062–6070. https://doi.org/10.1523/JNEUROSCI.22-14-06062.2002

7. Centonze, D., Siracusano, A., Calabresi, P., & Bernardi, G. (2004). The Project for a Scientific Psychology (1895): A Freudian anticipation of LTP-memory connection theory. Brain Research Reviews, 46(3), 310–314. https://doi.org/10.1016/j.brainresrev.2004.07.006

8. Chen, H., Yang, G. Z. X., Getachew, H., Acosta, C., Sierra Sánchez, C., & Konofagou, E. E. (2016). Focused ultrasound-enhanced intranasal brain delivery of brain-derived neurotrophic factor. *Scientific reports*, *6*(1), 28599. https://doi.org/10.1038/srep28599

9. Citri, A., & Malenka, R. C. (2008). Synaptic plasticity: Multiple forms, functions, and mechanisms. Neuropsychopharmacology, 33(1), 18–41. https://doi.org/10.1038/sj.npp.1301559

10. Collingridge, G. L., Pineau, S., Howland, J. G., & Wang, Y. T. (2010). Long-term depression in the CNS. Nature Reviews Neuroscience, 11(7), 459–473. https://doi.org/10.1038/nrn2867

11. Cotman, C. W., & Berchtold, N. C. (2002). Exercise: a behavioral intervention to enhance brain health and plasticity. Trends in Neurosciences, 25(6), 295–301. https://doi.org/10.1016/S0166-2236(02)02143-4

12. Coutinho, J., Perrone-McGovern, K., & Gonçalves, Ó. (2017). The Use of Neuroimaging Methodology in Counselling Psychology Research: Promises, Pitfalls, and Recommendations. Canadian Journal of Counselling and Psychotherapy (Online), 51(4), 327–348. https://cjc-rcc.ucalgary.ca/article/view/61169

13. Cramer, S. C., Sur, M., Dobkin, B. H., O'Brien, C., Sanger, T. D., Trojanowski, J. Q., … & Vinogradov, S. (2011). Harnessing neuroplasticity for clinical applications. Brain, 134(6), 1591–1609. https://doi.org/10.1093/brain/awr039

14. Dan, Y., & Poo, M. M. (2006). Spike timing-dependent plasticity: from synapse to perception. Physiological Reviews, 86(3), 1033–1048. https://doi.org/10.1152/physrev.00030.2005

15. DeFelipe, J. (2006). Brain plasticity and mental processes: Cajal again. Nature Reviews Neuroscience, 7(10), 811–817. https://doi.org/10.1038/nrn2005

16. Dempsey, C., & Sawtell, N. B. (2016). The timing is right for cerebellar learning. *Neuron*, 92(5), 931–933. https://doi.org/10.1152/physrev.00030.2005

17. Draganski, B., Gaser, C., Busch, V., Schuierer, G., Bogdahn, U., & May, A. (2004). Changes in grey matter induced by training – newly honed juggling skills are transient and selective structural brain changes. Nature, 427(6972), 311–312. https://doi.org/10.1038/427311a

18. Du, X., Yu, C., Hu, Z. Y., & Zhou, D. S. (2017). Commentary: Methamphetamine abuse impairs motor cortical plasticity and function—frontiers *in human neuroscience*, *11*, 562. https://doi.org/10.3389/fnhum.2017.00562

19. Duman, R. S., Aghajanian, G. K., Sanacora, G., & Krystal, J. H. (2016). Synaptic plasticity and depression: new insights from stress

and rapid-acting antidepressants. Nature Medicine, 22(3), 238–249. https://doi.org/10.1038/nm.4050

20. Erickson, K. I., Voss, M. W., Prakash, R. S., Basak, C., Szabo, A., Chaddock, L., … & Kramer, A. F. (2011). Exercise training increases the size of the hippocampus and improves memory. Proceedings of the National Academy of Sciences, 108(7), 3017–3022. https://doi.org/10.1073/pnas.1015950108
21. Fields, R. D. (2015). A new mechanism of nervous system plasticity: activity-dependent myelination. Nature Reviews Neuroscience, 16(12), 756–767. https://doi.org/10.1038/nrn4023
22. Fotuhi, M., Do, D., & Jack, C. (2016). Modifiable factors alter the size of the hippocampus with aging. Nature Reviews Neurology, 8(4), 189–202. https://doi.org/10.1038/nrneurol.2012.27
23. Fuchs, E., & Flügge, G. (2014). Adult neuroplasticity: more than 40 years of research. *Neural plasticity*, *2014*(1), 541870. https://doi.org/10.1155/2014/541870
24. Gijselaers, J., De Groot, R., & Kirschner, P. A. (2015). Biological lifestyle factors related to cognition and learning performance in adults in distance education. https://core.ac.uk/download/pdf/55538754.pdf
25. Gil-Mohapel, J. (2016). Mechanisms And Assessment Of Hippocampal-dependent Learning And Memory In Rodents. International Journal of Medical and Biological Frontiers, 22(2), 107–134.
26. Gire, D. H., Kapoor, V., Arrighi-Allisan, A., Seminara, A., & Murthy, V. N. (2013). Mice develop efficient strategies for foraging and navigation using complex natural stimuli. Neuron, 80(2), 451–465. https://doi.org/ 10.1016/j.cub.2016.03.040
27. Goldhorn, L., Wilhelm, T., & Spatz, V. (2023). Growth Mindset In Physics: Students' Beliefs About Learning Physics In Middle School And How To Foster A Subject Specific Growth Mindset. *International Journal of Innovation in Science and Mathematics Education*, *31*(4). https://doi.org/10.30722/IJISME.31.04.002
28. Gómez-Pinilla, F. (2008). Brain foods: The effects of nutrients on brain function. Nature Reviews Neuroscience, 9(7), 568–578. https://doi.org/10.1038/nrn2421

29. Gould, E., Reeves, A. J., Fallah, M., Tanapat, P., Gross, C. G., & Fuchs, E. (1999). Hippocampal neurogenesis in adult Old World primates. Proceedings of the National Academy of Sciences, 96(9), 5263–5267. https://doi.org/10.1073/pnas.96.9.5263

30. Hassabis, D., Kumaran, D., Summerfield, C., & Botvinick, M. (2017). Neuroscience-inspired artificial intelligence. Neuron, 95(2), 245–258. https://doi.org/10.1016/j.neuron.2017.06.011

31. Hebb, D. O. (1949). The organization of behavior: A neuropsychological theory. New York: Wiley. https://doi.org/10.1016/s0361-9230(99)00182-3

32. Herculano-Houzel, S. (2009). The human brain in numbers: a linearly scaled-up primate brain. Frontiers in Human Neuroscience, 3, 31. https://doi.org/10.3389/neuro.09.031.2009

33. Herculano-Houzel, S., Avelino-de-Souza, K., Neves, K., Porfírio, J., Messeder, D., Mattos Feijó, L., … & Manger, P. R. (2014). The elephant brain in numbers. Frontiers in neuroanatomy, 8, 46. https://doi.org/10.3389/fnana.2014.00046

34. Holtmaat, A., & Svoboda, K. (2009). Experience-dependent structural synaptic plasticity in the mammalian brain. Nature Reviews Neuroscience, 10(9), 647–658. https://doi.org/10.1038/nrn2699

35. Hubel, D. H., & Wiesel, T. N. (1970). The period of susceptibility to the physiological effects of unilateral eye closure in kittens. The Journal of Physiology, 206(2), 419–436. https://doi.org/10.1113/jphysiol.1970.sp009022

36. Iyer, K. K. (2018). Sleep, wake, and critical brain states: corollaries from brain dynamics—Frontiers in Neuroscience, 12, 948. https://doi.org/ 10.3389/fnins.2018.00948

37. Jamal, G. (2022). Sleep Quality, Life Satisfaction, and Psychological Well-being in Indian College Students. Indian Journal of Health and Wellbeing, 13(3), 307–316.

38. James, W. (1890). The principles of psychology. New York: Henry Holt and Company. https://doi.org/10.4324/9781912282494

39. Jochumsen, M., Signal, N., Nedergaard, R. W., Taylor, D., Haavik, H., & Niazi, I. K. (2015). Induction of long-term depression-like plasticity by pairings of motor imagination and peripheral electrical

stimulation. *Frontiers in Human Neuroscience*, *9*, 644. https://doi.org/10.3389/fnhum.2015.00644

40. Kempermann, G., Song, H., & Gage, F. H. (2015). Neurogenesis in the adult hippocampus. *Cold Spring Harbor perspectives in biology*, 7(9), a018812. https://doi.org/10.1002/0470870834.ch14
41. Kolb, B., & Gibb, R. (2011). Brain plasticity and behavior in the developing brain. Journal of the Canadian Academy of Child and Adolescent Psychiatry, 20(4), 265.
42. Lee, L. (2016). The neuroscience of fear and anxiety: A primer for clinicians. https://digitalcommons.pepperdine.edu/etd/1113
43. Lin, L., Bi, H., Wang, X., & Shi, X. (2023). Impact of Platelet-Rich Fibrin Combined with Silver Nanoparticle Dressing on Healing Time and Therapeutic Efficacy of Chronic Refractory Wounds. Alternative Therapies in Health and Medicine, 29(6), 264–267.
44. Loh, K. K., & Kanai, R. (2016). How has the internet reshaped human cognition? The Neuroscientist, 22(5), 506–520. https://doi.org/10.1177/1073858415595005
45. Lu, B., Nagappan, G., Guan, X., Nathan, P. J., & Wren, P. (2014). BDNF-based synaptic repair as a disease-modifying strategy for neurodegenerative diseases. Nature Reviews Neuroscience, 14(6), 401–416. https://doi.org/10.1038/nrn3505
46. Maguire, E. A., Gadian, D. G., Johnsrude, I. S., Good, C. D., Ashburner, J., Frackowiak, R. S., & Frith, C. D. (2000). Navigation-related structural change in the hippocampi of taxi drivers. Proceedings of the National Academy of Sciences, 97(8), 4398–4403. https://doi.org/10.1073/pnas.070039597
47. Malenka, R. C., & Bear, M. F. (2004). LTP and LTD: an embarrassment of riches. Neuron, 44(1), 5–21. https://doi.org/ 10.1016/j.neuron.2004.09.012
48. Mateos-Aparicio, P., & Rodríguez-Moreno, A. (2019). The impact of studying brain plasticity. Frontiers in Cellular Neuroscience, 13(February), 1–5. https://doi.org/10.3389/fncel.2019.00066
49. McEwen, B. S., & Morrison, J. H. (2013). The brain on stress: vulnerability and plasticity of the prefrontal cortex over the life course. Neuron, 79(1), 16–29. https://doi.org/ 10.1016/j.neuron.2013.06.028

50. Merzenich, M. M., Jenkins, W. M., Johnston, P., Schreiner, C., Miller, S. L., & Tallal, P. (1996). Temporal processing deficits of language-learning impaired children ameliorated by training. Science, 271(5245), 77–81. https://doi.org/10.1126/science.271.5245.77

51. Merzenich, M. M., Nelson, R. J., Stryker, M. P., Cynader, M. S., Schoppmann, A., & Zook, J. M. (1984). Somatosensory cortical map changes following digit amputation in adult monkeys. Journal of Comparative Neurology, 224(4), 591–605. https://doi.org/10.1002/cne.902240408

52. Mishra, J., & Gazzaley, A. (2014). Harnessing the neuroplastic potential of the human brain & the future of cognitive rehabilitation. *Frontiers in human neuroscience*, *8*, 218. https://doi.org/10.3389/fnhum.2014.00218

53. Nudo, R. J. (2013). Recovery after brain injury: mechanisms and principles. *Frontiers in human neuroscience*, *7*, 887. https://doi.org/10.3389/fnhum.2013.00887

54. Paolicelli, R. C., Bolasco, G., Pagani, F., Maggi, L., Scianni, M., Panzanelli, P., … & Gross, C. T. (2011). Synaptic pruning by microglia is necessary for normal brain development. Science, 333(6048), 1456–1458.

55. Rasch, B., & Born, J. (2013). About sleep's role in memory. Physiological Reviews, 93(2), 681–766. https://doi.org/10.1152/physrev.00032.2012

56. Scheidwasser, N., Faggella, M., Kozlova, E., & Sandi, C. (2021). Commentary: The Risky Closed Economy: A Holistic, Longitudinal Approach to Studying Fear and Anxiety in Rodents. *Frontiers in Behavioral Neuroscience, 15*, 664941. https://doi.org/10.3389/fnbeh.2021.664941

57. Smit, T. (2020). *Glial cells and neuronal function in Alzheimer's disease.* https://hdl.handle.net/11245.1/01f56fd0-fe11-4698-a86e-09bb34aefe33

58. Sweatt, J. D. (2013). The emerging field of neuroepigenetics. Neuron, 80(3), 624–632. https://doi.org/10.1016/j.neuron.2013.10.023

59. Tononi, G., & Cirelli, C. (2014). Sleep and the price of plasticity: from synaptic and cellular homeostasis to memory consolidation and integration. Neuron, 81(1), 12–34. https://doi.org/10.1016/j.neuron.2013.12.025

60. Van Praag, H., Kempermann, G., & Gage, F. H. (2000). Neural consequences of environmental enrichment. Nature Reviews Neuroscience, 1(3), 191–198. https://doi.org/10.1038/35044558

61. Werker, J. F., & Hensch, T. K. (2015). Critical periods in speech perception: new directions. Annual Review of Psychology, 66, 173–196. https://doi.org/10.1146/annurev-psych-010814-015104

62. Zatorre, R. J., Fields, R. D., & Johansen-Berg, H. (2012). Plasticity in gray and white: Neuroimaging changes in brain structure during learning. Nature Neuroscience, 15(4), 528–536. https://doi.org/10.1038/nn.3045

Cellular and Molecular Mechanisms Underlying Neuroplasticity

Zeynep Gül

1. Introduction to Neuroplasticity

This chapter aims to examine the cellular and molecular mechanisms driving neuroplasticity comprehensively. The discussion begins by defining the fundamental processes involved, such as synaptic remodeling, axonal guidance, and dendritic spine dynamics, and then continues with an examination of the roles of neurons and glial cells. Following this, we explore the molecular pathways that regulate synaptic plasticity, including neurotransmitter systems, signaling cascades, and intracellular processes.

To give a holistic view, the chapter also highlights advanced tools and techniques used in neuroplasticity research, such as optogenetics and CRISPR technology, which have revolutionized our ability to study and manipulate neural circuits. Finally, the implications of neuroplasticity for therapeutic interventions in neurological disorders are discussed, providing insights into potential future applications.

1.1. Importance of Cellular and Molecular Mechanisms in Neuroplasticity

Neuroplasticity relies on two essential processes. They are the strengthening (long-term potentiation, LTP) and weakening (long-term depression, LTD) of synaptic connections. These processes enable neural networks to function both stably and flexibly. Hebbian plasticity boosts synaptic strength based on experience. Homeostatic plasticity keeps the network stable by balancing these changes. Together, these mechanisms create metaplasticity. It is the capacity to adapt synaptic modifications (Li et al., 2019).

At the molecular level, neurotransmitter systems and signaling pathways support neuroplasticity. At the cellular level, it is driven by synaptic remodeling and changes in dendritic spines (Zhong & Gerges, 2010; Whitney & Peigneux, 2020).

2. Cellular Components Involved in Neuroplasticity

2.1. Neurons and Their Structure

Neurons are the basic building blocks of the nervous system. They vary significantly in their molecules, shapes, and connections. This diversity affects their ability to process inputs and produce outputs (Friedlander & Sherman, 1981; Zeng & Sanes, 2017). At the molecular level, neurons use different gene expressions. They do this to set their functions and chemical signals. Each neuron type is characterized by specific genes and proteins, influencing their responses to external and internal signals (Zeng & Sanes, 2017).

Structurally, neurons comprise a soma (cell body), dendrites, and axons. Dendrites branch out to receive and process signals. Axons send these signals to other cells. The dendritic and axonal structures vary based on the neuron's location and function. For instance, neurons that transmit signals rapidly have long axons and myelin sheaths. Those with localized functions have short axons and dense, branched dendrites (Friedlander & Sherman, 1981).

In terms of connectivity, neurons communicate with other cells through synaptic connections. Some neurons work in local circuits, while others connect different brain regions with long-range projections. These connections shape how neurons work and how they connect (Luo, 2021).

Neuronal differentiation involves a combination of temporal and spatial regulations. Temporal factors guide progenitor cells to specific neuronal types. Transcription factors direct the molecular processes involved. Spatial signals modulate differentiation via chemical and physical cues in the cellular environment (Sagner et al., 2020; Park et al., 2017).

The structural and functional diversity of neurons is essential for the nervous system's core functions. Also, glial cells are more than supportive structures. They regulate neuronal development, function, and plasticity (Allen & Barres, 2009; Barres, 2008).

2.1.1. Axonal Growth and Guidance Mechanisms

Axonal guidance is critical for nervous system development. Growth cones detect and respond to environmental cues, regulating their movements. Guidance molecules, like netrins, slits, semaphorins, and ephrins, are vital. They direct axonal growth by changing actin and microtubule structures in growth cones. These changes facilitate the formation of complex neural circuits (Dickson, 2002; Stoeckli, 2018).

Guidance molecules send attractive or repulsive signals. They steer axons to their targets or away from specific areas. This process varies among neuron populations depending on their environment and developmental stage. Receptor families that guide cells have changing expressions. This is regulated by transcription, translation, protein interactions, and mRNA transport (Stoeckli, 2018).

RNA-based mechanisms also play a significant role in growth cones. Environmental cues trigger localized translation of specific RNAs in growth cones, modulating the proteome and boosting its responsiveness. For instance, netrin and Robo signaling pathways are guidance systems. They are conserved across vertebrates and invertebrates. Some mechanisms contribute to evolutionary diversity (Shigeoka et al., 2013; Dickson, 2002).

2.1.2. Dendritic Spine Dynamics and Remodeling

Dendritic spines, housing most excitatory synapses, are essential for synaptic plasticity. Spine dynamics, including the formation and loss of spines, directly influence synaptic remodeling. Short-lived spines without PSD-95 protein contribute little to long-term neural circuit changes. In contrast, longer-lived spines with PSD-95 are vital for synaptic remodeling (Berry & Nedivi, 2017).

Synaptic activity alters dendritic spine shape. This changes synaptic strength and aids learning and memory. More prominent spines are linked to stable, long-term synapses. Smaller spines are transient connections (Kasai et al., 2010).

The actin cytoskeleton is vital for spine morphology regulation. Actin polymerization and depolymerization underpin the formation and stabilization of synaptic connections. Synaptic receptors, such as NMDA and metabotropic glutamate, interact with actin to shape dendritic spines. Molecules like Rho GTPases and ephrin receptors regulate spine dynamics (Ethell & Pasquale, 2005).

Studies have found "hotspots" on dendritic branches with clustered spine turnover. These support new synaptic connections during learning and help remodel circuits (Frank et al., 2018).

2.2. Glial Cells and Their Role in Neuroplasticity

Glial cells play a crucial role in neural development, plasticity, and repair. These cells are active regulators of neuroplasticity rather than merely supportive structures. During development, radial glial cells are neuronal progenitors. They later turn into astrocyte-like cells in the adult brain. These cells support neurogenesis (Kriegstein & Álvarez-Buylla, 2009).

Extrinsic factors, like matrix components and neurotrophic factors, promote the conversion of glial cells into neurogenic cells. Intrinsic mechanisms control neurogenesis. They include genomic regulatory programs and transcription factors (Falk & Götz, 2017).

2.2.1. Astrocytes: Supporting Synaptic Plasticity

Astrocytes regulate synaptic plasticity. They interact with synapses to promote their formation, stabilization, and elimination. Gliotransmitter release and calcium signaling enable two-way communication with neurons. They support learning and memory by adjusting synaptic connections (Chung et al., 2015).

Astrocytes influence synaptic plasticity by releasing gliotransmitters. These include D-serine, which activates NMDA receptors, and ATP, which modulates neurotransmitter release. Calcium signaling lets astrocytes regulate the uptake and recycling of neurotransmitter metabolites. This maintains circuit stability (Pirttimaki & Parri, 2013).

2.2.2. Microglial Synaptic Pruning

Microglia prune synapses by removing weak, unused ones during development and learning. This process optimizes neural circuits and preserves crucial synapses. It is guided by the complement system's signaling pathways like C1q and CR3 (Schafer et al., 2012).

2.3. Synapses as the Primary Sites of Plasticity

Synaptic plasticity refers to activity-dependent changes in synaptic function and structure. Mechanisms like LTP and depression LTD strengthen or weaken synapses, respectively. They insert or remove AMPA receptors from the postsynaptic membrane (Abraham, 2008).

2.3.1. Mechanisms of Short-Term and Long-Term Synaptic Plasticity

Short-term plasticity involves brief changes in synapses, like facilitation and depression. They are critical for fast information processing. Long-term plasticity, via LTP and LTD, modifies synapses. It is crucial for learning and memory (Deperrois & Graupner, 2019).

LTP is driven by calcium influx through NMDA receptors. This activates calcium-calmodulin-dependent protein kinase II (CaMKII), which promotes AMPA receptor insertion. LTD, conversely, removes AMPA receptors via protein phosphatase activation, maintaining synaptic balance and stability (Abraham, 2008).

As synapses serve as the primary sites of plasticity, their structural and functional changes are tightly regulated by underlying molecular mechanisms. These molecular processes, including activating signaling pathways, receptor dynamics, and ion channel modulation, are crucial for driving short-term and long-term synaptic changes. Understanding these molecular mechanisms provides insight into how synaptic connections are strengthened, weakened, or maintained, ultimately contributing to the adaptive capacity of neural circuits. The following sections examine the intricate molecular pathways underlying these processes, including LTP and LTD, and emphasize their relevance to neuroplasticity.

3. Molecular Mechanisms of Synaptic Plasticity

3.1. Long-Term Potentiation (LTP) and Long-Term Depression (LTD)

3.1.1. Signaling Pathways of LTP and LTD

LTP and LTD start when NMDA-type glutamate receptors are activated. They let calcium into the postsynaptic neuron. High calcium levels trigger synaptic strengthening (LTP). Low-level, sustained calcium influx weakens synapses (LTD). LTP inserts AMPA receptors into the postsynaptic membrane. This enhances synaptic strength. The autophosphorylation of CaMKII aids this process. It is also aided by the phosphorylation of GluA1 subunits at S831 and S845. These reactions speed up the incorporation of AMPA receptors into the membrane (Lüscher & Malenka, 2012; Purkey & Dell'Acqua, 2020).

LTD, on the other hand, is triggered by low, sustained calcium elevations. These activate phosphatases like PP1 and calcineurin. These phosphatases facilitate the endocytosis of AMPA receptors, reducing synaptic strength (Sanderson et al., 2012). The GluN2A subunit plays a critical role in the rapid induction of LTP. The GluN2B subunit supports LTD through prolonged calcium signaling (Ge & Wang, 2023).

3.1.2. Roles of Calcium, Kinases, and Phosphatases

Calcium plays a central role in initiating both LTP and LTD. High-intensity calcium influx through NMDA receptors activates CaMKII, which promotes synaptic plasticity. The autophosphorylation of CaMKII is essential for the long-term maintenance of LTP, as it increases the phosphorylation of the GluA1 subunit of AMPA receptors, enhancing synaptic strength (Purkey & Dell'Acqua, 2020).

In LTD, low-intensity calcium influx activates phosphatases like calcineurin and PP1. They dephosphorylate AMPA receptors, causing synaptic weakening. This phosphatase activity triggers receptor endocytosis, reducing synaptic strength and helping maintain balance (Sanderson et al., 2012; Cepeda-Prado et al., 2021).

3.2. Role of Ion Channels and Receptors in Synaptic Plasticity

3.2.1. Modulation of NMDA and AMPA Receptors

Due to their calcium permeability, NMDA receptors are critical for initiating LTP and LTD. Subunits GluN2A and GluN2B determine the receptor's roles in synaptic plasticity. GluN2B subunits enable LTD by providing prolonged calcium signals. GluN2A subunits enable rapid calcium signaling needed for LTP induction. Also, the GluN1 subunit helps stabilize and improve NMDA receptors (Paoletti et al., 2013).

AMPA receptors are vital for maintaining synaptic plasticity. Phosphorylation of the GluR1 subunit at S845 and S831 boosts AMPA receptor integration into the postsynaptic membrane, increasing synaptic strength. Dynamic processes such as lateral diffusion and receptor internalization further shape AMPA receptor behavior and plasticity (Makino & Malinow, 2009; Purkey & Dell'Acqua, 2020).

3.2.2. Dynamics of Ion Channels and Their Effects on Synaptic Activity

Voltage-dependent calcium channels (VDCCs) are vital for synaptic plasticity. They help induce and maintain it. These channels regulate calcium entry in response to high-frequency signals. They work with NMDA receptors to enhance synaptic strength (Menigoz et al., 2015). L-type calcium channels are most active during postsynaptic depolarization. They improve synaptic transmission and optimize network function (Sanderson et al., 2012).

TRPM4 channels work with NMDA receptors to aid synaptic plasticity. They do this by regulating calcium levels in postsynaptic cells and helping to induce LTP. The phosphorylation of TRPM4 channels boosts their activity. It also activates long-term signaling pathways that strengthen synapses (Menigoz et al., 2015). Conversely, calcium-permeable AMPA receptors (CP-AMPARs) modulate synaptic depression during LTD. They maintain synaptic homeostasis by responding to low-frequency stimulation (Purkey & Dell'Acqua, 2020).

3.3. Neurotransmitter Systems and Their Impact on Plasticity

3.3.1. Glutamatergic and GABAergic Transmission

Glutamatergic transmission is central to synaptic strengthening and learning processes. Glutamate binds to NMDA and AMPA receptors, regulating synaptic plasticity. NMDA receptor subunits GluN1 and GluN2B play a crucial role in LTP. GluN2B boosts calcium permeability. GluN1 stabilizes receptor complexes (Paoletti et al., 2013). AMPA receptors, especially the GluR1 subunit, help strengthen synapses. They do this via phosphorylation and lateral diffusion (Makino & Malinow, 2009; Zhang et al., 2021).

GABAergic transmission regulates the inhibitory aspects of synaptic plasticity. GABA-A receptors mediate postsynaptic chloride influx, promoting cellular hyperpolarization and synaptic depression. GABA-B receptors inhibit presynaptic glutamate release. This maintains synaptic balance and optimizes plasticity (Lüscher & Malenka, 2012). The glutamatergic and GABAergic systems fine-tune learning and memory (Mitsushima et al., 2011).

3.3.2. Dopaminergic and Other Modulatory Systems

Dopamine regulates synaptic plasticity via D1 and D2 receptors. D1 receptors enhance LTP by activating the cAMP-PKA pathway, which increases NMDA receptor phosphorylation. D2 receptors, however, facilitate LTD by lowering cAMP levels. This is especially true in striatal synapses (Surmeier et al., 2007; Nakano et al., 2010).

Presynaptically, dopamine modulates synaptic depression by inhibiting glutamate release. D2 receptors suppress vesicle recycling, promoting LTD processes (Schultz, 2002). Dopamine enhances NMDA receptor sensitivity, strengthening glutamatergic synapses (Paoletti et al., 2013).

3.3.3. Role of Neurotrophic Factors: BDNF

Brain-derived neurotrophic factor (BDNF) plays a vital role in:

- Synaptic plasticity
- Synaptic density
- Neuronal survival
- Neurogenesis

BDNF activates TrkB receptors. This stimulates signaling pathways like PI3K/Akt, MAPK/ERK, and PLC-γ. These pathways support synaptic adaptation and neuronal growth (Numakawa et al., 2018; Moya-Alvarado et al., 2024). These pathways regulate dendritic branching, synaptic efficacy, and neuronal differentiation.

Presynaptically, BDNF influences vesicle recycling and neurotransmitter release through TrkB receptors. It enhances synaptic strength by modulating NMDA and AMPA receptors. It affects their phosphorylation and trafficking (Vanevski & Xu, 2013). BDNF activates the MAPK/ERK pathway. This supports long-distance synaptic adaptation and dendritic growth (Moya-Alvarado et al., 2024).

BDNF regulates gene expression by phosphorylating CREB. This forms the molecular basis for learning and memory. It improves mitochondrial functions, supports energy balance, and boosts synaptic density (Vanevski & Xu, 2013). Disruptions in BDNF-TrkB signaling are associated with synaptic dysfunction and neuronal loss in neurodegenerative diseases such as Alzheimer's. Pharmacological TrkB agonists are being explored as potential treatments for these disorders (Ahmed et al., 2020; He et al., 2023).

4. Intracellular Signaling Pathways in Neuroplasticity

4.1. Calcium Signaling and Its Impact on Plasticity

4.1.1. Calcium Sources and Intracellular Dynamics

Calcium (Ca^{2+}) signaling is one of the pivotal mechanisms underlying neuroplasticity. During synaptic activity, NMDA receptors, L-type VGCCs, and the endoplasmic reticulum allow calcium to enter the cell. NMDA receptors play a critical role in synaptic strengthening and LTP (Hagerstown & Bading, 2011).

L-type calcium channels create Ca^{2+} signals. These signals regulate synaptic plasticity by activating kinases like CaMKII and modulating gene expression (Wayman et al., 2011). Calcium release from the endoplasmic reticulum is vital for neurons, fine-tuning their functions (Toth et al., 2016).

Calcium signaling affects gene expression and synaptic plasticity. Its timing is crucial. For instance, the duration of the calcium signal often determines transcription factor activation. This timing is usually controlled by the simultaneous activation of NMDA receptors and L-type calcium channels (Fields et al., 2005).

Calcium signals have spatial specificity, which leads to different functions in various cellular regions. Calcium waves can spread throughout the cell, but calcium microdomains regulate localized processes. This specificity comes from a strategic distribution of calcium stores and the spatial organization of calcium channels within the cell (Thomas et al., 1996).

4.1.2. Calcium-Dependent Transcription Factors

Calcium regulates gene expression by activating transcription factors, like CREB. CREB phosphorylation occurs through CaMKII and the MAPK cascade (Grewal et al., 2000).

CREB is vital for transcribing early response genes (e.g., c-Fos, Arc). These genes play a crucial role in neuroplasticity (Balu & Coyle, 2018). They also boost levels of neurotrophic factors like BDNF. This supports long-term neuron function and plasticity (Mohammadi et al., 2018).

4.2. Protein Kinases and Phosphatases in Synaptic Modulation

Protein kinases and phosphatases function as major regulators of synaptic modulation. Protein kinases, like those in the MAPK cascade, promote genes for synaptic plasticity (Wiegert & Bading, 2011). Conversely, phosphatases act as negative regulators, maintaining signaling balance.

4.2.1. Role of PP2A Isoforms

Protein phosphatase 2A (PP2A) functions as a critical enzyme. It is a serine/threonine phosphatase that regulates synaptic plasticity. PP2A isoforms, such as Bβ2 and Bβ1, differ in cellular localization and function. For instance, the

Bβ2 isoform localizes to mitochondria. It influences mitochondrial dynamics and energy metabolism during synaptic activity, vital for dendritic development and synapse formation (Dickey & Strack, 2011).

Alternative splicing of the Bβ2 isoform enables it to fulfill unique roles in cellular signaling. When targeted to mitochondria, this isoform can accelerate neuronal apoptosis, contributing to neurodegenerative processes (Dagda et al., 2003).

4.2.2. Mechanisms of MKPs and Other Phosphatases

MAPK phosphatases (MKPs) are particularly important for the negative regulation of ERK1/2. MKP1 and MKP3 isoforms have different cellular localizations and interact with target proteins (Kar et al., 2010). For instance, MKP3 inhibits the ERK pathway, limiting synaptic plasticity, while MKP1 regulates inflammatory cytokines.

Phosphatases have isoform-specific roles. They help regulate synaptic plasticity in time and space. This explains their ability to influence various genetic targets and adapt cells (Sontag et al., 2012).

4.3. Transcriptional Regulation and Gene Expression in Neuroplasticity

Gene transcription is regulated by calcium signaling and the MAPK cascade. For example, synaptic activity activates the ERK signaling pathway. It enhances gene expression and supports synaptic plasticity (Peng et al., 2010).

CREB activation regulates early response genes like c-Fos and Arc. They are critical for long-term memory and learning (Miyashita et al., 2018).

5. Epigenetic Regulation in Neuroplasticity

5.1. Epigenetic Mechanisms: DNA Methylation and Histone Modification

5.1.1. Methylation's Influence on Gene Expression

DNA methylation is a fundamental mechanism of epigenetic regulation that modulates gene expression. This process mainly occurs at CpG islands. Increased methylation in promoter regions inhibits transcription factor binding, resulting in gene silencing. DNA methylation triggers histone modifications and chromatin compaction via methyl-CpG-binding proteins, promoting heterochromatin formation and long-term gene control (Jones & Wolffe, 1999; Ellis et al., 2009).

Histone modifications complement DNA methylation as an additional mechanism regulating gene expression. Histone acetylation relaxes chromatin and boosts transcription, while histone deacetylation condenses chromatin and silences genes (Gregory et al., 2001).

5.1.2. Histone Acetylation and Chromatin Remodeling

Histone acetylation plays a critical role in the epigenetic regulation of gene expression. Adding acetyl groups to lysine residues on histone proteins relaxes chromatin, making DNA more accessible to the transcriptional machinery. This process is linked to euchromatin, a loosely packed chromatin state that promotes transcription (Kingston & Narlikar, 1999).

Chromatin remodeling plays a central role in regulating gene expression. ATP-dependent chromatin remodeling complexes and histone acetylation mediate this process. Histone acetyltransferases (HATs) and deacetylases (HDACs) are major contributors to this regulation. (Ellis et al., 2009; Basavarajappa & Subbanna, 2024).

5.1.3. Epigenetic Regulation of BDNF in Response to Environmental Stimuli

Epigenetic mechanisms regulate the BDNF gene. These mechanisms respond to environmental stimuli and physiological processes. DNA methylation and histone acetylation are critical for controlling BDNF gene expression. Stressful environments change histone acetylation at BDNF promoter regions, affecting neuronal plasticity and related processes (Fuchikami et al., 2009).

5.2. Role of Non-coding RNAs in Plasticity

Non-coding RNAs (ncRNAs) function as major regulators. They affect neuroplasticity, cognition, and neuropsychiatric disorders. MicroRNAs (miRNAs) and long non-coding RNAs (lncRNAs) represent prominent types of ncRNAs. They are involved in neurological processes (Spadaro & Bredy, 2012; Qureshi & Mehler, 2012).

Epigenetic mechanisms regulate neuroplasticity and gene expression. They include DNA methylation, histone modifications, and ncRNAs. These linked processes are vital. They affect synaptic plasticity, learning, and memory (Miller et al., 2008; Rudenko & Tsai, 2014).

6. Role of Mitochondrial Dynamics in Neuroplasticity

6.1. Mitochondrial Function and Energy Metabolism in Plasticity

Mitochondria are essential for energy production, ROS regulation, calcium balance, and cell survival. In the nervous system, they function as major regulators of neurons. They control energy production, neurotransmitter metabolism, and growth. They also manage synaptic remodeling and axonal and dendritic growth. Beyond supplying energy, mitochondria ensure cells work. They balance ROS and calcium levels (Lin & Luo, 2019; Cheng et al., 2010).

Mitochondria are dynamic organelles that keep their shape and function through fusion, fission, transport, and mitophagy. Fusion mixes mitochondrial contents and optimizes function under stress. Fission isolates damaged mitochondria for removal by mitophagy. Proteins like PINK1 and Parkin are vital to mitophagy and key to mitochondrial quality control (V. V. Laar & Berman, 2009).

In neurons, mitochondria move to meet energy demands. They shift between synaptic terminals, dendrites, cell bodies, and axons. This mobility is essential for meeting the energetic requirements of neuronal communication (Sheng & Cai, 2012; Blagov et al., 2022).

6.2. Mitochondrial Quality Control Mechanisms

Mitophagy is vital for mitochondrial quality control. It removes damaged or faulty mitochondria through autophagy. This process is particularly vital in neurons, which have high energy demands. Mitophagy, linked to mitochondrial fusion and fission, ensures their integrity and energy metabolism (Cai & Jeong, 2020).

Mitochondrial dysfunction is an early sign of neurodegenerative diseases. It is linked to energy metabolism problems. In Alzheimer's, failed mitophagy causes synaptic losses and other issues. This occurs despite high but ineffective OXPHOS activity (Han et al., 2021; Feng et al., 2019).

6.3. Mitochondria-Dependent Pathways in Synaptic Function

Mitochondrial dynamics play a crucial role in regulating energy metabolism, maintaining cell health, and supporting neuronal function. Events like fusion, fission, transport, and mitophagy are vital for preserving mitochondrial balance and enabling adaptation to stress (Chen & Chan, 2009).

Mitochondrial dysfunction harms energy production and cell health. It is a factor in neurodegenerative diseases like Alzheimer's and Parkinson's. It also affects synaptic function. Understanding these processes provides therapeutic potential for managing such diseases (Flannery & Trushina, 2019).

7. Endocannabinoid System and Its Role in Neuroplasticity

7.1. Endocannabinoid Signaling in Synaptic Modulation

The endocannabinoid (eCB) system is a lipid-based neurotransmitter system essential for learning, memory, synaptic transmission, and plasticity. eCBs are released from postsynaptic neurons and travel back to presynaptic CB1 receptors, where they regulate neurotransmitter release. This mechanism influences both short—and long-term synaptic modulation. Additionally, eCBs support synaptic plasticity by inhibiting glutamate and GABA. (Piette et al., 2020; Ohno-Shosaku et al., 2012).

During LTP and LTD, eCBs act via TRPV1 channels and CB1 receptors. eCB signaling also affects glial cells' and neurons' interactions. It influences neuroinflammation and neuronal network activity. Astroglial eCB signaling regulates neuroinflammation and synaptic homeostasis (Piette et al., 2020; Young & Denovan-Wright, 2022).

The eCB system affects neuroplasticity. Its role in this is linked to stress, addiction, Alzheimer's, and Parkinson's diseases. These links make the eCB system a good target for treating neurological diseases (Piette et al., 2020; Ohno-Shosaku et al., 2012).

7.2. Interactions Between Endocannabinoids and Neurotransmitter Systems

The eCB system is vital in modulating dopamine (DA) neurotransmission. PCBs released from postsynaptic neurons bind to presynaptic CB1 receptors, regulating dopamine release. This modulation affects reward, motivation, learning, and motor control. In dopaminergic centers like the VTA and substantia nigra, CB1 receptor activation can either boost or reduce dopamine transmission (Micale et al., 2023; Rodriguez de Fonseca et al., 2004).

Interactions with glutamatergic and GABAergic systems indirectly modulate dopaminergic signaling. At glutamatergic synapses, eCBs enhance LTP. At GABAergic synapses, they inhibit, maintaining balance. These cross-system interactions are vital for neurotransmitter balance (Katona & Freund, 2012; Micale et al., 2023).

The ECB and dopamine systems interact. eCB signaling targets addiction, schizophrenia, and Parkinson's disease (Micale et al., 2023; Rodriguez de Fonseca et al., 2004).

7.3. Therapeutic Implications of Endocannabinoid Regulation in Neuroplasticity

7.3.1. Role of Receptors

CB1 receptors are in the central nervous system. They are involved in learning, memory, motor control, and synaptic balance. CB2 receptors, expressed on immune cells, play a significant role in suppressing neuroinflammation. In Alzheimer's, CB2 activation reduces neuroinflammation and regulates microglia (Pacher et al., 2006; Oddi et al., 2019).

7.3.2. Role of Enzymes in the Endocannabinoid System

Enzymes such as FAAH and MAGL regulate eCB metabolism. Inhibitors of these enzymes provide neuroprotection and support synaptic plasticity. FAAH inhibitors reduce neuroinflammation and β-amyloid buildup. MAGL inhibitors boost the survival of dopaminergic neurons in Parkinson's disease (Ren et al., 2020; Van Der Stelt & Di Marzo, 2003).

7.3.3. Synaptic Plasticity and Neurogenesis

At glutamatergic synapses, CB1 receptor activation supports LTP. It underpins the neurobiology of learning and memory. CB1 receptors modulate glutamate release, playing a critical role in these processes (Oddi et al., 2019).

Both CB1 and CB2 receptors boost hippocampal plasticity. They support synaptic adaptation by promoting adult neurogenesis. In the hippocampus, the eCB system affects neurogenesis. It is tied to learning and stress responses (Zhao et al., 2015).

The eCB system plays a vital role in regulating neuroplasticity and synaptic transmission. It also controls learning, memory, and motor skills. eCBs help regulate synapses and maintain balance. They do this through retrograde signaling and interactions with neurotransmitters. These include dopamine, glutamate, and GABA (Lovinger & Mathur, 2012).

Research on the eCB system may help treat some brain disorders. These include Alzheimer's, Parkinson's, depression, and addiction. There is great promise in selectively modulating CB1, CB2, FAAH, and MAGL. It may lead to new therapies that boost neuroplasticity and protect the brain (Ren et al., 2020).

8. Autophagy and Neuroplasticity

8.1. Mechanisms of Autophagy in Neural Cells

Autophagy represents a fundamental process in eukaryotic cells. It degrades long-lived proteins and organelles. Autophagosomes, with double membranes, transport cytoplasmic components to lysosomes. There, lysosomal enzymes recycle them. This process is vital for cell function and stress adaptation (Glick et al., 2010; He & Klionsky, 2009).

Autophagy activates in response to cellular stresses and environmental stimuli. These include amino acid deprivation, reduced growth factors, ER stress, and infections. At the molecular level, autophagy is regulated by the mTOR (mammalian target of rapamycin) pathway. It suppresses autophagy in nutrient-rich conditions. Conversely, activated during energy stress, AMPK (AMP-activated protein kinase) promotes autophagy. Key molecules, like Beclin1 and the PI3K complex, are vital for autophagosome formation (Glick et al., 2010; He & Klionsky, 2009).

In neural cells, autophagy is vital. It keeps post-mitotic neurons, long-lived cells, alive and functional. Autophagy in neurons keeps cells healthy. It clears damaged proteins and organelles. In chronic neurodegenerative diseases, autophagy helps. It reduces protein buildup and damaged organelles. This protects the brain. However, excessive autophagy in acute neural disorders can worsen stress and kill cells (Peker & Gozuacik, 2020; Puyal et al., 2012).

Autophagy is vital in embryonic development. It maintains cellular order and supports organ formation. Ambra1, a critical regulator of autophagy through Beclin1, has been shown to be indispensable. Ambra1 deficiency in mice causes severe neural tube defects, autophagy issues, and ubiquitin buildup (Fimia et al., 2007).

8.2. Autophagy's Role in Synaptic Maintenance and Remodeling

Autophagy is vital for regulating synaptic components and sustaining synaptic plasticity. Synaptic autophagy selectively degrades synaptic proteins, vesicles, receptors, and mitochondria. It supports synaptic remodeling and neurotransmission (Liang, 2019; Lieberman & Sulzer, 2019).

Dysfunctional synaptic autophagy is linked to neurodegenerative diseases, like Alzheimer's and Parkinson's. Additionally, secretory autophagy plays a critical role in processes like trans-synaptic signaling. Proteins linked to secretory autophagy, such as Snap29, Sec22, and Rab8, play a crucial role in synaptic remodeling. It depends on activity (Chang et al., 2023; Lieberman & Sulzer, 2019).

8.3. BDNF's Role in Modulating Autophagic Processes

Brain-derived neurotrophic Factor (BDNF) regulates synaptic plasticity, which is key to memory processes. BDNF suppresses autophagy via signaling pathways like TrkB and PI3K/Akt, enhancing synaptic plasticity and memory formation. Under fasting conditions, autophagy is downregulated in forebrain regions, facilitating synaptic remodeling processes. However, BDNF deficiency disrupts synapses and raises autophagy (Nikoletopoulou et al., 2017).

Autophagy emerges as a critical mechanism in regulating neuroplasticity and synaptic function. It is vital for cellular function in health and disease. Synaptic component regulation through autophagy supports synaptic plasticity and memory processes. Dysfunctions in autophagy mechanisms significantly contribute to the pathogenesis of neurodegenerative processes. Neurotrophic factors like BDNF play a central role. They regulate autophagy, which is vital for synaptic plasticity and brain health. These findings suggest that if we understand autophagy, we could develop new treatments for brain disorders (Glick et al., 2010; He & Klionsky, 2009; Nikoletopoulou et al., 2017; Peker & Gozuacik, 2020).

9. Advanced Tools and Techniques in Neuroplasticity Research

9.1. Optogenetic Approaches in Neuroplasticity

9.1.1. Basics of Optogenetics and Its Mechanisms

Optogenetics is a groundbreaking neuroscience technology. It allows precise control of neurons using light-sensitive proteins (opsins) created by genes. Opsins can be introduced into specific neurons, allowing researchers to modulate their activity using light. They can then investigate neuroplasticity and synaptic remodeling (Beyeler et al., 2014; Deubner et al., 2019).

Opsins like Channelrhodopsin-2 (ChR2), Halorhodopsin (NpHR), and Archaerhodopsin-3 (Arch) are pivotal in optogenetics. ChR2 is a blue-light-activated cation channel. It depolarizes neurons. NpHR and Arch are activated by yellow and green light, respectively. They hyperpolarize neurons and inhibit their activity. Recent advancements include red-shifted opsins. They improve tissue penetration and allow studies in deeper brain areas (Mikhailova et al., 2017; Prosseda & Sun, 2022).

9.1.2. Applications of Optogenetics in Studying Synaptic Plasticity

Optogenetics serves as a powerful tool for studying synaptic plasticity and neural circuits. Tools like SynaptoPAC raise presynaptic cAMP levels using blue light. This regulates transmitter release and induces presynaptic LTP (Oldani et al., 2020). Similarly, optogenetic tools like photoactivatable CaMKII (paCaMKII) enable a detailed study of the mechanisms behind synaptic plasticity (Shibata et al., 2021).

They have been used to study learning and memory. They provide insights into synaptic plasticity and neuronal groups in memory, including acquisition, consolidation, and recall (Beyeler et al., 2014; Xie et al., 2013).

9.1.3. Advantages and Limitations of Optogenetic Methods in Neuroplasticity

Optogenetics allows precise control of neurons in time and space, which helps us better understand synaptic plasticity. However, blue light has poor tissue penetration, and red-shifted opsins pose technical challenges. These limitations are significant (Prosseda & Sun, 2022). Also, ethical and regulatory frameworks must ensure safety for potential human use (Shen et al., 2020).

9.2. CRISPR Technology and Gene Editing in Neuroplasticity

9.2.1. Overview of CRISPR-Cas9 and Its Mechanism

CRISPR-Cas9 is a versatile gene-editing tool derived from bacterial defense mechanisms. It uses a guide RNA (gRNA) to direct the Cas9 endonuclease to specific DNA sequences, creating double-strand breaks (DSBs). These breaks enable precise genetic modifications. They allow fine control over gene expression (Jiang & Doudna, 2017; Palermo et al., 2016).

High-fidelity Cas9 variants and RNA-targeting CRISPR-Cas13 systems improve CRISPR tech. They reduce off-target effects and enhance precision in complex systems (Zhang et al., 2014; Wang et al., 2016).

9.2.2. CRISPR Applications in Modulating Genes Related to Neuroplasticity

CRISPR has played a pivotal role in studying genes involved in synaptic plasticity and neurotransmission. Tools like dCas9 can regulate genes like BDNF, NMDA receptors, and TrkB, revealing how learning, memory, and synapses change (Duke et al., 2020; Leal et al., 2014).

CRISPR activator systems (CRISPRa) can boost gene expression. They have helped us learn about neurotransmission and neuroplasticity (Di Maria & Moindrot, 2020).

9.2.3. Epigenetic and Alternative CRISPR Systems

Beyond gene editing, CRISPR-Cas9 enables epigenetic changes. It uses dCas9 to regulate histone modifications and DNA methylation, without cutting the genes. Alternative CRISPR systems like Cas12 and Cas13 expand the toolbox. They allow precise DNA and RNA modifications (Duke et al., 2020; Zhang et al., 2014).

9.2.4. Ethical and Technical Considerations in CRISPR-Based Neuroplasticity Research

Off-target effects and ethical issues, like misuse in cognitive enhancement, show a need for regulations. Advances in precision tools and ethics are critical. They ensure the safe use of CRISPR in neuroplasticity research (Brokowski & Adli, 2020; Reyes & Lanner, 2017).

10. Conclusion

This chapter has explored the cellular and molecular mechanisms that underpin neuroplasticity, emphasizing the roles of neurons, glial cells, synaptic plasticity, intracellular signaling pathways, and epigenetic regulation. It is evident that neuroplasticity is a dynamic and multifaceted process involving changes at the synaptic, cellular, and molecular levels, which together shape the brain's adaptability and function.

One of the most important insights from this chapter is the critical role of synapses as primary sites of plasticity, driven by mechanisms such as LTP and LTD. These processes are influenced by calcium signaling, neurotransmitter systems, and protein kinases, which regulate synaptic strength and contribute to learning and memory. The chapter also highlighted the importance of glial cells, particularly astrocytes and microglia, in supporting and modulating synaptic plasticity, as well as the significant impact of mitochondrial dynamics and autophagy on neuronal health and function.

The chapter further discussed cutting-edge research tools such as optogenetics and CRISPR technology, which have revolutionized our understanding of neuroplasticity by allowing precise manipulation of neural circuits and gene expression. These advancements have paved the way for potential therapeutic interventions in neurological disorders, offering new hope for conditions like Alzheimer's disease, Parkinson's disease, and schizophrenia.

Future research directions in neuroplasticity should focus on unraveling the complex interactions between various molecular pathways and their impact on long-term brain adaptation. Additionally, understanding the role of environmental factors, such as diet, physical activity, and stress, in influencing neuroplasticity could provide valuable insights into promoting brain health and resilience. Moreover, translating findings from basic research into clinical applications remains a significant challenge, requiring further investigation into how neuroplastic mechanisms can be harnessed for therapeutic purposes.

In conclusion, the study of neuroplasticity continues to evolve, offering promising avenues for enhancing cognitive function, improving mental health, and developing novel treatments for brain-related disorders. By integrating knowledge from molecular biology, neuroscience, and clinical practice, researchers can unlock the full potential of neuroplasticity to transform our understanding of the human brain and its remarkable capacity for change.

11. Takeaways

1. Neuroplasticity Defined:
   Neuroplasticity is the brain's ability to reorganize itself in response to stimuli. It involves short- and long-term changes, both functional and structural. These include synaptogenesis and axonal branching.
   It plays a crucial role in learning, memory, and rehabilitation processes.
2. Cellular Components in Neuroplasticity:
   Neurons have diverse structures: soma, dendrites, and axons. They play a vital role in processing and transmitting information.
   Axonal growth and dendritic spine dynamics are essential processes. They are vital for synaptic remodeling and forming functional neural circuits.
3. Role of Glial Cells:
   Astrocytes and microglia regulate synaptic plasticity. They influence synapse formation, stabilization, and pruning.
   Oligodendrocytes contribute by modulating myelination, which impacts signal transmission efficiency and synaptic stability.
4. Molecular Mechanisms:
   Mechanisms like LTP and LTD drive synaptic plasticity. They are mediated through NMDA and AMPA receptor modulation.
   Calcium signaling, protein kinases, and transcription factors like CREB play a central role. They translate synaptic activity into long-term cellular changes.
5. Epigenetic Regulation:
   DNA methylation, histone acetylation, and non-coding RNAs serve as critical regulators. They modulate gene expression to sustain neuroplasticity.
   BDNF is a fundamental epigenetically regulated factor key for synaptic adaptation and memory.
6. Mitochondrial Dynamics:
   Mitochondria supply energy, regulate calcium and balance ROS. These are vital for synaptic function and neuronal health.

Dysregulation in mitochondrial processes, such as mitophagy, is linked to neurodegenerative diseases.

7. Endocannabinoid System:
   The system affects synaptic transmission and plasticity via retrograde signaling. This has implications for learning, stress response, and neuroprotection.
   CB1 and CB2 receptors are targets for treating Alzheimer's and Parkinson's.
8. Autophagy in Neural Health:
   Autophagy clears damaged organelles and protein aggregates. It supports synaptic remodeling and maintains cellular balance.
   Dysregulated autophagy is implicated in neurodegenerative diseases but also offers therapeutic potential.
9. Advanced Tools in Neuroplasticity Research:
   Optogenetics allows precise control of neural activity, enabling in-depth studies of synaptic plasticity and circuit dynamics.
   CRISPR Technology: It enables targeted gene editing and epigenetic changes. This helps study the genetics of neuroplasticity and test new therapies.
10. Therapeutic Potential:
    Neuroplasticity research may lead to new treatments for brain disorders. Optogenetics and CRISPR hold promise for personalized medicine. They could help manage neurodegenerative diseases and boost cognitive function.

Abbreviations

AMPA: α-amino-3-hydroxy-5-methyl-4-isoxazolepropionic acid
ATP: Adenosine triphosphate
BDNF: Brain-derived neurotrophic factor
CaMKII: Calcium-calmodulin-dependent protein kinase II
CB1: Cannabinoid receptor 1
CB2: Cannabinoid receptor 2
ChR2: Channelrhodopsin-2
cAMP: Cyclic adenosine monophosphate
cAMP-PKA pathway: Cyclic adenosine monophosphate-protein kinase A pathway

Cas12: CRISPR-associated protein 12
Cas13: CRISPR-associated protein 13
CR3: Complement receptor 3
CREB: cAMP response element-binding protein
CRISPR-Cas9: Clustered regularly interspaced short palindromic repeats-CRISPR-associated protein 9
CpG islands: Regions of DNA with a high frequency of CpG sites (cytosine-phosphate-guanine)
CP-AMPARs: Calcium-permeable AMPA receptors
C1q: Complement component 1q
DA: Dopamine
DNA: Deoxyribonucleic acid
DSBs: Double-strand breaks
eCB: Endocannabinoid
ER: Endoplasmic reticulum
ERK1/2: Extracellular signal-regulated kinases 1 and 2
FAAH: Fatty acid amide hydrolase
GABA-A: Gamma-aminobutyric acid type A receptor
GABA-B: Gamma-aminobutyric acid type B receptor
GluN1: Glutamate receptor N1
GluN2A: Glutamate receptor N2A subunit
GluN2B: Glutamate receptor N2B subunit
GluR1: Glutamate receptor 1
HDACs: Histone deacetylases
HATs: Histone acetyltransferases
lncRNAs: Long non-coding RNAs
LTD: Long-term depression
LTP: Long-term potentiation
MAGL: Monoacylglycerol lipase
MAPK/ERK: Mitogen-activated protein kinase/extracellular signal-regulated kinase
MAPK phosphatases (MKPs): Mitogen-activated protein kinase phosphatases
MKP1: Mitogen-activated protein kinase phosphatase 1
MKP3: Mitogen-activated protein kinase phosphatase 3
miRNAs: MicroRNAs
mTOR: Mammalian target of rapamycin

ncRNAs: Non-coding RNAs
NpHR: Halorhodopsin
OXPHOS: Oxidative phosphorylation
paCaMKII: Photoactivatable calcium-calmodulin-dependent protein kinase II
PCBs: Polychlorinated biphenyls
PINK1: PTEN-induced kinase 1
PLC-γ: Phospholipase C-gamma
PP1: Protein phosphatase 1
PP2A: Protein phosphatase 2A
PP2A isoforms: Protein phosphatase 2A isoforms, such as Bβ2 and Bβ1
PSD-95: Postsynaptic density protein 95
RNA: Ribonucleic acid
ROS: Reactive oxygen species
SynaptoPAC: Synapse-targeted photoactivatable adenylyl cyclase
TRPM4: Transient receptor potential melastatin 4
TRPV1: Transient receptor potential vanilloid 1
TrkB: Tropomyosin receptor kinase B
VDCCs: Voltage-dependent calcium channels
VTA: Ventral tegmental area

Bibliography

1. Abraham, W. C. (2008). Metaplasticity: Tuning synapses and networks for plasticity. Nature Reviews Neuroscience, 9(5), 387–399. https://doi.org/10.1038/nrn2356
2. Allen, N. J., & Barres, B. A. (2009). Neuroscience: Glia - More than just brain glue. Nature, 457(7230), 675–677. https://doi.org/10.1038/457675a
3. Ahmed, S., Kwatra, M., Gawali, B., Panda, S. R., & Naidu, V. (2020). Potential role of TrkB agonist in neuronal survival by promoting CREB/BDNF and PI3K/Akt signaling in vitro and in vivo model of 3-nitropropionic acid (3-NP)-induced neuronal death. Apoptosis, 26(1), 52–70. https://doi.org/10.1007/s10495-020-01645-x
4. Barker, A. J., & Ullian, E. M. (2010). Astrocytes and synaptic plasticity. The Neuroscientist, 16(1), 40–50. https://doi.org/10.1177/1073858409339215

5. Barres, B. A. (2008). The mystery and magic of glia: A perspective on their roles in health and disease. Neuron, 60(3), 430–440. https://doi.org/10.1016/j.neuron.2008.10.013

6. Basavarajappa, B. S., & Subbanna, S. (2024). Unlocking the epigenetic symphony: Histone acetylation's impact on neurobehavioral change in neurodegenerative disorders. Epigenomics. Advance online publication. https://doi.org/10.2217/epi-2023-0428

7. Balu, D. T., & Coyle, J. T. (2018). Altered CREB binding to activity-dependent genes in serine racemase deficient mice, a mouse model of schizophrenia. ACS Chemical Neuroscience, 9(9), 2205–2209. https://doi.org/10.1021/acschemneuro.7b00404

8. Benson, D. L., Schnapp, L. M., Shapiro, L., & Huntley, G. W. (2000). Making memories stick: Cell-adhesion molecules in synaptic plasticity. Trends in Cell Biology, 10(11), 473–482. https://doi.org/10.1016/S0962-8924(00)01838-9

9. Berry, K. P., & Nedivi, E. (2017). Spine dynamics: Are they all the same? Neuron, 96(1), 43–55. https://doi.org/10.1016/j.neuron.2017.08.008

10. Beyeler, A., Eckhardt, C. A., & Tye, K. M. (2014). Deciphering memory function with optogenetics. Progress in Molecular Biology and Translational Science, 122, 341–390. https://doi.org/10.1016/B978-0-12-420170-5.00012-X

11. Blagov, A. V., Grechko, A. V., Nikiforov, N. G., Borisov, E. E., Sadykhov, N. K., & Orekhov, A. N. (2022). Role of impaired mitochondrial dynamics processes in the pathogenesis of Alzheimer's disease. International Journal of Molecular Sciences, 23(13), Article 6954. https://doi.org/10.3390/ijms23136954

12. 1Bosworth, A. P., & Allen, N. J. (2017). The diverse actions of astrocytes during synaptic development. Current Opinion in Neurobiology, 47, 38–43. https://doi.org/10.1016/j.conb.2017.08.017

13. Brokowski, C., & Adli, M. (2020). Ethical considerations in therapeutic clinical trials involving novel human germline-editing technology. The CRISPR Journal, 3(1), 18–26. https://doi.org/10.1089/crispr.2019.0051

14. Cai, Q., & Jeong, Y. Y. (2020). Mitophagy in Alzheimer's disease and other age-related neurodegenerative diseases. Cells, 9(1), Article 150. https://doi.org/10.3390/cells9010150

15. Castillo, P. E. (2012). Presynaptic LTP and LTD of excitatory and inhibitory synapses. Cold Spring Harbor Perspectives in Biology, 4(2), a005728. https://doi.org/10.1101/cshperspect.a005728

16. Cepeda-Prado, E., Khodaie, B., Quiceno, G. D., Beythien, S., Edelmann, E., & Lessmann, V. (2021). Calcium-permeable AMPA receptors mediate timing-dependent LTP elicited by low repeat coincident pre- and postsynaptic activity at Schaffer collateral-CA1 synapses. Cerebral Cortex, 32(8), 1682–1695. https://doi.org/10.1093/cercor/bhab306

17. Chang, Y. C., Gao, Y., Lee, J. Y., Langen, J., & Chang, K. T. (2023). Identification of secretory autophagy as a novel mechanism modulating activity-induced synaptic remodeling. bioRxiv. https://doi.org/10.1101/2023.10.06.560931

18. Chen, H., & Chan, D. C. (2009). Mitochondrial dynamics—fusion, fission, movement, and mitophagy—in neurodegenerative diseases. Human Molecular Genetics, 18(R2), R169–R176. https://doi.org/10.1093/hmg/ddp326

19. Chen, H., Tang, A. H., & Blanpied, T. A. (2018). Subsynaptic spatial organization as a regulator of synaptic strength and plasticity. Current Opinion in Neurobiology, 51, 147–153. https://doi.org/10.1016/j.conb.2018.05.004

20. Cheng, A., Hou, Y., & Mattson, M. P. (2010). Mitochondria and neuroplasticity. ASN Neuro, 2(3), AN20100019. https://doi.org/10.1042/AN20100019

21. Chung, W. S., Allen, N. J., & Eroglu, C. (2015). Astrocytes control synapse formation, function, and elimination. Cold Spring Harbor Perspectives in Biology, 7(9), a020370. https://doi.org/10.1101/cshperspect.a020370

22. Colucci-D'Amato, L., Speranza, L., & Volpicelli, F. (2020). Neurotrophic factor BDNF, physiological functions and therapeutic potential in depression, neurodegeneration and brain cancer.

International Journal of Molecular Sciences, 21(20), Article 7777. https://doi.org/10.3390/ijms21207777

23. Dagda, R. K., Zaucha, J. A., Wadzinski, B. E., & Strack, S. (2003). A developmentally regulated, neuron-specific splice variant of the variable subunit Bβ targets protein phosphatase 2A to mitochondria and modulates apoptosis. Journal of Biological Chemistry, 278(27), 24976–24985. https://doi.org/10.1074/jbc.M302832200
24. Deisseroth, K. (2011). Optogenetics. Nature Methods, 8(1), 26–29. https://doi.org/10.1038/nmeth.f.324
25. Deperrois, N., & Graupner, M. (2019). Short-term depression and long-term plasticity together tune sensitive range of synaptic plasticity. PLoS Computational Biology, 16(9), e1008265. https://doi.org/10.1371/journal.pcbi.1008265
26. Deubner, J., Coulon, P., & Diester, I. (2019). Optogenetic approaches to study the mammalian brain. Current Opinion in Structural Biology, 57, 157–163. https://doi.org/10.1016/j.sbi.2019.04.003
27. Dickson, B. J. (2002). Molecular mechanisms of axon guidance. Science, 298(5600), 1959–1964. https://doi.org/10.1126/science.1072165
28. Dickey, A. S., & Strack, S. (2011). PKA/AKAP1 and PP2A/Bβ2 regulate neuronal morphogenesis via Drp1 phosphorylation and mitochondrial bioenergetics. The Journal of Neuroscience, 31(44), 15716–15726. https://doi.org/10.1523/JNEUROSCI.3159-11.2011
29. Di Maria, V., Moindrot, M., Ryde, M., Bono, A., Quintino, L., & Ledri, M. (2020). Development and validation of CRISPR activator systems for overexpression of CB1 receptors in neurons. Frontiers in Molecular Neuroscience, 13, Article 168. https://doi.org/10.3389/fnmol.2020.00168
30. Duke, C. G., Bach, S. V., Revanna, J. S., Sultan, F. A., Southern, N. T., Davis, M. N., Carullo, N. V. N., Bauman, A. J., Phillips, R. A., & Day, J. J. (2020). An improved CRISPR/dCas9 interference tool for neuronal gene suppression. Frontiers in Genome Editing, 2, 9. https://doi.org/10.3389/fgeed.2020.00009
31. Dickey, A. S., & Strack, S. (2011). PKA/AKAP1 and PP2A/Bβ2 regulate neuronal morphogenesis via Drp1 phosphorylation and mitochondrial bioenergetics. Journal of Neuroscience, 31(44), 15716–15726. https://doi.org/10.1523/JNEUROSCI.3159-11.2011

32. Ellis, L., Atadja, P. W., & Johnstone, R. W. (2009). Epigenetics in cancer: Targeting chromatin modifications. Molecular Cancer Therapeutics, 8(6), 1409–1420. https://doi.org/10.1158/1535-7163.MCT-08-0860

33. Ethell, I. M., & Pasquale, E. B. (2005). Molecular mechanisms of dendritic spine development and remodeling. Progress in Neurobiology, 75(3), 161–205. https://doi.org/10.1016/j.pneurobio.2005.02.003

34. Falk, S., & Götz, M. (2017). Glial control of neurogenesis. Current Opinion in Neurobiology, 47, 188–195. https://doi.org/10.1016/j.conb.2017.10.025

35. Fields, R. D., Lee, P. R., & Cohen, J. E. (2005). Temporal integration of intracellular Ca2+ signaling networks in regulating gene expression by action potentials. Cell Calcium, 37(5), 433–442. https://doi.org/10.1016/j.ceca.2005.01.011

36. Fields, R. D. (2015). A new mechanism of nervous system plasticity: Activity-dependent myelination. Nature Reviews Neuroscience, 16(12), 756–767. https://doi.org/10.1038/nrn4023

37. Fimia, G. M., Stoykova, A., Romagnoli, A., Giunta, L., Di Bartolomeo, S., Nardacci, R., … & Cecconi, F. (2007). Ambra1 regulates autophagy and development of the nervous system. Nature, 447(7148), 1121–1125. https://doi.org/10.1038/nature05925

38. Flannery, P. J., & Trushina, E. (2019). Mitochondrial dynamics and transport in Alzheimer's disease. Molecular and Cellular Neuroscience, 98, 109–120. https://doi.org/10.1016/j.mcn.2019.06.009

39. Frank, A. C., Huang, S., Zhou, M., Gdalyahu, A., Kastellakis, G., Silva, T. K., Lu, E., Wen, X., Poirazi, P., Trachtenberg, J. T., & Silva, A. J. (2018). Hotspots of dendritic spine turnover facilitate clustered spine addition and learning and memory. Nature Communications, 9, Article 422. https://doi.org/10.1038/s41467-017-02751-2

40. Friedlander, M., & Sherman, S. (1981). Morphology of physiologically identified neurons. Trends in Neurosciences, 4(6), 211–214. https://doi.org/10.1016/0166-2236(81)90067-9

41. Fuchikami, M., Morinobu, S., Kurata, A., Yamamoto, S., & Yamawaki, S. (2009). Single immobilization stress differentially alters the expression profile of transcripts of the brain-derived

neurotrophic factor (BDNF) gene and histone acetylation at its promoters in the rat hippocampus. The International Journal of Neuropsychopharmacology, 12(1), 73–82. https://doi.org/10.1017/S1461145708008997

42. Feng, S. T., Wang, Z. Z., Yuan, Y., Wang, X. L., Sun, H. M., Chen, N., & Zhang, Y. (2019). Dynamin-related protein 1: A protein critical for mitochondrial fission, mitophagy, and neuronal death in Parkinson's disease. Pharmacological Research, 145, 104553. https://doi.org/10.1016/j.phrs.2019.104553
43. Ge, Y., & Wang, Y. T. (2023). GluN2B-containing NMDARs in the mammalian brain: Pharmacology, physiology, and pathology. Frontiers in Molecular Neuroscience, 16, Article 1190324. https://doi.org/10.3389/fnmol.2023.1190324
44. Gclick, D., Barth, S., & Macleod, K. F. (2010). Autophagy: Cellular and molecular mechanisms. The Journal of Pathology, 221(1), 3–12. https://doi.org/10.1002/path.2697
45. Gkogkas, C., Sonenberg, N., & Costa-Mattioli, M. (2010). Translational control mechanisms in long-lasting synaptic plasticity and memory. The Journal of Biological Chemistry, 285(43), 31913–31917. https://doi.org/10.1074/jbc.R110.154476
46. Gregory, P. D., Wagner, K., & Hörz, W. (2001). Histone acetylation and chromatin remodeling. Experimental Cell Research, 265(2), 195–202. https://doi.org/10.1006/excr.2001.5187
47. Grewal, S. S., Fass, D. M., Yao, H., Ellig, C. L., Goodman, R. H., & Stork, P. J. S. (2000). Calcium and cAMP signals differentially regulate cAMP-responsive element-binding protein function via a Rap1-extracellular signal-regulated kinase pathway. The Journal of Biological Chemistry, 275(44), 34433–34441. https://doi.org/10.1074/jbc.M004728200
48. Guru, A., Post, R. J., Ho, Y.-Y., & Warden, M. R. (2015). Making sense of optogenetics. International Journal of Neuropsychopharmacology, 18(11), pyv079. https://doi.org/10.1093/ijnp/pyv079
49. Han, S., Zhang, M., Jeong, Y. Y., Margolis, D. J., & Cai, Q. (2021). The role of mitophagy in the regulation of mitochondrial energetic status in neurons. Autophagy, 17(12), 4182–4201. https://doi.org/10.1080/15548627.2021.1907167

50. Hansel, C., & Bear, M. (2008). LTD – Synaptic depression and memory storage. Progress in Brain Research, 4, 327–365. https://doi.org/10.1016/B978-012370509-9.00017-6

51. Harris, K. P., & Littleton, J. T. (2015). Experimental methods for examining synaptic plasticity in Drosophila. Cold Spring Harbor Protocols, 2012(2), 162–173. https://doi.org/10.1101/pdb.top067785

52. He, C., & Klionsky, D. J. (2009). Regulation mechanisms and signaling pathways of autophagy. Annual Review of Genetics, 43(1), 67–93. https://doi.org/10.1146/annurev-genet-102808-114910

53. He, K., Nie, L., Dai, Z., Li, S., Yang, X., & Liu, G. (2023). The role of TrkB receptor signaling in Alzheimer's disease. Human Brain. https://doi.org/10.37819/hb.4.1788

54. Ho, V. M., Lee, J. A., & Martin, K. C. (2011). The cell biology of synaptic plasticity. Science, 334(6056), 623–628. https://doi.org/10.1126/science.1209236

55. Hagenston, A. M., & Bading, H. (2011). Calcium signaling in synapse-to-nucleus communication. Cold Spring Harbor Perspectives in Biology, 3(11), a004564. https://doi.org/10.1101/cshperspect.a004564

56. Incontro, S., Asensio, C. S., & Nicoll, R. A. (2017). Dissecting the role of synaptic proteins with CRISPR. In Genome Editing in Neurosciences (pp. 51–62). Springer. https://doi.org/10.1007/978-3-319-60192-2_5

57. Jiang, F., & Doudna, J. (2017). CRISPR-Cas9 structures and mechanisms. Annual Review of Biophysics, 46, 505–529. https://doi.org/10.1146/annurev-biophys-062215-010822

58. Jones, P. L., & Wolffe, A. P. (1999). Relationships between chromatin organization and DNA methylation in determining gene expression. Seminars in Cancer Biology, 9(5), 339–347. https://doi.org/10.1006/scbi.1999.0134x

59. Kar, S., Ukil, A., Sharma, G., & Das, P. K. (2010). MAPK-directed phosphatases preferentially regulate pro- and anti-inflammatory cytokines in experimental visceral leishmaniasis: Involvement of distinct protein kinase C isoforms. Journal of Leukocyte Biology, 88(1), 9–20. https://doi.org/10.1189/jlb.0909644

60. Kasai, H., Fukuda, M., Watanabe, S., Hayashi-Takagi, A., & Noguchi, J. (2010). Structural dynamics of dendritic spines in memory and cognition. Trends in Neurosciences, 33(3), 121–129. https://doi.org/10.1016/j.tins.2010.01.001

61. Katona, I., & Freund, T. F. (2012). Multiple functions of endocannabinoid signaling in the brain. Annual Review of Neuroscience, 35(1), 529–558. https://doi.org/10.1146/annurev-neuro-062111-150420

62. Khakh, B. S., & Sofroniew, M. V. (2015). Diversity of astrocyte functions and phenotypes in neural circuits. Nature Neuroscience, 18(7), 942–952. https://doi.org/10.1038/nn.4043

63. Kingston, R. E., & Narlikar, G. J. (1999). ATP-dependent remodeling and acetylation as regulators of chromatin fluidity. Genes & Development, 13(18), 2339–2352. https://doi.org/10.1101/gad.13.18.2339

64. Kriegstein, A., & Álvarez-Buylla, A. (2009). The glial nature of embryonic and adult neural stem cells. Annual Review of Neuroscience, 32, 149–184. https://doi.org/10.1146/annurev.neuro.051508.135600

65. Leal, G., Comprido, D., & Duarte, C. (2014). BDNF-induced local protein synthesis and synaptic plasticity. Neuropharmacology, 76, 639–656. https://doi.org/10.1016/j.neuropharm.2013.04.005

66. Lee, H., Takamiya, K., Han, J. S., et al. (2003). Phosphorylation of the AMPA receptor GluR1 subunit is required for synaptic plasticity and spatial memory retention. Cell, 112(5), 631–643. https://doi.org/10.1016/S0092-8674(03)00122-3

67. Li, J., Park, E., Zhong, L., & Chen, L. (2019). Homeostatic synaptic plasticity as a metaplasticity mechanism—A molecular and cellular perspective. Current Opinion in Neurobiology, 54, 44–53. https://doi.org/10.1016/j.conb.2018.08.010

68. Liang, Y. T. (2019). Emerging concepts and functions of autophagy as a regulator of synaptic components and plasticity. Cells, 8(1), 34. https://doi.org/10.3390/cells8010034

69. Liberman, O. J., & Sulzer, D. (2019). The synaptic autophagy cycle. Journal of Molecular Biology. https://doi.org/10.1016/j.jmb.2019.12.028

70. Lin, F., & Luo, S. (2019). Mitochondria in neurodegenerative diseases. CNS Neuroscience & Therapeutics, 25(10), 813–815. https://doi.org/10.1111/cns.13183

71. Liu, J., Dietz, K., DeLoyht, J. M., Pedre, X., Kelkar, D., Kaur, J., Vialou, V., Lobo, M. K., Dietz, D. M., Nestler, E. J., Dupree, J., & Casaccia, P. (2012). Impaired adult myelination in the prefrontal cortex of socially isolated mice. Nature Neuroscience, 15(11), 1621–1623. https://doi.org/10.1038/nn.3263

72. Luo, L. (2021). Architectures of neuronal circuits. Science, 373, 610–620. https://doi.org/10.1126/science.abg7285

73. Lüscher, C., & Malenka, R. C. (2012). NMDA receptor-dependent long-term potentiation and long-term depression. Cold Spring Harbor Perspectives in Biology, 4(6), a005710. https://doi.org/10.1101/cshperspect.a005710

74. Lovinger, D. M., & Mathur, B. N. (2011). Endocannabinoids in striatal plasticity. Parkinsonism & Related Disorders, 17(Suppl 1), S10–S13. https://doi.org/10.1016/S1353-8020(11)70041-4

75. Makino, H., & Malinow, R. (2009). AMPA receptor incorporation into synapses during LTP: The role of lateral movement and exocytosis. Neuron, 64(3), 381–390. https://doi.org/10.1016/j.neuron.2009.08.035

76. Maffei, A. (2011). The many forms and functions of long-term plasticity at GABAergic synapses. Neural Plasticity, 2011, Article 254724. https://doi.org/10.1155/2011/254724

77. McKenzie, I. A., Ohayon, D., Li, H., Paes de Faria, J., Emery, B., Tohyama, K., & Richardson, W. D. (2014). Motor skill learning requires active central myelination. Science, 346(6207), 318–322. https://doi.org/10.1126/science.1254960

78. Mei, Y., & Zhang, F. (2012). Molecular tools and approaches for optogenetics. Biological Psychiatry, 71(12), 1033–1038. https://doi.org/10.1016/j.biopsych.2012.02.019

79. Merlin, S., & Vidyasagar, T. (2023). Optogenetics in primate cortical networks. Frontiers in Neuroanatomy, 17, Article 1193949. https://doi.org/10.3389/fnana.2023.1193949

80. Memi, F., Ntokou, A., & Papangeli, I. (2018). CRISPR/Cas9 gene-editing: Research technologies, clinical applications, and ethical considerations. Seminars in Perinatology, 42(8), 487–500. https://doi.org/10.1053/j.semperi.2018.09.003
81. Menigoz, A., Ahmed, T., Sabanov, V., Philippaert, K., Pinto, S., Kerselaers, S., … & Balschun, D. (2015). TRPM4-dependent post-synaptic depolarization is essential for the induction of NMDA receptor-dependent LTP in CA1 hippocampal neurons. Pflugers Archiv, 468, 593–607. https://doi.org/10.1007/s00424-015-1764-7
82. Micale, V., Ruggiero, A. M., Di Marzo, V., & Drago, F. (2023). Crosstalk between the endocannabinoid and midbrain dopaminergic systems: Implication in dopamine dysregulation. Frontiers in Behavioral Neuroscience, 17, Article 1137957. https://doi.org/10.3389/fnbeh.2023.1137957
83. Miller, C. A., Gavin, C. F., White, J. A., Parrish, R. R., Honasoge, A., Yancey, C. R., Rivera, H. S., & Sweatt, J. D. (2008). DNA methylation and histone acetylation work in concert to regulate memory formation and synaptic plasticity. Neurobiology of Learning and Memory, 89(4), 599–603. https://doi.org/10.1016/j.nlm.2007.07.016
84. Mikhailova, M. A., Deal, A. L., Budygin, E. A., & Gainetdinov, R. R. (2017). Optogenetics: Applications in neurobiology. Biological Communications, 62(4), 261–271. https://doi.org/10.21638/11701/spbu03.2017.405
85. Mitsushima, D., Ishihara, K., Sano, A., & Takahashi, T. (2011). Contextual learning requires synaptic AMPA receptor delivery in the hippocampus. Proceedings of the National Academy of Sciences, 108(30), 12503–12508. https://doi.org/10.1073/pnas.1104558108
86. Miyashita, T., Kikuchi, E., Horiuchi, J., & Saitoe, M. (2018). Long-term memory engram cells are established by c-Fos/CREB transcriptional cycling. Cell Reports, 25(10), 2716–2728.e3. https://doi.org/10.1016/j.celrep.2018.11.022
87. Mohammadi, A., Amooeian, V. G., & Rashidi, E. (2018). Dysfunction in brain-derived neurotrophic factor signaling pathway and susceptibility to schizophrenia, Parkinson's, and Alzheimer's diseases. Current Gene Therapy, 18(1), 45–63. https://doi.org/10.2174/1566523218666180302163029

88. Moya-Alvarado, G., Valero-Peña, X., Aguirre-Soto, A., Bustos, F. J., Lazo, O. M., & Bronfman, F. (2024). The PLC-γ-Ca2+ pathway regulates axonal TrkB endocytosis and is required for long-distance propagation of BDNF signaling. Frontiers in Molecular Neuroscience, 17. https://doi.org/10.3389/fnmol.2024.1009404

89. Nakano, T., Doi, T., Yoshimoto, J., & Doya, K. (2010). A kinetic model of dopamine- and calcium-dependent striatal synaptic plasticity. PLoS Computational Biology, 6(7), e1000670. https://doi.org/10.1371/journal.pcbi.1000670

90. Nicoll, R. A. (2017). A brief history of long-term potentiation. Neuron, 93(2), 281–290. https://doi.org/10.1016/j.neuron.2016.12.015

91. Nikoletopoulou, V., Sidiropoulou, K., Kallergi, E., Dalezios, Y., & Tavernarakis, N. (2017). Modulation of autophagy by BDNF underlies synaptic plasticity. Cell Metabolism, 26(1), 230–242.e5. https://doi.org/10.1016/j.cmet.2017.06.005

92. Nugent, F. S., & Kauer, J. (2008). LTP of GABAergic synapses in the ventral tegmental area and beyond. The Journal of Physiology, 586(1), 1471–1478. https://doi.org/10.1113/jphysiol.2007.148098

93. Numakawa, T., Odaka, H., & Adachi, N. (2018). Actions of brain-derived neurotrophic factor in the neurogenesis and neuronal function, and its involvement in the pathophysiology of brain diseases. International Journal of Molecular Sciences, 19(11), Article 3650. https://doi.org/10.3390/ijms19113650

94. Oddi, S., Scipioni, L., & Maccarrone, M. (2019). The endocannabinoid system and adult neurogenesis: A focused review. Current Opinion in Pharmacology, 50, 25–32. https://doi.org/10.1016/j.coph.2019.11.002

95. Ohno-Shosaku, T., Tanimura, A., Hashimotodani, Y., & Kano, M. (2012). Endocannabinoids and retrograde modulation of synaptic transmission. The Neuroscientist, 18(2), 119–132. https://doi.org/10.1177/1073858410397377

96. Oldani, S., Moreno-Velasquez, L., Stumpf, A., Rosenmund, C., Schmitz, D., & Rost, B. R. (2020). SynaptoPAC is an optogenetic tool for inducing presynaptic plasticity. Journal of Neurochemistry, 156(3), 324–336. https://doi.org/10.1111/jnc.15210

97. Ortega, J. M., Genç, Ö., & Davis, G. W. (2018). Molecular mechanisms that stabilize short-term synaptic plasticity during presynaptic homeostatic plasticity. eLife, 7, Article e40385. https://doi.org/10.7554/eLife.40385

98. Palermo, G., Miao, Y., Walker, R. C., Jinek, M., & McCammon, J. A. (2016). Striking plasticity of CRISPR-Cas9 and key role of non-target DNA, as revealed by molecular simulations. ACS Central Science, 2(9), 756–763. https://doi.org/10.1021/acscentsci.6b00218

99. Pacher, P., Batkai, S., & Kunos, G. (2006). The endocannabinoid system is an emerging target of pharmacotherapy. Pharmacological Reviews, 58(3), 389–462. https://doi.org/10.1124/pr.58.3.2

100. Paixão, S., & Klein, R. (2010). Neuron–astrocyte communication and synaptic plasticity. Current Opinion in Neurobiology, 20(4), 466–473. https://doi.org/10.1016/j.conb.2010.04.008

101. Panatier, A., Theodosis, D. T., Mothet, J.-P., Touquet, B., Pollegioni, L., Poulain, D. A., & Oliet, S. H. R. (2006). Glia-derived D-serine controls NMDA receptor activity and synaptic memory. Cell, 125(4), 775–784. https://doi.org/10.1016/j.cell.2006.02.051

102. Pannasch, U., Derangeon, M., & Rouach, N. (2011). Astroglial networks scale synaptic activity and plasticity. Proceedings of the National Academy of Sciences, 108(19), 8467–8472. https://doi.org/10.1073/pnas.1016650108

103. Park, S., Choi, K., Kim, D., Kim, W., Lee, D., Kim, H., Hyun, H., Lim, K., Kim, J.-W., Kim, Y.-R., & Kim, J. (2017). Controlled extracellular topographical and chemical cues for the acceleration of neuronal development. Journal of Industrial and Engineering Chemistry, 61, 65–70. https://doi.org/10.1016/j.jiec.2017.12.001

104. Paoletti, P., Bellone, C., & Zhou, Q. (2013). NMDA receptor subunit diversity impacts receptor properties, synaptic plasticity, and disease. Nature Reviews Neuroscience, 14(6), 383–400. https://doi.org/10.1038/nrn3504

105. Peker, N., & Gozuacik, D. (2020). Autophagy is a cellular stress response mechanism in the nervous system. Journal of Molecular Biology, 432(8), 2560–2580. https://doi.org/10.1016/j.jmb.2020.01.017

106. Peng, S., Zhang, Y., Zhang, J., Wang, H., & Ren, B. (2010). ERK in learning and memory: A review of recent research. International

Journal of Molecular Sciences, 11(1), 222–232. https://doi.org/10.3390/ijms11010222

107. Perea, G., & Araque, A. (2005). Glial calcium signaling and neuron-glia communication. Cell Calcium, 38(3–4), 375–382. https://doi.org/10.1016/j.ceca.2005.06.015

108. Piette, C., Cui, Y., Gervasi, N., & Venance, L. (2020). Lights on endocannabinoid-mediated synaptic potentiation. Frontiers in Molecular Neuroscience, 13, Article 132. https://doi.org/10.3389/fnmol.2020.00132

109. Pirttimaki, T., & Parri, H. R. (2013). Astrocyte plasticity: Implications for synaptic and neuronal activity. The Neuroscientist, 20(1), 79–91. https://doi.org/10.1177/1073858413504999

110. Prosseda, P. P., Tran, M., Kowal, T., Wang, B., & Sun, Y. (2022). Advances in ophthalmic optogenetics: Approaches and applications. Biomolecules, 12(2), Article 269. https://doi.org/10.3390/biom12020269

111. Puyal, J., Ginet, V., Grishchuk, Y., Truttmann, A. C., & Clarke, P. G. H. (2012). Neuronal autophagy mediates life and death: Contrasting roles in chronic neurodegenerative and acute neural disorders. The Neuroscientist, 18(3), 224–236. https://doi.org/10.1177/1073858411404948

112. Purkey, A. M., & Dell'Acqua, M. L. (2020). Phosphorylation-dependent regulation of calcium-permeable AMPA receptors during hippocampal synaptic plasticity. Frontiers in Synaptic Neuroscience, 12, Article 8. https://doi.org/10.3389/fnsyn.2020.00008

113. Qureshi, I. A., & Mehler, M. F. (2012). Emerging roles of non-coding RNAs in brain evolution, development, plasticity, and disease. Nature Reviews Neuroscience, 13(8), 528–541. https://doi.org/10.1038/nrn3234

114. Ren, S.-Y., Wang, Z.-Z., Zhang, Y., & Chen, N. (2020). Potential application of endocannabinoid system agents in neuropsychiatric and neurodegenerative diseases—focusing on FAAH/MAGL inhibitors. Acta Pharmacologica Sinica, 41(10), 1263–1271. https://doi.org/10.1038/s41401-020-0385-7

115. Rodriguez de Fonseca, F., Navarro, M., Gomez, R., Escuredo, L., Nava, F., Fu, J., Murillo-Rodriguez, E., Giuffrida, A., LoVerme, J., Gaetani, S., Kathuria, S., & Piomelli, D. (2004). The endocannabinoid

system in the basal ganglia and mesolimbic reward system: Implications for neurological and psychiatric disorders. European Journal of Pharmacology, 480(1–3), 133–150. https://doi.org/10.1016/j.ejphar.2003.08.101

116. Rudenko, A., & Tsai, L. (2014). Epigenetic modifications in the nervous system and their impact upon cognitive impairments. Neuropharmacology, 80, 70–82. https://doi.org/10.1016/j.neuropharm.2014.01.043

117. Reyes, Á. P., & Lanner, F. (2017). Towards a CRISPR view of early human development: Applications, limitations and ethical concerns of genome editing in human embryos. Development, 144(1), 3–7. https://doi.org/10.1242/dev.139683

118. Sanderson, J. L., Gorski, J. A., Gibson, E. S., Lam, P., Freund, R. K., Chick, W. S., & Dell'Acqua, M. L. (2012). AKAP150-anchored calcineurin regulates synaptic plasticity by limiting synaptic incorporation of Ca^{2+}-permeable AMPA receptors. The Journal of Neuroscience, 32(43), 15036–15052. https://doi.org/10.1523/JNEUROSCI.3326-12.2012

119. Sagner, A., Zhang, I., Watson, T., Lázaro, J., Melchionda, M., & Briscoe, J. (2020). Temporal patterning of the central nervous system by a shared transcription factor code. bioRxiv. https://doi.org/10.1101/2020.11.10.376491

120. Sanes, J. R., & Lichtman, J. W. (1999). Development of the vertebrate neuromuscular junction. Annual Review of Neuroscience, 22(1), 389–442. https://doi.org/10.1146/annurev.neuro.22.1.389

121. Sathler, M. F., Khatri, L., Roberts, J. P., Schmidt, I. G., Zaytseva, A., Kubrusly, R. C. C., Ziff, E. B., & Kim, S. (2020). Phosphorylation of the AMPA receptor subunit GluA1 regulates clathrin-mediated receptor internalization. bioRxiv. https://doi.org/10.1101/2020.12.03.410258

122. Schafer, D. P., Lehrman, E. K., & Stevens, B. (2012). The "quad-partite" synapse: Microglia-synapse interactions in the developing and mature CNS. Glia, 61(1), 24–36. https://doi.org/10.1002/glia.22389

123. Schultz, W. (2002). Getting formal with dopamine and reward. Neuron, 36(2), 241–263. https://doi.org/10.1016/S0896-6273(02)00967-4

124. Shen, Y., Campbell, R. E., Côté, D. C., & Paquet, M.-E. (2020). Challenges for therapeutic applications of opsin-based optogenetic tools in humans. Frontiers in Neural Circuits, 14, Article 41. https://doi.org/10.3389/fncir.2020.00041

125. Sheng, Z.-H., & Cai, Q. (2012). Mitochondrial transport in neurons: Impact on synaptic homeostasis and neurodegeneration. Nature Reviews Neuroscience, 13(2), 77–93. https://doi.org/10.1038/nrn3156

126. Shibata, A. C. E., Ueda, H. H., Eto, K., Onda, M., Sato, A., Ohba, T., Nabekura, J., & Murakoshi, H. (2021). Photoactivatable CaMKII induces synaptic plasticity in single synapses. Nature Communications, 12, Article 21025. https://doi.org/10.1038/s41467-021-21025-6

127. Shigeoka, T., Lu, B., & Holt, C. (2013). RNA-based mechanisms underlying axon guidance. The Journal of Cell Biology, 202(7), 991–999. https://doi.org/10.1083/jcb.201305139

128. Sontag, J.-M., Nunbhakdi-Craig, V., White, C. L., Halpain, S., & Sontag, E. (2012). The protein phosphatase PP2A/Bα binds to the microtubule-associated proteins Tau and MAP2 at a motif also recognized by the kinase Fyn: Implications for tauopathies. The Journal of Biological Chemistry, 287(18), 14984–14993. https://doi.org/10.1074/jbc.M111.338681

129. Spadaro, B., & Bredy, T. W. (2012). The emerging role of non-coding RNA in neural plasticity, cognitive function, and neuropsychiatric disorders. Frontiers in Genetics, 3, Article 132. https://doi.org/10.3389/fgene.2012.00132

130. Stoeckli, E. (2018). Understanding axon guidance: Are we nearly there yet? Development, 145(9), dev151415. https://doi.org/10.1242/dev.151415

131. Straub, C., Granger, A. J., Saulnier, J. L., & Sabatini, B. L. (2014). CRISPR/Cas9-mediated gene knock-down in post-mitotic neurons. PLOS ONE, 9(8), e105584. https://doi.org/10.1371/journal.pone.0105584

132. Surmeier, D. J., Ding, J., Day, M., Wang, Z., & Shen, W. (2007). D1 and D2 dopamine-receptor modulation of striatal glutamatergic signaling in striatal medium spiny neurons. Trends in Neurosciences, 30(5), 228–235. https://doi.org/10.1016/j.tins.2007.03.008

133. Thomas, A. P., Bird, G. S. J., Hajnóczky, G., Robb-Gaspers, L. D., & Putney, J. W. (1996). Spatial and temporal aspects of cellular calcium signaling. The FASEB Journal, 10(12), 1505–1517. https://doi.org/10.1096/fasebj.10.13.8940296

134. Toth, A. B., Shum, A. K., & Prakriya, M. (2016). Regulation of neurogenesis by calcium signaling. Cell Calcium, 59(2–3), 124–134. https://doi.org/10.1016/j.ceca.2016.02.011

135. Tremblay, M. È., Stevens, B., & Sierra, A. (2011). The role of microglia in the healthy brain. Journal of Neuroscience, 31(45), 16064–16069. https://doi.org/10.1523/JNEUROSCI.4158-11.2011

136. Van der Stelt, M., & Di Marzo, V. (2003). The endocannabinoid system in the basal ganglia and the mesolimbic reward system: Implications for neurological and psychiatric disorders. European Journal of Pharmacology, 480(1–3), 133–150. https://doi.org/10.1016/j.ejphar.2003.08.101

137. Vanevski, F., & Xu, B. (2013). Molecular and neural bases underlying BDNF's roles in controlling body weight. Frontiers in Neuroscience, 7, Article 37. https://doi.org/10.3389/fnins.2013.00037

138. Van Laar, V. S., & Berman, S. B. (2009). Mitochondrial dynamics in Parkinson's disease. Experimental Neurology, 218(2), 247–256. https://doi.org/10.1016/j.expneurol.2009.03.019

139. Wang, H., La Russa, M., & Qi, L. S. (2016). CRISPR/Cas9 in genome editing and beyond. Annual Review of Biochemistry, 85, 227–264. https://doi.org/10.1146/annurev-biochem-060815-014607

140. Wayman, G. A., Tokumitsu, H., Davare, M. A., & Soderling, T. R. (2011). Analysis of CaM-kinase signaling in cells. Cell Calcium, 50(1), 1–8. https://doi.org/10.1016/j.ceca.2011.02.007

141. Whalley, K. (2007). Synaptic plasticity: Balancing LTP and LTD. Nature Reviews Neuroscience, 8, 249–249. https://doi.org/10.1038/nrn2123

142. Whitney, S., & Peigneux, P. (2020). Post-learning micro- and macro-structural neuroplasticity changes with time and sleep. Biochemical Pharmacology, 114, Article 369. https://doi.org/10.1016/j.bcp.2020.114369

143. Wright-Jin, E. C., & Gutmann, D. H. (2019). Microglia as dynamic cellular mediators of brain function. Trends in Neurosciences, 42(12), 943–956. https://doi.org/10.1016/j.tins.2019.08.013

144. Wu, Y., Liu, Q., Guo, B., Ye, F., Ge, J., & Xue, L. (2020). BDNF activates postsynaptic TrkB receptors to induce endocannabinoid release, inhibiting presynaptic calcium influx at a calyx-type synapse. The Journal of Neuroscience, 40, 8070–8087. https://doi.org/10.1523/JNEUROSCI.2838-19.2020

145. Wiegert, J. S., & Bading, H. (2011). Activity-dependent calcium signaling and ERK-MAP kinases in neurons: A link to structural plasticity of the nucleus and gene transcription regulation. Cell Calcium, 49(5), 296–305. https://doi.org/10.1016/j.ceca.2010.11.009

146. Xie, Yf., Jackson, M. & MacDonald, J. Optogenetics and synaptic plasticity. Acta Pharmacol Sin 34, 1381–1385 (2013). https://doi.org/10.1038/aps.2013.150

147. Young, A. P., & Denovan-Wright, E. M. (2022). The dynamic role of microglia and the endocannabinoid system in neuroinflammation. Frontiers in Pharmacology, 12, Article 806417. https://doi.org/10.3389/fphar.2021.806417

148. Zhang, F., Wen, Y., & Guo, X. (2014). CRISPR/Cas9 for genome editing: Progress, implications, and challenges. Human Molecular Genetics, 23(R1), R40–R46. https://doi.org/10.1093/hmg/ddu125

149. Zhang, J.-M., Wang, H.-K., Ye, C.-Q., Ge, W., Chen, Y., Jiang, Z.-L., Wu, C.-P., Poo, M.-M., & Duan, S. (2003). ATP released by astrocytes mediates glutamatergic activity-dependent heterosynaptic suppression. Neuron, 40(5), 971–982. https://doi.org/10.1016/s0896-6273(03)00717-7

150. Zhang, D., Watson, J. F., Matthews, P. M., Cais, O., & Greger, I. H. (2021). Gating and modulation of a hetero-octameric AMPA glutamate receptor. Nature, 594(7863), 454–458. https://doi.org/10.1038/s41586-021-03613-0

151. Zhang, X., Liu, Y., Hong, X. et al. NG2 glia-derived GABA release tunes inhibitory synapses and contributes to stress-induced anxiety. Nat Commun 12, 5740 (2021). https://doi.org/10.1038/s41467-021-25956-y

152. Zhao, L., Yeh, M. L.-W., & Levine, E. (2015). Role for endogenous BDNF in endocannabinoid-mediated long-term depression at neocortical inhibitory synapses. eNeuro, 2(1), ENEURO.0029-14.2015. https://doi.org/10.1523/ENEURO.0029-14.2015

153. Zeng, H., & Sanes, J. R. (2017). Neuronal cell-type classification: Challenges, opportunities, and the path forward. Nature Reviews Neuroscience, 18(9), 530–546. https://doi.org/10.1038/nrn.2017.85

154. Zhong, L., & Gerges, N. Z. (2010). Neurogranin and synaptic plasticity balance. Communicative & Integrative Biology, 3(4), 340–342. https://doi.org/10.4161/cib.3.4.11763

155. Zucker, R. S., & Regehr, W. G. (2002). Short-term synaptic plasticity. Annual Review of Physiology, 64, 355–405. https://doi.org/10.1146/annurev.physiol.64.092501.114547

Neural Renaissance: A Journey Through Learning and Neuroplasticity

Asena Himmetoğlu Padgett

1. Introduction to Neuroplasticity and Learning

Neuroplasticity allows the brain to adapt by forming and changing pathways based on experiences, learning, and the environment (Marzola et al., 2023). Through learning, synaptic connections are strengthened, supporting memory formation (Mateos-Aparicio & Rodríguez-Moreno, 2019). At the same time, the weaker or less active connections may possibly be pruned to conserve energy. This process, called synaptic plasticity, is vital for learning and understanding brain function.

Neuroplasticity also contributes to the recovery process following brain injuries. For instance, stroke patients often regain motor skills as healthy brain areas take over lost functions. This recovery process is known as functional neuroplasticity and will be discussed in later chapters (Gaviria et al., 2020).

At this stage, neuroplasticity is understood to comprise multiple types; structural neuroplasticity, one of these types, differs from functional neuroplasticity, as will be elaborated in subsequent sections. This chapter will outline the different types of neuroplasticity and examine how it influences emotions and thinking by establishing its connection to learning (Gaviria et al., 2020). We'll examine how long-term illness, trauma, or stress can alter the brain, affecting emotions and thinking. Furthermore, we will see how positive environments can counteract negative effects by encouraging beneficial brain changes.

Understanding the brain's learning, adapting, and healing processes is vital for cognitive health. Learning fosters neuroplasticity, which alters the brain by changing connections and creating new pathways (Marzola et al., 2023). Thus, understanding the learning process is key to grasping neuroplasticity. Mastering skills like playing an instrument or learning a language enhances cognitive abilities (Bialystok & Poarch, 2015). Regular practice strengthens connections, boosting memory and processing. Neuroplasticity drives these changes, making learning ongoing. It fosters skill development and adaptation, showing that the brain evolves with experience. Learning and neuroplasticity support each other. And, learning also involves breaking weak connections and moving beyond old patterns (Shors et al., 2014). This process supports the idea of "use it or lose it."

This chapter will explore the link between neuroplasticity and learning, building on the previous chapter. We will focus on the effects of neuroplasticity on brain structure and lifelong learning. We will examine the connections, types, and changes involved.

First, we will define neuroplasticity and show how it alters the brain based on experiences. We will highlight its role in forming connections and aiding learning.

Next, we will discuss how neuroplasticity supports skill acquisition and associated brain changes. We will also investigate the relationship between learning and neuroplasticity in older adults, along with learning disorders and their effects on neuroplasticity. Furthermore, we will explore how technology has transformed learning and its implications for neuroplasticity. Finally, we will analyze how stress and sleep affect learning and neuroplasticity. This chapter aims to illuminate our learning journey and the brain's influence on our capacity to learn.

2. Learning and Types of Neuroplasticity

This section covers neuroplasticity types. It is vital for learning, memory, and motor skills. We'll explore how neuroplasticity types manifest, along with the differences between structural and functional neuroplasticity.

Synaptic plasticity, a part of structural neuroplasticity, underpins learning. It describes how connections between neurons strengthen or weaken with experience. Key mechanisms include long-term potentiation (LTP) and long-term depression (LTD). LTP enhances synaptic transmission, aiding learning. Consider it this way: regular practice, like daily painting, strengthens synaptic connections (Bliss & Cooke, 2011). While synaptic plasticity is crucial, it is just one aspect of structural neuroplasticity. Other factors exist and will be discussed later.

Structural neuroplasticity might be broader than synaptic plasticity alone. It could involve mechanisms like dendritic branching, axon sprouting, and neuronal migration (Menon & Gupton, 2018). These processes help the brain adapt, learn, and maintain flexibility throughout life.

Let us discuss dendritic arborization. This process makes dendrites longer and more complex. Dendritic arborization boosts dendritic branching, allowing neurons to form more synapses. This boost is especially true during learning and new experiences (Jan & Jan, 2010). With the formation of more synapses, the brain's function improves. Each task initially requires more synapses. Therefore, learning occurs faster in neurons with more branching. For this reason, we need to strengthen synaptic transmission to boost brain power.

Now, let us move to axonal sprouting. Neurons create new axon branches when damaged or when synapses weaken (Marshall & Farah, 2021). Axonal sprouting allows the axon branches to connect with other neurons. Moreover, keeps a balance between neural plasticity and learning. Axonal sprouting is vital when learning new skills or compensating for lost functions. The process is key to recovery after brain injuries.

Neuronal migration is the movement of new neurons to their final spots, mainly during brain development. It happens mostly in the developing brain, but some evidence shows it continues in adults. This process is vital for learning and memory (Ghashghaei et al., 2007). New neurons join circuits in the hippocampus, boosting learning and memory. In adults, REM sleep triggers neuronal migration, increasing brain flexibility. This process helps adapt or recover brain functions (Meerlo et al., 2009). We've seen that structural neuroplasticity is more than just synaptic changes. Thus, learning neuroplasticity clears up a myth that structural neuroplasticity is just synaptic changes. Neuroplasticity also includes functional neuroplasticity, which is often overlooked. As a result, many mistakenly interpret "neuroplasticity" as referring solely to structural changes. A complete discussion needs to cover both structural and functional aspects.

Functional neuroplasticity allows one brain area to take over for another if damaged. This takeover process helps the brain reorganize and learn new skills, vital for injury recovery (Zotey et al., 2023). For example, stroke patients can regain lost functions thanks to this ability. It shows the brain's capacity to adapt and learn through experience, even after damage.

Both structural and functional neuroplasticity work together in learning, highlighting how our abilities grow and foster lifelong learning (Zotey et al., 2023). Structural plasticity involves changing synaptic connections, while functional plasticity adjusts neural pathways. This adaptability helps the brain meet changing demands that improves learning efficiency and continuity. The brain keeps enhancing its functions by tackling new challenges.

3. The Role of Neurotransmitters and Neurotrophic Factors in Learning and Neuroplasticity

The effects of neurotrophic factors and neurotransmitters on neuroplasticity and learning occur through various mechanisms. This chapter will explain these occurrences through six fundamental neurotransmitters and neurotrophic factors.

3.1. Neurotransmitters

Neurotransmitters regulate synaptic transmission and synaptic plasticity in specific ways. Each neurotransmitter plays a specific role in regulating synaptic transmission and synaptic plasticity (Teleanu et al., 2022b). Key neurotransmitters include glutamate, gamma-aminobutyric acid (GABA), dopamine, serotonin, acetylcholine, and norepinephrine. We will explore each of these neurotransmitters, starting with glutamate.

3.1.1. Glutamate

The amino acid glutamate is essential in all brain functions, like learning and memory formation. Due to its central role in synaptic transmission, glutamate forms the basis of neuroplasticity processes such as long-term potentiation (LTP).

Where glutamate operates at the synapse and transduces the signals through the intracellular domains of these membrane-bound receptors, other molecules like AMPA and NMDA receptors at the cell surface get activated (Mattson, 2009). The potentiation of AMPA receptors, one of the glutamate receptors in the brain, is thought to play a critical role in several areas, such as respective learning. AMPA receptors are ionotropic receptors because they are bound by glutamate, and their regulation of the direct flow of ions is simultaneous. The primary electrical responses that trigger synaptic transmission are based upon this ion flow. These early electrical changes are triggered by AMPA receptors, which facilitate the acceleration of synaptic transmission by enabling potassium (K^+) ions to pass through the membrane and sodium (Na^+) ions to rush into the cell quickly (Brown et al., 2022). In addition to accelerating synaptic signaling, this ion influx initiates a cascade of biophysical processes that boost synaptic functional strength for the time being.

One of the key processes that triggers transient mechanisms of synaptic plasticity is these early electrical changes. Learning and memory are founded on temporarily strengthening existing synaptic connections caused by the sudden influx of ions. For example, the AMPA receptors are crucial for learning and memory via their participation in synaptic plasticity, including LTP and LTD. These receptors have a tetrameric structure, and the specific arrangement of subunits, such as GluA1-4, determines their function. Each subunit within the receptor contributes to its unique properties and activities (Brown et al., 2022). AMPA receptors increase the speed of synaptic transmission (they are some of the fastest transmitting receptors) by making the postsynaptic cell membrane more permeable to sodium and potassium, which increases the potential for LTP and recording transition (participating in learning processes) duality at synapses (Lüscher, 2012).

On the other hand, NMDA receptors are ionotropic receptors located in nerve cells and activated by glutamate. This activation opens Ca^{2+} channels through which calcium ions flow into the cell and cause LTP when glutamate binds (Youn et al., 2013). Glutamate is an excitatory neurotransmitter that fires the postsynaptic cell with which it connects. For the NMDA receptor to be activated, simultaneous glutamate release and slight depolarization are required to remove the Mg^{2+} block from the ion channel. This process reduces EPSPs in the NMDA receptor. Thus, the blockade is removed, and the calcium ions can move into the cell without reserves (Espinosa & Kavalali, 2009).

The calcium entering the cell triggers a host of enzymes and synaptic plasticity signaling pathways. These pathways include enzymes and holoenzymes, for example, Protein Kinase C (PKC), CaMKII (Calcium/Calmodulin-Dependent Protein Kinase II), adenylate cyclase, Protein Kinase A (PKA), nitric oxide synthase (NOS) (Mohanan et al., 2022). Calcium-activated PKC supports the initialization and maintenance of synaptic transmission by facilitating the strengthening and elongation of synapses. So far, behaviors indicating LTP have been abolished by using inhibitors of the CaMKII enzymes, signifying their role as crucial controllers of LTP necessary for synaptic plasticity and brain function. As an AgAGP function, calcium can also activate adenylate cyclase to heighten intracellular cAMP levels and thus mediate synaptic modifications via PKA (Wei et al., 2006). NMDA also activates nitric oxide synthase (NOS), which, in the presence of calcium, leads to the synthesis of nitric oxide, a widely studied contributor to synaptic plasticity. Such activation is known to cause synaptic plasticity, which strengthens synaptic transmission (LTP), leading to structural and functional changes at synapses (Mohanan et al., 2022). Cyclic AMP and PKA increase NMDAR activity, polymerizing the actin cytoskeleton. This landmark effect is characteristic of LTP and forever provides something more solid to anchor learning and memory (Abel & Nguyen, 2014). The problem is that too much glutamate activity can be toxic, and, as expected, this has been connected to neurodegenerative diseases.

3.1.2. Gamma-Aminobutyric Acid (GABA)

The brain's principal inhibitory neurotransmitter, gamma-aminobutyric acid (GABA), is essential to the preservation of synaptic equilibrium. This critical role in synaptic equilibrium is due to stabilizing inhibitory systems, of which GABA is a significant component, as well as excitatory signals influencing the formation of synapses. It induces hyperpolarization of neurons through its receptors, GABA_A and GABA_B (Lambert, 2007). The GABA_A receptor permits chloride ions to rush into the cell. That means when GABA binds, it opens up the chloride channels, increasing the negative charge inside the cell, thereby making it more negatively charged inside the membrane (Excitatory et al., 1997). This negative charge reduces the likelihood of neurons generating action potentials, inhibiting synaptic transmission, and preventing excessive excitation.

On the other hand, the GABA_B receptor increases the efflux of potassium (K^+) ions from the cell via G-protein-coupled signaling pathways. This mechanism supports hyperpolarization by further decreasing the neuron's membrane potential. The efflux of potassium ions raises the threshold for neuronal excitation, thus preventing synaptic overexcitability (Erunuma, 2018). These regulatory effects help maintain the stability of synaptic connections and prevent excessive activity, enhancing the effectiveness of learning processes and preventing the formation of abnormal synaptic connections.

3.1.3. Dopamine

Dopamine is an essential neurotransmitter involved in the reward and motivation processes of the brain. It also plays a role in neuronal plasticity, which is responsible for supporting learning processes, ritualized and minimal activities, and even learning cheats. Dopamine operates in reward centers like the ventral tegmental area (VTA) and nucleus accumbens (NAc), and there are five dopamine receptors: D1, D2, D3, D4, and D5 (Juárez Olguín et al., 2016). The activation of D1 receptors encourages the synaptic potentiating effects of dopamine. Upon binding a ligand, these receptors couple to stimulatory G proteins that activate adenylate cyclase to increase cAMP levels. cAMP is an important intracellular signaling molecule that indirectly activates PKA. It mediates the phosphorylation of glutamate receptors and other synapse-related proteins, enhancing synaptic transmission by promoting stronger connections between neurons (Jones-Tabah et al., 2022). On the other hand, D5 receptors work long-term by regulating dopamine effects over time and promoting synaptic reinforcement through increasing cAMP levels, influencing learning and memory processes (Sibley & Neve, 2007).

In contrast, D2 receptors play an inhibitory role by inhibiting adenylate cyclase and reducing cAMP levels, which contributes to the regulation of synaptic transmission (Christopher P Ford, 2014). This mechanism controls excessive dopamine release. D3 receptors regulate reward anticipation and motivational processes, enhancing reward-oriented learning (Kiss et al., 2021). D4 receptors are involved in attention and cognitive control processes; dopamine binding to these receptors aids in regulating cognitive functions (Woods, 2010). Generally, synaptic strengthening is promoted by releasing dopamine through the D1 and D5 receptors. When dopamine binds to D1 and D5 receptors, it activates adenylate cyclase, thus increasing cAMP levels. cAMP triggers the activation of PKA, which phosphorylates glutamate receptors along with other synaptic proteins (Speranza et al., 2021). This activation ensures further strengthening of synaptic connections and facilitates the process of LTP. Further, dopamine potentiates the opening of NMDA receptors and allows the entry of calcium ions into the cell, thus supporting synaptic plasticity and LTP. These actions of dopamine, coupled with mechanisms of reward and reinforcement processes, enhance learning and maximize motivation through rewards.

Increased dopamine levels during reinforcement promote learning and heighten the anticipation of rewards. In turn, individuals exert more effort toward rewards and goals (Speranza et al., 2021). While dopamine deficiency leads to disturbances in motor functions like Parkinson's disease and emotional disturbances like depression, in cases of Parkinson's disease, it is reduced levels of dopamine that produce difficulties in motor control (Ramesh & Arachchige, 2023). In depression, an imbalance in dopamine may be related to a loss of motivation. The effective functioning of dopamine is critical for both motor and emotional balance.

3.1.4. Serotonin

Serotonin is a neurotransmitter that persists as essential for mood regulation and cognitive adaptability. The numerous 5-HT receptor subtypes, such as the 5-HT1A, 5-HT2, 5-HT3, and 5-HT4 subtypes, have distinct effects in distinct neural circuits and are responsible for the decisive actions of serotonin within the brain (Carhart-Harris & Nutt, 2017).

Serotonin plays a vital role in enabling the individual to maintain attention, select appropriate behavior, and adapt to changes in his or her environment via the modulation of neuron excitability in higher-order cognitive brain regions such as the prefrontal cortex. It modulates limbic system activity to balance emotional responses and optimize cognitive processes (Janet et al., 2023). For instance, serotonin binds to 5-HT1A receptors in the prefrontal cortex, exciting the adenylate cyclase that increases the intracellular levels of cyclic adenosine monophosphate—the rise of cAMP results in a succeeding activation of protein kinase A, PKA. It activates the cAMP response element-binding protein, CREB, which increases gene expression, promoting synaptic plasticity. Thus, the 5-HT1A receptors reinforce the inhibitory effects in the prefrontal cortex, reduce anxiety, and extend attention. The 5-HT2A receptors reinforce neuronal excitability that may contribute to learning and adaptation to changing conditions of the environment (Wang et al., 2018).

In addition, serotonin's action through binding to 5-HT4 receptors activates intracellular mechanisms that enhance the learning and memory processes. Indeed, the 5-HT4 receptors enhance adenylate cyclase activity by raising the levels of cAMP. The increased level of cAMP leads to the activation of PKA and, hence, CREB, supporting the expression of genes involved in synaptic plasticity and memory. In this manner, such a mechanism strengthens learning and memory processes and facilitates adaptation to environmental changes (Wang et al., 2018).

Serotonin gates the activity in a brain region, including the prefrontal cortex and striatum. Besides, it interacts with the dopamine and norepinephrine systems such that the processes of reward and motivation are well-balanced in maintaining appropriate behavioral choices. At the molecular level, serotonin fluctuations within the neural circuits affect neuronal plasticity and synaptic integrity, impacting learning and adaptation processes (Bang et al., 2020). These interactions enable the individual to respond appropriately to environmental variations and direct attention. An optimal level of serotonin thus facilitates decision-making in a balanced manner rather than acting impulsively based on immediate emotions. Low concentrations of serotonin, on the contrary, manifest as depression and anxiety, which decrease cognitive flexibility and hinder learning. Serotonin positively influences learning and memory since it supports synaptic plasticity. Due to the results of such processes, serotonin leads to the activation of new protein synthesis and structural strengthening of synaptic contacts (Dobielska et al., 2022; Jenkins et al., 2016). Serotonin supports such an essential form of synaptic plasticity as long-term potentiation, by which synaptic contacts become more prominent and more robust. LTP represents the cellular basis of learning and memory, and serotonin contributes to this process due to its control of the entry of calcium ions through the NMDA and AMPA receptors (Lüscher, 2012). The receptors mentioned earlier lie at the heart of LTP formation because they provide for lasting changes in synaptic contacts.

Accordingly, the changes in serotonin levels directly interfere with cognitive flexibility and learning processes. On the other hand, optimal serotonin levels facilitate synaptic plasticity, enhance the ability to change in response to new information, and reduce the negative consequences of stress. Deficient and extremely high levels of serotonin can disturb this process and limit learning. A sufficient serotonin level contributes to neuroplasticity since psychiatric diseases like depression and anxiety can cause less, and individuals can be more productive in their emotional and cognitive spheres. Serotonin operates this way in the human brain; however, because of the presence of depression and the interference of the serotonin system, neuroplasticity is reduced (Kraus et al., 2017). This reduction also reduces the ability to carry out learning processes. On the cellular level, synaptic connectivity, in particular, is interfered with by depression. At the same time, the rate at which neurogenesis occurs slows down and eventually lessens the sensitivity of serotonin receptors in synapses. Thus, serotonin is being highly administered to increase the levels of serotonin for the treatment of depression (Thompson et al., 2015). The quantity of serotonin in synaptic clefts increases due to the effect of SSRIs. Higher levels of serotonin support neuroplasticity by increasing the time serotonin binds to 5-HT receptors, which strengthens synaptic relationships between neurons, improves communication abilities, normalizes learning processes, and accelerates neurogenesis in the hippocampus.

Higher levels of serotonin, in turn, contribute to the reversal of cognitive deficits associated with depression and allow an improvement in learning ability (Jenkins et al., 2016). These would be the effects of SSRIs that reverse the decreases in neuroplasticity associated with depression and support cellular and molecular processes underlying learning.

3.1.5. Acetylcoline

Acetylcholine is one of the major neurotransmitters that participate in the processes of learning and neuroplasticity. Acetylcholine provides for signal transmission between neurons on the molecular level. It is synthesized within the neurons from choline and acetyl-CoA and stored in the synaptic vesicles. When an action potential electrical signal occurs in the neuron, it releases acetylcholine into the synapse. The latter acts on nicotinic and muscarinic receptors in the postsynaptic cell (Picciotto et al., 2012). Protein receptors regulate ion channels; the ion channel opens when acetylcholine binds to the receptor. When the ion channel opens, positive ions—sodium ions (Na^+)—enter the cell, which starts an electrical signal. This electrical alteration can excite the cell, which enables signals to be sent to the subsequent neuron and starts neuroplasticity processes that allow learning (Owens & Tanner, 2017). Electrical response significantly contributes to synaptic plasticity, particularly when learning is involved. There is a relation between electrical response and long-term potentiation, which refers to the process used to strengthen connections between neurons; an important role is also played in learning and memory development. Through muscarinic receptors, acetylcholine induces the release of intracellular second messenger molecules such as calcium. Once calcium ions enter the cell, they initiate all those biochemical processes that will eventually result in the structural and functional reinforcement of the synapse, both of which improve the process of learning and retrieval of already learned information (Selway et al., 2012).

Similarly, acetylcholine is a neurotransmitter mediating communications between all those parts of the brain involved in learning and memory, such as the hippocampus and prefrontal cortex. With higher amounts present in these regions, synaptic plasticity increases to allow the gain of new information and storage with greater ease. The electrical response triggered by the opening of ion channels is sustained by the transient effect of acetylcholine in the synaptic cleft. The enzyme acetylcholinesterase rapidly hydrolyzes acetylcholine, thus terminating the signal (Voglis & Tavernarakis, 2006). Such a balance prevents over-excitation of neurons and ensures that learning processes go on unabated in a controlled manner.

3.1.6. Norepinephrine

As a critical neurotransmitter, norepinephrine is required to control cognitive functions, including attention and memory, for adaptive responses to environmental stimuli, and for aiding people to cope with changing conditions. Its effects are mediated by regulating neuronal excitability via α-adrenergic and β-adrenergic receptors. The stimulation of postsynaptic neurons by α1-adrenergic receptors activates the PLC enzyme to release the second messengers IP3 and DAG (Perez, 2021). With this process, intracellular calcium levels rise, enhancing neuronal excitability and amplifying attentional processes. On the other hand, α2-adrenergic receptors have been shown to depress the release of norepinephrine in presynaptic neurons through negative feedback, thereby inhibiting excessive excitation and balancing attentional processes. Norepinephrine influences learning and memory processes through β-adrenergic receptors (Maletic et al., 2017). These receptors activate adenylate cyclase on the cell membrane, increasing intracellular levels of cyclic adenosine monophosphate (cAMP). The increased level of cAMP initiates PKA, a significant mediator of intracellular signaling cascades in the regulation of synaptic plasticity. PKA mediates the phosphorylation and activation of transcription factors like CREB, enhancing gene expression required for synaptic plasticity and long-term memory (Wang et al., 2018). Norepinephrine acts on synaptic plasticity due to stress or arousal by enhancing amygdala and prefrontal cortex neuronal excitability. High norepinephrine levels enable the adaptation of the learning process according to environmental demands and can briefly accelerate learning through the actions of α1 and β receptors (Berridge & Spencer, 2016). During sustained high activity, however, α2 receptors regulate the level of norepinephrine to protect neuroplasticity against overexcitement. Therefore, this narrow balance optimizes attentive learning and adaptation to stressful situations.

3.2. Neurotrophic Factors

Neurotrophic factors are essential biochemical compounds that enhance neuroplasticity, learning process enhancement, and neurotransmitter function. Neurotrophic factors are proteins essential in neuron growth, differentiation, survival, and functionality. They assist significantly in guiding the development of nerve cells, strengthening synaptic connections, and repairing damaged neural tissue. Neurotrophic factors represent a crucial substrate for developing persistent and functional interconnections between neurons that constitute the substrate for learning, memory, and neuroplasticity (Bathina & Das, 2015).

In sharp contrast to neurotransmitters, which act upon relatively fast transmitter-secretion mechanisms, significant actions of neurotrophic factors are related to the regulation of long-term cell survival and function. Neurotransmitters represent chemical messengers through which fast signal transmission occurs between neurons and through which transient, short-term responses are elicited. For example, the serotonin neurotransmitter transmits signals across two neurons in milliseconds (Eiden et al., 2022). Neurotrophic factors influence the structure and function of nerve cells over extended periods, often days or weeks.

Thus, neurotransmitters participate in transmitting the signal, while neurotrophic factors are concerned with the health and functions of nerve cells. In light of the cellular and molecular processes by which neurotrophic factors encourage growth, differentiation, survival, and functionality of neurons, neurotrophic factors such as brain-derived neurotrophic factor (BDNF), glial cell-derived neurotrophic factor (GDNF), and nerve growth factor (NGF) carry out their many roles through the binding to specific receptors and the resultant activation of cellular signaling pathways (Huang & Reichardt, 2001). For instance, the binding of BDNF with its receptor, TrkB (tropomyosin receptor kinase B), may trigger several intracellular signaling cascades that include PI3K/Akt, MAPK/ERK, and PLC-γ. This cascade, sets off the gene expression chain of genes concerned with neuron survival, protein synthesis, and synaptic plasticity (Bathina & Das, 2015; Gupta et al., 2013). GDNF utilizes the same receptors, RET and GFRα1, to exert support for the protection and repair of motor neurons and dopaminergic neurons. NGF exerts its activity through binding with the TrkA receptor, whose activation sets off intracellular signaling for the survival and growth of sensory and peripheral nerve cells (Ibáñez et al., 2020). On a molecular level, these factors maintain synaptic plasticity for the long-term functionality of neurons. For example, BDNF strengthens neuronal connections that underpin learning and memory by supporting synaptic strengthening and LTP. Activation of BDNF enhances the expression of genes regulating synaptic plasticity through the phosphorylation of protein kinases and transcription factors, including cAMP response element-binding protein (CREB) (Bathina & Das, 2015).

On the other hand, GDNF and NGF strengthen neuronal connections but are mainly involved in repairing damaged or degenerating nerve cells. While neurotransmitters participate in the transmission of rapid signals, neurotrophic factors often take part in long-term processes that regulate cellular health, growth, and synaptic plasticity to support nervous system activities. The molecular basis for this distinction includes neurotrophic factors eliciting the signaling pathways that regulate gene expression and promote long-term neuronal survival and repair (Voglis & Tavernarakis, 2006). Lowered BDNF levels in the brain, as in the case of Alzheimer's disease, characterize reduced synaptic plasticity and lowered learning capabilities. On the other hand, higher levels of BDNF during the treatment of Alzheimer's have been supportive of neuronal survival, thereby rejuvenating neuroplasticity and improving learning abilities. Elevation of BDNF strengthens synapses in the hippocampus (Ng et al., 2021). It contributes to generating new nerve cells, helping to decelerate the course of Alzheimer's disease and preserve memory and learning abilities. BDNF also contributes to restructuring learning processes through improved communication between brain cells. The second crucial neurotrophic factor in the maintenance of learning processes under the treatment of Alzheimer's is NGF (Ng et al., 2021; Schindowski et al., 2008). NGF significantly supports the health of cholinergic neurons, and in response to insult to such neurons from Alzheimer's disease, NGF supports their survival through the slowdown of neuronal death. This support maintains synaptic plasticity, facilitates the formation of new synapses, and strengthens existing synaptic contacts (Schindowski et al., 2008). In this manner, this neurotrophic support reinforces learning capacities, keeps learning and memory functions, and contributes to slowing disease progression.

4. Neuroplasticity Across the Lifespan and Critical Periods in Learning and Development

Critical periods describe times in an individual's life when they are best able to learn specific skills and are usually confined to early childhood. These are the times when our brain development is high and capable of more neural plasticity. The classic example is the critical period for language learning, which runs from infancy through early childhood, when the brain is optimized to absorb a particular language (Gariépy et al., 2020). During this period, the brain reaches its peak synaptic density and learning potential, functioning like a sponge that can rapidly absorb new experiences and information. This absorption behavior explains why children can easily pick up a new language. Critical periods have a temporal window of effective operation related to neurofunctional and neuroanatomical development and organization of brain regions (Laland et al., 2008).

Through childhood, the brain is highly plastic; that is, neural connections are rapidly reorganized and changed. Essential structures for language development include a dorsal stream with Broca's area and a ventral stream with Wernicke's area. The dorsal stream comprises the posterior superior temporal gyrus (pSTG) and Broca's area and feeds into the motor processing and production of language. This stream develops very early in childhood and is essential for acquiring the motor dimensions of language, both speech and writing. Synaptic connections are strengthened according to activity in support of these motor functions.

The **ventral stream** is associated with the recognition and comprehension of objects, with the superior temporal gyrus playing a role in this process. In contrast, the **Wernicke's area** plays an important role in understanding word meaning and the syntactic structure of language.(Aguilar-Mediavilla et al., 2022). The ventral stream also develops rapidly during infancy and early childhood, enabling the comprehension of the meaning of language and its essential elements. During this period, the connectivity among different brain parts is most robust, establishing neural networks that process syntactic and semantic language features. Critical periods of development are marked by the intensive formation of synaptic connectivity between neurons at the cellular and molecular levels. They comprise more dynamic processes of synaptogenesis, the formation of new synaptic connections through synaptic pruning. Synaptogenesis refers to forming new synaptic connections, while synaptic pruning is about eliminating the less useful ones and strengthening the neuronal connections for more efficient brain functioning. These mechanisms maximize the brain's capacity for learning a language (Gariépy et al., 2020). At the end of this period, neuronal plasticity begins to decline, and the synaptic links become stable. Learning a new language takes more time and effort in adulthood since plasticity decreases with age. Integrating new information in existing synaptic networks becomes more complex, slowing the language-learning process and making it less fluent (Steber & Rossi, 2021). While the phonetic and grammatical rules may be learned, with an increase in age, it becomes more complicated, and the rate at which learning is acquired diminishes. Though brain plasticity does decrease after these critical periods, capabilities for learning and adaptation do not disappear; new knowledge and language skills are acquired, but with increased effort and time abilities (Gariépy et al., 2020; Steber & Rossi, 2021). Similarly, there are critical periods in acquiring motor skills, visual processing, musical skills, and sporting skills.

Motor skills development includes coordination between the brain and the muscles, often developing quickly during childhood (Sutapa et al., 2021). The development occurs as a product of collaboration between the brain's motor cortex and subcortical structures. The most sensitive age range for motor skills usually lies between 2 and 6 years of age (Tierney & Nelson, 2009). The most rapid and effective motor skill learning occurs in this critical age window. From the neurological viewpoint, it mainly relies on the motor cortex, basal ganglia, and cerebellum. In itself, the motor cortex is responsible for planning and controlling movement. Thus, the motor cortex forms the very basis of motor skill development (Shi & Feng, 2022).

Throughout childhood, synaptic connections in the motor cortex grow rapidly, which in turn facilitates the learning of motor skills. The development of motor activities is based on the basal ganglia, which take responsibility for regulating and coordinating movements. In contrast, the cerebellum cares about the precision and balance of movements. As all of these parts mature fast during this period, we can state that it can be considered a reason for the rapid development of motor skills in childhood (Meier et al., 2008). Neurophysiological development in acquiring or learning motor skills is based on synaptic plasticity. Synaptic connections are more flexible in early childhood, enabling quicker acquisition of motor skills. As a result, synaptic connections are strengthened when learning a motor skill to ensure that the movements will be done accurately and effectively (Tierney & Nelson, 2009).

Moreover, motor memory formation also develops fast in the early childhood stage; the more children repeat complicated movements, the more decisive such actions are in motor memory. Both synaptogenesis and synaptic pruning at the cellular level support the acquisition of motor skills. More extreme synaptogenesis occurs during childhood and accounts for the persistence of learned motor skills. Synaptic pruning supports effective learning of motor skills and movements. Consequently, most critical periods for acquiring motor skills fall between 2 and 6 years. During this period, all the mentioned parts of the brain undergo rapid maturation; synaptic connections with high levels of plasticity are strengthened, and motor memory formation is supported (Tierney & Nelson, 2009).

It is well known that the proper functioning of the visual cortex is mainly responsible for visual processing. Portions of the visual cortex, particularly the primary visual cortex, V1, and its higher-order areas, V2, V3, V4, and V5, lie in the occipital lobe and play essential roles in processing visual information (Siu & Murphy, 2018). Experiments by Hubel and Wiesel showed that their development is adversely affected if these areas do not receive visual stimulation at an early age (Wurtz, 2009). Within these critical periods, the organization and synaptic connections in the visual cortex develop rapidly, a prerequisite for processing visual signals (Tierney & Nelson, 2009). For example, V1 is essential in analyzing simple visual features such as lines and edges; these elementary analysis procedures can be interfered with due to the lack of visual experience. Neurophysiologically, the processes of visual processing have to do with synaptic plasticity and the strengthening of neuronal connections. Synaptic connections are established in significant intensities during the critical periods, and the visual cortex shows a high degree of plasticity. Inadequate visual stimulation during this tender age leads to insufficient synaptic connection formation, weakening the already formed. Such a situation can interfere with processes of synaptic pruning in the visual cortex, impair the proper processing of visual information, and cause irreparable damage to the function of vision. In visual processing, there are heavy formations of synaptic connections between neurons. In these critical periods of the visual system, synaptic contacts are made in profuse numbers and shaped by visual experiences (Siu & Murphy, 2018). Thus, visual processing directly relates to synaptic connections and neuronal organization between different brain regions. The functional development of the visual cortex and synaptic connections among neurons is significant from infancy to early childhood, around approximately six years of age. During this period, the brain maximizes its capacity to process visual stimuli, making it meaningful and providing enough visual experiences.

Deficiencies in the input of vision during critical periods may lead to irrecoverable losses in visual performance (Siu & Murphy, 2018). Therefore, ensuring appropriate visual experiences within this period allows for the normal maturation of the visual system to establish the optimum possible visual processing abilities.

Developing musical skills under the influence of brain plasticity and synaptic connections allows one to learn rhythm, melody, and harmony. Rhythm is the understanding of the arrangement of sounds over time; melody refers to the sequential arrangement of notes, whereas harmony involves the simultaneous combination of different notes (Dumont et al., 2017). Musical skills will be processed by the music-related brain areas and developed by consolidating synaptic connections within the same regions. Children develop musical skills rapidly and effectively, particularly during their early ages. (Hayes et al., 2003). Research into musical competencies in the early years also indicates that high brain flexibility is associated with this process and that musical training is evident. According to other studies, much of the processing related to rhythm and melody has involved the premotor cortex, motor cortex, and temporal lobe. Critical periods during childhood provide the means for the fast acquisition of fundamental musical skills like rhythm and melody. Rhythm typically starts to develop around age 2 when a child can imitate simple rhythms (Bailey & Penhune, 2013). Melody and harmony manifest clear development between the ages of 4 and 6, where children can differentiate between melodic sequences and start learning simple harmonic structures (Bailey & Penhune, 2013). Neuroanatomically, the reinforcement of connections among various brain regions is essential for the learning of musical skills. Notably, the brain regions responsible for linguistic and musical aspects, such as Broca's and Wernicke's, become crucial for establishing musical synapses (Dumont et al., 2017).

The brain regions for sensory-motor integration and reward systems are also crucial in acquiring and applying musical skills. Synaptogenesis and synaptic pruning processes are at the heart of musical skill development. During critical periods, the neurons connect intensively; these processes occur most efficiently during peak brain plasticity. This intense learning process, experienced in the early ages of taking musical training, supports forming these synaptic connections in children. On the other hand, this process becomes difficult after these critical periods because synaptic connections stabilize and need tremendous effort for changes. Critical periods allow the acquisition of rhythm, melody, and harmony quickly during childhood. With appropriate musical experiences provided during these periods, the establishment of musical skills is considerably enhanced at an optimum. It helps to solder synaptic connections between different parts of the brain. Sporting skills are attributed to physical coordination, balance, and agility and are based on the interaction of the brain and muscles. These skills are quickly acquired, especially in childhood, and critical periods play an essential role (Sutapa et al., 2021). Critical periods refer to stages in life when a particular function or skill is most strongly supported by learning and development (Tierney & Nelson, 2009). Indeed, fast physical coordination and balance development occur in early childhood, especially between 0 and 7 years of age (Patel et al., 2017). This period is critical because a child can learn basic sporting skills more quickly and effectively. Children pick up motor skills faster at this stage and create a strong foundation in sporting skills. For sporting skills, agility and speed are more complicated motor skills that become significant. These skills typically emerge between the ages of 7 and 12.(Patel et al., 2017). Developing the abilities of sport and performance in this age group is easier because coordination between the brain and muscles is better established. Neuroanatomically, motor skills are controlled by brain regions such as the motor cortex, cerebellum, and basal ganglia. While the motor cortex coordinates the movements of muscles, the cerebellum maintains balance and precision of movement. The basal ganglia are also critical in planning and controlling movements. In childhood, the connections and functionality of these regions develop at an alarming rate to enable children to learn motor skills more effectively (Abraham & Collins, 2011). The critical period of life is when synaptic links between parts of the brain are enhanced, which is necessary for motor learning. The learning of motor skills is also directly associated with neurophysiological events. Synaptogenesis and synaptic pruning occur extensively during motor learning and skills development. Synaptic connections are strengthened quickly in childhood, which is essential for learning motor skills. The more children learn motor skills, the more synaptic connections are formed, the more intensely and effectively (Sutapa et al., 2021). Proper sports training and practice during these periods support the strengthening of synaptic connections and the development of motor skills.

Critical periods maximize any given learning opportunity with appropriate stimuli that result in better long-term cognitive development and success for the individual. However, it has to be considered that even though there is a critical developmental period, the brain remains, to a minimal extent, capable of changing and learning. Neuroplasticity may also be encouraged in older age with appropriate stimuli, thus positively maintaining and improving cognitive functions (Tierney & Nelson, 2009).

5. Neuroplasticity in Memory Formation and Learning

Memory formation is defined by the processes of elaboration, information processing, and retrieval. It has been described as an elaborate process comprising three basic steps: encoding, storage, and retrieval. Generally, whenever an environmental stimulus undergoes the encoding process, it usually undergoes significant associative processing with contexts of an emotional or cognitive nature (Mujawar et al., 2021). Learning and the establishment of memory depend on the interaction between different regions of the brain. The central regions comprise the prefrontal cortex, hippocampus, basal ganglia, cerebellum, and amygdala (Rusu & Pennartz, 2020). Because the hippocampus is such an essential region in integrating memories in humans, neural adjustments allow information to pass from short-term to long-term memory through neural adjustments that strengthen synaptic contacts. More often than not, learning within the hippocampus induces the formation of new synaptic connections among neurons and enhances new ones. It is further supported by mechanisms of synaptic plasticity, such as LTPs discussed previously (Mujawar et al., 2021; Rusu & Pennartz, 2020). LTP strengthens synaptic transmission via neurotransmitters like glutamate, promoting the ability of neurons to fire at each other and, in turn, conduct information. The prefrontal cortex mediates higher-order cognitive processes, including decision-making, problem-solving, and complex intellectual functions, with changes at the level of synaptic connectivity, which are fundamental for these processes (Hare, 2022). It also strengthens the pre-existing synapses in the prefrontal cortex, besides making new synapses, and allows it to communicate effectively by supporting information processing (Rusu & Pennartz, 2020). These changes are essential for learning complex tasks and give cognitive flexibility that enables one to better adapt to everyday challenges. The prefrontal cortex and the hippocampus are interconnected within a more extensive network that includes parts of the amygdala. Where the amygdala contributes to the formation of emotional memories, the basal ganglia and cerebellum play their role in motor memories and learning (Tang et al., 2022). In all, the interrelations between these regions synergize information processing and the formation of memories. Interactions between the prefrontal cortex and hippocampus, synaptic plasticity, neurotrophic factors, and cellular skeleton dynamics enrich the interactions underlying memory formation (Goto, 2022). Changes in morphology at the level of the dendritic spines are cellular correlates of learning and memory. These tiny projections on the dendritic surface, wherein synapses reside, participate in synaptic transmission and connectivity. Changes at the molecular structure level have been related to the motility of elements of the cellular skeleton, such as actin filaments of the dendritic spines (Goto, 2022).

These actin filaments participate in the cellular skeleton and have found their place in the dendrites of neurons, contributing more to the maintenance of the shape and structure of the dendritic spines. The composition of the dendritic cytoskeleton consists of actin filaments, microtubules, and intermediate filaments.

These skeletal elements provide the structural basis for neuronal morphology and afford dynamic regulation of synaptic connectivity. Alterations in the structure of dendritic spines correspond to changes in the cytoskeleton, which regulates synaptic formation (Goto, 2022; Hare, 2022). LTP not only strengthens the existing synapses but also adds new synapses to the dendritic spines. Along with these morphological changes that indicate stronger synaptic connections, alterations in the dynamics of actin filaments have also been observed. Generally speaking, quantitative changes in the number and structural features of dendritic spines reflect the dynamic nature of neuronal connections, hence, the flexibility of learning processes (Hare, 2022).

Neurotrophic factors, the foremost among them being the brain-derived neurotrophic factor (BDNF), allow synaptic plasticity and memory molecularly by facilitating synaptic strengthening and neuronal health. The effects of BDNF increase synaptic plasticity and improve communication among neurons, which enables synaptic changes during learning (Bathina & Das, 2015). The neurotrophic factors maintain neuronal health and increase synaptic structure stability by sustaining information processing and memory formation. These multilevel processes allow the dynamic adaptation and reorganization of brain functions. Neurotrophic factors accomplish this by both developing and strengthening the brain's ability to process information, further solidifying its capacity to adapt to new information and environmental changes. Hence, this stage is a pointer that molecular and cellular changes establish a structure and functional changes. These changes are crucial for effective information processing, storage, and retrieval. Synaptic plasticity, neurotrophic factors, and dendritic spine changes contribute to the effectiveness of the learning and memory development process. During the storage phase, information is deposited within the long-term memory (Rusu & Pennartz, 2020). Here, the hippocampus seems very important, especially in consolidating short-term to long-term memory by strengthening synaptic connections within the hippocampus and forming new ones.

Proposed mechanisms of synaptic plasticity underlying these processes are LTP and LTD, which enhance or reduce synaptic efficacy. LTP is the cellular phenomenon achieved through the neurotransmitter action of glutamate by increasing synaptic transmission, hence increasing synaptic efficiency to store information more effectively and laying down long-term memory traces. The stored information is reaccessed at retrieval when required. The organization of the restructured information at retrieval consists of the brain regions of the prefrontal cortex. It is complex, based on synergistic interactions among brain regions and continuous adaptations. Neurotrophic factors include BDNF, which plays the most relevant role in synaptic strengthening and healthy neuronal maintenance alongside cellular skeleton elements such as actin filaments.

6. Neuroplasticity and Emotional Regulation in Learning

Emotional regulation is the control, direction, and balancing of emotional responses. It enables one to modulate the intensity, duration, and expression of one's emotions to respond adequately to whatever situation may involve stress or other solid emotional feelings (Kadović et al., 2023). Emotional regulation development is contingent upon the interaction of different brain regions and plays a critical role in developing complex cognitive processes and social behaviors. The prefrontal cortex, especially the dorsolateral and the ventromedial prefrontal cortex, is centrally involved in emotion regulation. The prefrontal cortex is considered equally important in regulating responses to emotional stimuli through its connections with other limbic structures, such as the amygdala, in organizing responses associated with emotionally arousing events relevant to threat/stress. The anterior cingulate cortex informs and regulates the emotional state, while the hippocampus stores the memories of emotional events (Dixon et al., 2017). Synaptic connectivity between the structures enables processing and dealing with emotional information.

Since emotional states directly impact decision-making, attention, and memory processes, emotional regulation is at the core of different cognitive activities. It modulates, for instance, the capability to focus on tasks and to carry out reasoned decisions (Dixon et al., 2017). On the other hand, lousy emotion regulation may mean that an overactive amygdala impairs attention and cognitive performance. When emotions are well-regulated, particularly in stressful situations, they allow thoughts to flow more freely, fostering creativity in finding solutions to problems. The brain's emotional regulatory processes in memory and decision-making could also support neuroplasticity by establishing novel synaptic connections (Salzman & Fusi, 2010). Emotional regulation is an indispensable part of social relations. The recognition of others' emotional states and elaborating an appropriate response are possible because of the brain's mirror neuron system and medial prefrontal cortex (PFC). The interaction between the latter two underpins healthy communication, social harmony, and empathic understanding (He et al., 2018; Plata-Bello et al., 2023). The interplay between the insula, PFC, and amygdala plays a role in maintaining the emotional balance during the social interaction process (Denny et al., 2014). Without emotional regulation, misunderstandings and social conflicts will increase, hurting relationships with others (Seo et al., 2008).

On a molecular level, hormones and neurotransmitters are the neurophysiological mechanisms involved in emotional regulation. The most widely used substances in such processes include cortisol, serotonin, and dopamine (Eslinger et al., 2021). In this regard, dopamine controls motivation and reward-related emotional responses, and serotonin is responsible for general mood balance and maintaining emotional stability (Eslinger et al., 2021; Seo et al., 2008). High levels of stress chemicals may impair communications between the amygdala and the PFC, disrupting a person's behavioral control of their emotions (McEwen & Morrison, 2013). These chemical interfaces directly affect one's capacity for learning and memory, relationships, and even neuroplastic processes of the brain, including synaptic transmission (Arnsten, 2009).

Learning ability can be especially compromised by the over-activation of brain regions like the amygdala, which are vulnerable to poorly managed emotional stress and linked to neuroplastic alterations (Järvenoja et al., 2019; Martin & Oschner, 2016). The influence of emotional regulation on learning during development has often been described, starting from early childhood to old age. The children's brains are very neuroplastic and sensitive to outer stimulation (Marzola et al., 2023). Since those areas of the brain responsible for emotional regulation are still in the process of maturation during the early years, newborn babies have difficulty dealing with stressful situations or uncomfortable feelings. At this stage, the brain's limbic system, amygdala, and prefrontal cortex are still developing and extremely sensitive to external stimuli (He et al., 2018; Marzola et al., 2023; Salzman & Fusi, 2010). Consequently, newborns have difficulty managing their emotional responses and have ineffective coping mechanisms. Such brain structures develop appropriately if critical emotional support is provided through secure and supportive relationships with caregivers.

The caregiver can help the infants build emotional regulation capabilities by creating an ambient and reassuring environment; this soothes their stress responses and gives a feeling of security. Infants experience stress because they have not learned to control emotional arousal (Buhler-Wassmann & Hibel, 2021). The comforting and protective behavior of caregivers reduces stress hormone levels, including cortisol, allowing the baby to relax. In such a way, an infant's comfort and sense of trust are enhanced, helping him or her to regulate emotional responses better (Buhler-Wassmann & Hibel, 2021; Eslinger et al., 2021). In secure attachment relationships, the amygdala, which processes emotions, and the prefrontal cortex, which regulates emotions, both develop. The synaptic solid connection that is developed between these structures serves later in life to maintain emotional balance (Buhler-Wassmann & Hibel, 2021).

Additionally, in an environment with secure attachment relationships, synaptic connections are strengthened due to the brain's heightened neuroplastic potential. The newborn's brain receives positive signals for the soothing and regular responses of the carers. The stronger these neural cell connections in the brain are, the more the experience is repeated. More in support of that, such synaptic connectivity with structures implicated in emotional awareness includes the anterior cingulate cortex and the hippocampus implicated in memory (Kadović et al., 2023). This support of synaptic connectivity encourages the processing and regulation of emotional information. The strength of such synaptic connections enhances the capability of the brain to learn and, eventually, allows the newborn to handle increasingly complex emotional and cognitive tasks. On the bedrock of this constant support that caregivers provide, emotional regulatory capabilities are laid down as part and parcel of the developing brain (Buhler-Wassmann & Hibel, 2021; Giuliani et al., 2011).

Moreover, the brain's neuroplasticity is heightened where synaptic connections are reinforced. The baby's brain gets positive signals from the caregivers' soothing and predictable behavior. The more times an experience is repeated, the stronger the neural cell connections are in the brain (Etkin et al., 2015). This synaptic connection allows the structure of memory, such as the hippocampus, and the structure of emotional awareness, such as the anterior cingulate cortex, to process and control emotional information. The strength of these synaptic connections gives way to the brain's capacity for learning and, in time, allows the infant to progressively execute more challenging emotional and cognitive tasks (Gaviria et al., 2020). The secure base caregivers offer, to which emotional regulation skills are intrinsically linked, is ingrained in the growing brain. While neuroplastic capacity declines with age, emotional control abilities generally improve. Whereas learning is complex in older individuals due to age-related changes in brain structure, older individuals can regulate their emotional experiences better (Isaacowitz, 2022).

Emotional control at this stage may serve as support for learning processes. Neuroplastic processes support learning and maintenance of cognitive skills more when older adults can manage emotional stress. Neuroplasticity, therefore, affects learning and emotional control differently depending on the age and elasticity of an individual's brain.

7. Neuroplasticity and Aging in Learning

Neuroplasticity and aging affect the molecular, neuroanatomical, and neurophysiological levels of learning and adaptation processes. The brain's declining adaptability with advancing age has consequently hindered learning and cognitive performance. Neurotrophic factors and neurotransmitter systems determine neuroplasticity at the molecular level; however, with developing age, their chemical levels, such as BDNF, have reduced value that influences synaptic plasticity and learning capability. The neurochemical changes caused by the decline of neurotransmitters with age, including dopamine, acetylcholine, and serotonin, may disturb motivation, emotional functioning, and reward-based learning (Marzola et al., 2023).

On the other hand, aging-related enhancement in oxidative stress and inflammation disrupts brain functioning by destroying neuroplasticity. Oxidative stress is one of the main biological features significantly contributing to the decline in learning capabilities and neuroplasticity associated with aging. Excessive amounts of free radicals, or ROS, destroy cellular structures. However, adequate quantities are required for cellular action and energy production. The high energy requirement makes the brain highly vulnerable to oxidative stress (Kandlur et al., 2020). Due to the increased generation of ROS with advancement in age and deterioration in the antioxidant defense mechanisms of the brain cells, there is an impairment in learning and memory. This impairment also includes damage to cellular organelles, DNA, and synaptic proteins. Due to long-term exposure to excessive oxidative stress, LTP and synaptic function pathways are impaired. Learning becomes problematic since oxidative damage interferes with synaptic signaling and neurotransmitter release (Massaad & Klann, 2011). Also, oxidative stress impairs synaptic connections in the hippocampus and prefrontal cortex, thereby accelerating cognitive loss in older adults and problems with retrieving previously learned information and memory functions. Neurogenesis in the hippocampus decreases with age and is further suppressed by rising levels of oxidative stress. As neurogenesis is a vital constituent in learning and adaptation, the rise in oxidative stress levels may result in a decline in the ability of an individual to learn something new and eventually adapt to the changing world (Serrano & Klann, 2004).

Over-activation of immune cells, such as microglia and astrocytes, leads to nervous system injury or neurodegenerative diseases connected with neuroinflammation. Neuroinflammation denotes the inflammation of either the brain or the central nervous system. The interaction of high levels of oxidative stress, coupled with neuroinflammation, can lead to aging-related decreases in cognitive function and learning ability. Neuroinflammation can increase oxidative stress levels that insult the brain tissue. The presence of this vicious cycle suppresses neurogenesis and accelerates the course of neurodegenerative diseases. Increased oxidative stress can inhibit the antioxidant defense mechanisms of the brain, leading to an inflammatory response. Immune cells activated by free radicals initiate neuroinflammation. Neuroinflammation and oxidative stress can permanently cause damage to the neurological system by destroying cognitive capabilities (Teleanu et al., 2022a). Oxidative stress serves as one of the driving forces of cognitive decline in this respect. Chronic neuroinflammation might cause a long-standing decline in neurogenesis and subsequent impairment of brain functions.

In contrast, short-term and controlled inflammation might be necessary for the recovery of damaged brain tissue. Thus, neuroinflammation impairs hippocampal neurogenesis via interference with the signaling pathways critical for the proliferation and growth of new neurons. Such inflammatory cytokines as Tumor Necrosis Factor alpha (TNF-α)and Interleukin 6 (IL-6) impede, in turn, the proliferation and differentiation of neural stem cells, which are vital for neurogenesis. These processes could result from the propagation of reactive oxygen species or free radicals, leading to oxidative stress that might enhance inflammatory response and neuroinflammation (Teleanu et al., 2022a). Glial cells mainly include astrocytes and microglia that get activated by injury or stress to release inflammatory cytokines and other chemicals. Chronic neuroinflammation, however, eats away at synaptic connections and threatens neuronal health.

Despite all these adverse outcomes, the bottom line is that there is a way to reduce oxidative stress and neuroinflammation. Antioxidant nutrition, continued exercise, and mental activity potentiate brain health and neuroplasticity (Ciancarelli et al., 2023). Preserving learning ability can be further supported by the anti-inflammatory response of nutrition. The interplay between neuroinflammation and oxidative stress will be crucial in adequately managing aging-related cognitive impairments. It is associated with neuroanatomical changes at a structural level, including cortical thinning, white matter variation, and hippocampal degeneration (Serrano & Klann, 2004). Losses in synaptic connections and the brain volume indicate that such losses affect learning ability. The neurophysiological levels include the disturbances in the fundamentals of learning and memory processes, constituting LTD and LTP. Learning processing becomes more challenging as LTP decline impairs neurotransmitter release and synaptic connectivity.

Moreover, with aging, neurogenesis further declines, and synaptic density in the hippocampus is reduced, which makes the person less capable of learning and adapting (Kandel et al., 2021). Regular physical and mental exercises and lifelong learning stimulate the production of neurotrophic factors, which reinforce synapses. Diets rich in antioxidants maintain health in the brain, while social engagement positively influences neuroplasticity, enhancing the capability to learn. These are important in maintaining learning and adaptability into older age (Ciancarelli et al., 2023).

8. Neuroplasticity and Learning Disabilities

Some neurological disabilities, referred to as learning disabilities, affect selective areas of learning. Areas that are affected by learning disabilities include talking, writing, listening, thinking, mathematics, and science. For example, a learning disability, dyslexia, affects an individual's abilities with reading and understanding written language; dysgraphia affects an individual's writing and thoughts on paper, while dyscalculia deals with problems associated with arithmetic and an understanding of numbers. One of the other learning disabilities is auditory processing disability, or APD, where individuals have difficulty processing what they hear (Ramdhane, 2024). Learning difficulties are not linked to indolence or low intelligence Quotient (IQ); instead, they are a consequence of the difference in the functioning of each brain, as, generally, intellect is seldom affected by learning problems (Fletcher-Janzen & Reynolds, 2008; Ramdhane, 2024). These disabilities emanate from how the brain acquires and manages knowledge. Since neuroplasticity is the foundation of the capability of the brain to adapt or change, it makes learning possible due to synaptic density changes, the formation and removal of neuronal connections, and the remodeling of the brain's architecture. Insufficiencies in neuroplasticity are directly leading to the appearance of learning disabilities such as dyslexia, dyscalculia, and dysgraphia. The failures in synaptic strengthening and disruptions in neural transmissions are the primary roots of these disabilities (Battro et al., 2011; Galván, 2010). These conditions usually start from school and follow into adulthood. Along with instruction that enhances neuroplasticity, early diagnosis can diminish some of these effects and enhance the learning capability of an individual (Fletcher-Janzen & Reynolds, 2008). The relation of neuroplasticity to some of the learning disabilities will be described in more detail below.

8.1. Neuroplasticity and Dyslexia

Dyslexia is the most common learning disability that disrupts language and reading ability. Anatomically, studies of dyslexia have shown structural and functional differences in the brain's left inferior frontal gyrus (Broca's area) and left temporoparietal area in those affected by this disability. LTP generally involves strengthening synaptic relationships and is typically found across normal development in these regions, while dyslexia may diminish neuroplastic mechanisms (Battro et al., 2011). Moreover, people with dyslexia have shown lower white matter density and less connectivity between such cortical regions, including the corpus callosum, which enables inter-hemispheric transfer (Norton et al., 2015). The functions of neurotransmitters, including glutamate and GABA, demonstrate the neurochemical underpinnings of a relationship that might exist between dyslexia and neuroplasticity. The excitatory glutamate neurotransmitter triggers processes that increase synapses, while the inhibitory neurotransmitter GABA supports activities that decrease or stabilize synapses. Imbalances among these two neurotransmitters in dyslexic individuals may result in insufficient strengthening of synapses; hence, learning ability is impaired (Kershner, 2019). Neurophysiologically, aberrant shifts in the frequencies of brain waves have been observed; more specifically, disruptions in alpha and beta rhythms during phonological processing are thought to indicate defective cortical plasticity (Cainelli et al., 2023).

8.2. Neuroplasticity and Dyscalculia

Dyscalculia is a learning problem wherein a person's understanding and performance of mathematical activities are impaired (Battro et al., 2011). Neuroanatomical studies have implicated the parietal lobe, especially the intraparietal sulcus, in numerical processing. Dyscalculic individuals have demonstrated reduced synapse density and changes in cortical thickness. Such findings indicate an impairment in neuroplasticity processes, implying that neural networks underlying numerical functioning would never fully develop (Ramdhane, 2024). Dyscalculia can be attributed to learning difficulties on a molecular basis because of impaired synaptic plasticity. In neurons, calcium ions are considered the basis of synaptic plasticity. A fluctuation in intracellular calcium ion concentration can trigger synaptic strengthening, LTP, or synaptic weakening, LTD. Alterations in the balance of calcium ions may work to restrict neuroplasticity in dyscalculia, making the learning of numbers, in particular, very challenging (Iuculano et al., 2015; Kaufmann, 2008; McCaskey et al., 2020). Besides, dyscalculia is justified to be engendered by irregularities in the brain's dopaminergic system. Dopamine influences reward-motivation processes, enhancing the ability to learn. In dyscalculic subjects, this may inhibit the development of the mathematical ability mediocrity in the function of the above system (Iuculano et al., 2015; Ramdhane, 2024).

8.3. Neuroplasticity and Dysgraphia

One writing-related learning disability is termed dysgraphia. This neuroanatomically based disability involves the frontal lobe and premotor cortex, two brain areas that regulate an individual's motor components (Battro et al., 2011). The same regions control the use of fine motor skills during writing. Individuals with dysgraphia show insufficient synaptic connectivity in these regions. The flexibility of the motor cortex is one of the critical players in the development of writing ability. Writing repeatedly from neurons forms a motor memory through synaptic activity (Ramdhane, 2024). In dysgraphia situations, this mechanism may remain undeveloped, and writing difficulties could result from a lack of neuroplastic adaptation. The involvement of dopamine and acetylcholine systems in neuroplasticity accompanying dysgraphia can be discussed at a molecular level. Dopamine is critical for the initial learning of skills and the refinement of movements. People with dysgraphia will find writing more labored because their motor control systems are less effective due to the insufficient activity of the dopamine system. In contrast, acetylcholine enhances synaptic plasticity by modulating attention and learning processes. In cases of dysgraphia, malfunctioning of this neurotransmitter system may lead to inadequately strong synaptic connections that would interfere with the full development of motor skills (Rapp & McCloskey, 2018).

8.4. Neuroplasticity and Auditory Processing Disorder (APD)

People with Auditory Processing Disorder (APD) have problems understanding spoken language, especially when there is noise, because of abnormalities in how their brains process sounds. These abnormalities might cause learning issues, especially in skills based on audition, such as reading, language comprehension, and spoken instructions (Ramdhane, 2024). Children with APD might mix up similar sounds, impeding their phonetic learning and making deciphering more complex auditory information, understanding conversations, or perceiving words much more difficult. These difficulties may be followed by a decline in academic performance, especially in classes that demand high language proficiency (Moore, 2007). Neuroplasticity, however, helps in learning and somewhat compensates for APD. Targeted interventions and auditory training exercises enhance neuroplasticity processes, allowing the brain to improve and change its processing (Anderson & Kraus, 2013). This process is molecularly based on the changes in synaptic plasticity, which are generally believed to involve the strengthening or weakening of connections between nerve cells, which are called synapses. During auditory training exercises, the consistently used pathways in the brain strengthen synaptic connectivity by initiating LTP. In LTP, molecular and structural changes encourage signal transmission between neuronal cells. The levels of neurotransmitters, especially excitatory neurotransmitters, increase, as does the sensitivity of synaptic receptors, including the NMDA and AMPA receptors, when synaptic transmission is enhanced.

The brain also sends signals of chemicals, like BDNF, responsible for causing the nerve cells to grow, cling together, and forming new synaptic pathways. Such a signaling process could perhaps lead to the reorganization of the synaptic network, especially after extensive repeated training and auditory experiences (Ahmmed, 2020). The brain can change chemically by strengthening synaptic connections, making neurotransmitters and synaptic receptors more efficient, and enhancing the processing ability with auditory training. Because of neuroplasticity, a person with APD can strengthen auditory pathways and develop alternative ways of processing the sounds, enhancing learning capability. In the case of APD, early and frequent treatment is required to enhance neuroplasticity; for example, auditory training or speech therapy can make new learning more effective. Although APD seriously affects learning, neuroplasticity offers the opportunity for compensation and recovery by rewiring the pathways in the brain through specific therapies (Alanazi, 2023).

To sum up, brain neuroplasticity deficits are related to intimate learning disabilities such as dyslexia, dyscalculia, and dysgraphia. Deficiency in synaptic strength in various brain parts, chemical imbalances within neurotransmitters, and deficiency in neural communication have caused these disabilities. However, early therapies that support neuroplasticity can improve those problems and enable the person to recreate their learning capability.

9. Stress, Learning, and Neuroplasticity

Stress is a complex physiological and psychological reaction to a perceived danger or challenging situation. On the neurobiological side, stress results from an interaction among the hypothalamus, the pituitary gland, and the adrenal glands (Godoy et al., 2018). When danger is detected, it causes the hypothalamus to produce CRH (Corticotropin-Releasing Hormone), which orders the pituitary gland to produce ACTH (Adrenocorticotropic Hormone). Subsequently, the latter instructs the adrenal glands to produce cortisol and other stress hormones (Chu et al., 2019). Psychologically, stress has been linked with a person's trepidation and apprehension in reaction to environmental stimuli that can affect their mood and thought processes. Acute stress is defined as events occurring all of a sudden and being transient. Physically, this involves high blood pressure, rapid breathing, musculoskeletal tension, and an increased heart rate.

In response to a life-threatening condition, the body rapidly delivers nutrients and oxygen while slowing digestion (Godoy et al., 2018). Chronic stress, however, is a long-lasting condition and consists of some adverse physiological consequences. Long-term stress impairs immunity due to the high levels of cortisol.

The excess amount of cortisol can cause many health disabilities. Apart from these, metabolic changes, impaired memory and learning abilities, as well as the onset of psychological disorders, such as anxiety disorders, may also originate from them (Montgomery, 2024). Stress is associated with a complex psychological and neurological influence that impacts the general health of an individual. Acute stress may trigger an adaptive response in the short run; however, chronic stress can give rise to long-term health disabilities (Brivio et al., 2020). Stress management and learning various coping techniques are, therefore, crucial to being in a healthy state. Let us now consider the connection between stress, the learning process, and neuroplasticity.

In that state, stress exerts interference upon the balancing ability of the brain at both molecular and anatomical levels; it interferes negatively with learning and neuroplasticity. One of the significant effects of stress is that it heightens cortisol production within the body. Cortisol, through its harmful effects on hippocampal neurons, decreases the quantity of BDNF produced within this area. BDNF significantly supports the development, maintenance, and creation of new connections between nerve cells (Brivio et al., 2020). Stress-induced reduction in the level of BDNF interferes with the ability of neurons to make synaptic connections; hence, learning becomes difficult (Miao et al., 2020). Besides this, glutamate release due to stress in excessive amounts causes damage to the nerve cells and results in an imbalance in the nerve cells due to over-excitation (Popoli et al., 2013). The GABA pathway is also significantly affected by stress. GABA synthesis and release may be reduced during acute stress conditions, maintaining a balance in neuronal excitation as it is an inhibitory neurotransmitter in the brain (Brivio et al., 2020; Hepsomali et al., 2020). It causes a disturbance in the system that regulates the excitation of the brain, which results in much higher activity than usual. Owing to this imbalance, there is a deficit in cognitive and learning processes (Córdova et al., 2023; Vogel & Schwabe, 2016). Chronic stress has been found to change the neuroanatomical structure of the hippocampus, prefrontal cortex, and amygdala, among other brain structures. Hippocampal volume may be reduced under prolonged stress along with neuronal loss while simultaneously weakening synaptic connections in the prefrontal cortex, which may affect executive processes such as planning and decision-making (Montgomery, 2024). Stress is known to increase the amygdala activity, the part of the brain that deals with emotional response, making it more difficult to control emotions. It also makes learning and remembering particularly challenging (McEwen et al., 2015). Neurotransmitter systems, typically dopamine, serotonin, and norepinephrine, become out of balance during stressful periods and impede the brain from carrying on with its work. The increased level of cortisol during stressful times impedes the release of dopamine. Lower dopamine levels due to stress negatively impact learning capacity and motivation because dopamine is vital in reward, motivation, and learning processes (Hepsomali et al., 2020; Mora et al., 2012; Popoli et al., 2013).

Similarly, stress lowers the levels of serotonin in the body. Low levels of this neurotransmitter are characterized by emotional instability, anxiety, and sadness, factors that impede one's ability to concentrate or learn. Serotonin level is crucial in regulating mood and maintaining emotional stability (Den Buuse & Hale, 2019; Mora et al., 2012). Severe stress results in failure of a decline in the level of norepinephrine, which is the neurotransmitter involved in release during the activation of the body's response to stress (Mora et al., 2012; Morilak et al., 2005). This response could develop into a perpetual state of arousal, leading to overactivity in the nervous system, disrupting learning, and making one unable to focus. Stress-induced imbalances within these neurotransmitter systems significantly impair general learning and cognitive functioning. By agitating these systems, stress may lower motivation and prevent the positive reinforcement required for learning. Further decreases in serotonin create emotional instability, which further worsens the adverse effects of stress on learning.

10. Sleep, Learning, and Neuroplasticity

Sleep helps brain functions related to learning and memory; it is related to neuroplasticity, which means that the importance of sleep is amplified. When we sleep in bed, we are not just sleeping; our brains summarize information and strengthen and process during sleep. These very crucial functions of remembering and learning are well-supported by sleep. These very crucial functions of remembering and learning are well-supported by sleep. Sleep consolidates and transforms the daytime experience by evoking deep and intricate systems that cement knowledge (Maquet, 2001). Other studies have shown how sleep interacts with learning: sleep onset is a time of ordering, synthesizing, and consolidating new knowledge within the brain (Wong et al., 2013). The backbone of the latter processes, which occur whenever one is resting, is considered the default mode network (DMN). DMN controls the internal mental processes that a person passes through, memories, and thoughts to link or understand newly learned information easily (Vidal-Pineiro et al., 2014). The DMN plays a vital role in digesting daily gained experiences through sleep. For example, DMN is active during REM sleep and supports the transfer of information to memory (Di et al., 2024). A change in this direction is particularly effective in enhancing one's creative thinking and problem-solving potential (Poerio et al., 2017).

Further, DMN synthesizes and solidifies the knowledge gained during the day, which assists in achieving detailed insight into things during slow-wave sleep (Wamsley & Stickgold, 2011). Enhanced DMN functioning means enhanced cognitive flexibility in the brain. On the other hand, comprehension and application of newly learned information are also improved (Safron et al., 2022). Therefore, sleep supports learning and maintains normal brain functions, especially the DMN. On the other hand, sleep deprivation can damage all these cognitive functions and interfere with this network, slowing down the learning process (Mi et al., 2023). The most critical role the DMN probably plays in further explaining the interrelationship of sleep and learning is that adequate sleep is vital for maintaining good mental health. The two stages of sleep, REM and NREM, support different forms of learning. While NREM sleep supports memory consolidation, REM sleep enhances creativity. Adequate sleep bolsters neuroplasticity, or the brain's ability to synthesize knowledge acquired during the day to consolidate it further and enhance synaptic connectivity. The synaptic homeostasis hypothesis holds that essential connections are strengthened during sleep, and unnecessary connections are pruned (Sippel et al., 2020). These strengthened and pruned connections have been particularly shown during REM and slow-wave sleep; the latter phase has changes associated with novel knowledge to render information stable over an extended period. In a more general interpretation, sleep strengthens the performance of brain regions, such as the prefrontal cortex, which is involved in cognitive control and decision-making. Well-qualified sleep heightens concentration and, therefore, improves academic results.

Experimental evidence shows that well-rested individuals possess a better learning and memorizing capacity than those who do not get enough sleep. In other words, sleep, neuroplasticity, and learning are related, so this capability is guaranteed to be maximum (Kavanau & Ph, 2000; Ono et al., 2020). Qualified sleep promotes brain activities and solidifies information for permanence even before and after periods of learning. Sleeping better will be necessary for cognitive enhancement in learning abilities (Maha Putra et al., 2023).

11. The Future of Learning and Neuroplasticity Research

Researchers on neuroplasticity and learning have developed landmark findings about the nature of brain activity as a dynamic process. Neuroplasticity has thoroughly refuted the traditional view that brain development stabilizes at a certain age.. Experiences, learning, trauma, and other events cause neural connections to shift and change moment by moment (Voss et al., 2017). This aspect is brought out by the brain's ability to reorganize itself through learning. This reorganization ability makes the brain change; hence, it is paramount in acquiring new knowledge and skills, and equally important in the recovery from head trauma. Current research on neuroplasticity concerning learning provides new lessons necessary to maximize people's ability to learn at all stages of life. Investigations into the critical periods of learning processes, especially, show that an individual is better off mastering specific skills at a particular age (Marzola et al., 2023). While these are the critical periods for the most significant changes, the brain can still evolve after these times, but at a more gradual rate and shallow in breadth.

The acquisition of knowledge, understanding, or experience is widely recognized as a phenomenon involving morphological changes in the brain. These changes indicate that learning and neuroplasticity are closely related, as learning is a process that leads to the strengthening or weakening of synaptic connections in the brain, a fundamental principle of neuroplasticity (Gariépy et al., 2020). These changes, therefore, imply that learning brings about measurable neural effects. The relationship between neuroplasticity and learning can also explain individual differences. In this respect, neuronal connections in some people's brains are said to be directly related to their brains' flexibility and adaptability, enabling them to learn skills faster than others (Stee & Peigneux, 2021). The treatments directly affecting the brain are being researched increasingly, especially with new technology in this field. For example, neurofeedback accelerates the learning process through brain stimulation. The detailed investigation of the long-term effects of the methods on neuroplasticity holds excellent potential for improving learning skills (Bliss & Cooke, 2011).

Meanwhile, research into neuroplasticity offers a glimmer of hope for those who have struggled with learning disabilities. Neuroplasticity-based treatments target the reorganization of the brain as a way to alleviate neurodevelopmental disorders such as dyslexia and attention deficit hyperactivity disorder (ADHD) (Doyle, 2020). Such findings prove that individualized interventions possess immense potential within an educational context and further indicate that individuals can achieve their full learning potential.

Finally, neuroplasticity studies are leading us closer to a complete understanding of the dynamic nature of learning and the brain's adaptability to the learning process (Voss et al., 2017). What is going to be important in further research is how to maximize the capacity to learn in each individual and to provide more effective rehabilitation for individuals suffering from brain injury or neurodevelopmental problems.

12. Takeaways

1. Neuroplasticity comes in two flavors: structural and functional. While synaptic plasticity tends to be the most associated form of structural neuroplasticity, axonal sprouting, dendritic arborization, and neuronal migration are other forms.
2. The development, differentiation, survival, and functionality of nerve cells depend on neurotrophic substances. These elements fortify the synaptic connections that serve as the building blocks of memory and learning.
3. Neurotrophic factors control nerve cells' long-term health and function, whereas neurotransmitters facilitate fast signal transmission. For instance, by enhancing synaptic plasticity, substances like BDNF promote memory and learning capacity.
4. More specifically, our brains can learn and develop faster during childhood. These are the critical periods for acquiring skills and, most of all, language acquisition.

5. Synaptic development in childhood is fast, allowing learning motor abilities quickly. Learning is more difficult with age as synaptic plasticity decreases.
6. Early visual experiences are crucial in determining the course of maturation and refining processes in visual skills. Poor visual stimulation may bring about long-lasting visual deficits due to the weakening of synaptic connections.
7. Even though the brain's plasticity decreases with age, with appropriate stimuli, learning and adaptation continue even at later stages of life. This continuation forms a basis for maintaining or improving cognitive abilities.
8. Encoding, storage, and retrieval are the phases of forming memories. Many regions of the brain act together in both learning and memory.
9. The hippocampus strengthens synaptic connections to solidify short-term memory into long-term memory. Synaptic plasticity strengthens the synaptic connections between the neurons, which enables learning.
10. In controlling complex cognitive function, new interneural connections and strengthened old ones develop in the prefrontal cortex. This process forms the necessary changes for adapting life to its various challenges.
11. Synaptic plasticity dynamically controls neuronal connections. Both structural and functional changes in the brain give rise to efficient learning and memory processes.
12. A person with emotional regulation can better monitor and manage his or her emotions. This management helps them concentrate and make better decisions while performing mental tasks.
13. Neuroplasticity, involving memory and learning, might be supported by the brain's control systems of emotions. Emotional balance also aids in establishing new synaptic connections.
14. This neuroplasticity of the brain to outer inputs is very high at the beginning of life. Safe and supportive interactions positively influence the development of emotional control abilities.
15. With age, the ability to adapt and learn is altered negatively. Neuroinflammation and oxidative stress result in deficiencies in learning and memory functions.

16. Increased oxidative stress results in damage to cellular structure and disrupts the pathways of learning. This condition may reduce memory and learning in older individuals.
17. Frequent mental and physical activity and a diet rich in antioxidants enhance neuroplasticity, maintaining this organ's functions. Similarly, social connectedness while aging translates into better learning and adaptability.
18. Changes in brain function are due to whatever IQ level relates to learning problems. Thus, early diagnosis and treatment could increase synaptic changes, enabling the brain to learn.
19. Dyslexia, dyscalculia, and dysgraphia are learning disorders caused by neurotransmitter imbalance and poor synaptic connections. Training that encourages neuroplasticity can, however, reduce these problems.
20. Stress heightens cortisol excretion within the hippocampus, decreasing the amount of BDNF within it. This decrease obviously prevents new neural connections from sprouting, making learning difficult. Overstimulation and damage to the nerve cells also follow due to excessive glutamate release.
21. Chronic stress alters the morphology of the amygdala, prefrontal cortex, and hippocampal area, thus negatively influencing learning and memory, as well as emotional regulation. Learning potential is decreased as it interferes with motivation and cognitive processes.
22. Sleep is vital in pre- and post-learning, as it is related to neuroplasticity. Neuroplasticity helps in sleeping, learning, and memorizing processes. While sleeping, our brain consolidates, reinforces, and processes the knowledge learned during the day.
23. While NREM sleep is helpful for memory consolidation, REM sleep enhances a person's ability to be creative. REM and NREM sleep support different kinds of learning. Sleep has been proven to enhance neuroplasticity, which then serves to strengthen synaptic connections.
24. Long-term neuroplasticity effects are studied, and interventions like brain stimulation and neurofeedback are applied to accelerate learning. Neuroplasticity-based treatments appear promising in the remediation of neurodevelopmental disorders like dyslexia and ADHD.

Abbreviations

APD: Auditory processing disorder
ADHD: Attention deficit hyperactivity disorder
BDNF: Brain-derived neurotrophic factor
CRH: Corticotropin-releasing hormone
ACTH: Adrenocorticotropic hormone
LTD: Long-term depression
LTP: Long-term potentiation
GABA: Gamma-aminobutyric acid
PKC: Protein kinase C
CaMKII: Calcium/Calmodulin-dependent protein kinase II
cAMP: Cyclic adenosine monophosphate
PKA: Protein kinase A
NOS: Nitric oxide synthase
VTA: Ventral tegmental area
NAc: Nucleus accumbens
pSTG: Posterior superior temporal gyrus

Bibliography

1. Abel, T., & Nguyen, P. V. (2014). Regulation of hippocampus-dependent memory by cyclic AMP-dependent protein kinase. *NIH Public Access*, *23*(1), 1–7. <https://doi.org/10.1016/S0079-6123(07)00006-4>

2. Abraham, A., & Collins, D. (2011). Effective skill development: How should athletes' skills be developed. In *Performance Psychology* (Issue January 2011). <https://doi.org/10.1016/B978-0-443-06734-1.00015-8>

3. Aguilar-Mediavilla, E., Pérez-Pereira, M., Serrat-Sellabona, E., & Adrover-Roig, D. (2022). Introduction to Language Development in Children: Description to Detect and Prevent Language Difficulties. *Children*, *9*(3), 1–7. <https://doi.org/10.3390/children9030412>

4. Ahmmed, A. U. (2020). Auditory processing, co-morbidities, and parental report of sleep disturbance in children with auditory processing disorder (APD). *International Journal of Pediatric Otorhinolaryngology*, *135*(May), 110117. <https://doi.org/10.1016/j.ijporl.2020.110117>

5. Alanazi, A. A. (2023). Understanding Auditory Processing Disorder: A Narrative Review. *Saudi Journal of Medicine and Medical Sciences*, *11*(4), 275–282. <https://doi.org/10.4103/sjmms.sjmms_218_23>
6. Anderson, S., & Kraus, N. (2013). Auditory Training: Evidence for Neural Plasticity in Older Adults. *Perspectives on Hearing and Hearing Disorders Research and Diagnostics*, *17*(1), 37. <https://doi.org/10.1044/hhd17.1.37>
7. Arnsten, A. (Yale). (2009). stress signalling pathways and PFC. *Molecular Biology*, *26*(2), 148. <https://doi.org/10.1038/nrn2648>
8. Bailey, J. A., & Penhune, V. B. (2013). The relationship between the age of onset of musical training and rhythm synchronization performance: validation of sensitive period effects. *Frontiers in Neuroscience*, 7(November), 1–9. <https://doi.org/10.3389/fnins.2013.00227>
9. Bang, D., Kishida, K. T., Lohrenz, T., White, J. P., Laxton, A. W., Tatter, S. B., Fleming, S. M., & Montague, P. R. (2020). Sub-second Dopamine and Serotonin Signaling in Human Striatum during Perceptual Decision-Making. *Neuron*, *108*(5), 999-1010.e6. <https://doi.org/10.1016/j.neuron.2020.09.015>
10. Bathina, S., & Das, U. N. (2015). Brain-derived neurotrophic factor and its clinical Implications. *Archives of Medical Science*, *11*(6), 1164–1178. <https://doi.org/10.5114/aoms.2015.56342>
11. Battro, A. M., Dehaene, S., & Singer, W. J. (2011). *Human Neuroplasticity and Education* (Issue October 2010).
12. Berridge, C. W., & Spencer, R. C. (2016). Differential cognitive actions of norepinephrine a2 and a1 receptor signaling in the prefrontal cortex. *Brain Research*, *1641*, 189–196. <https://doi.org/10.1016/j.brainres.2015.11.024>
13. Bialystok, E., & Poarch, G. (2015). Language Experience Changes Language and Cognitive Ability. *NIH Public Access*, *17*(3), 1–13. <https://doi.org/10.1007/s11618-014-0491-8>
14. Bliss, T. V. P., & Cooke, S. F. (2011). Long-term potentiation and long-term depression: A clinical perspective. *Clinics*, *66*(SUPPL.1), 3–17. <https://doi.org/10.1590/S1807-59322011001300002>
15. Brivio, P., Sbrini, G., Riva, M. A., & Calabrese, F. (2020). Acute Stress Induces Cognitive Improvement in the Novel Object Recognition Task by Transiently Modulating Bdnf in the Prefrontal Cortex of Male

Rats. *Cellular and Molecular Neurobiology*, *40*(6), 1037–1047. <https://doi.org/10.1007/s10571-020-00793-7>

16. Brown, J. C., Higgins, E. S., & George, M. S. (2022). Synaptic Plasticity 101: The Story of the AMPA Receptor for the Brain Stimulation Practitioner. *Neuromodulation*, *25*(8), 1289–1298. <https://doi.org/10.1016/j.neurom.2021.09.003>
17. Buhler-Wassmann, A. C., & Hibel, L. C. (2021). Studying caregiver-infant co-regulation in dynamic, diverse cultural contexts: A call to action. *Infant Behavior and Development*, *64*, 1–29. <https://doi.org/10.1016/j.infbeh.2021.101586>
18. Cainelli, E., Vedovelli, L., Carretti, B., & Bisiacchi, P. (2023). EEG correlates of developmental dyslexia: a systematic review. *Annals of Dyslexia*, *73*(2), 184–213. <https://doi.org/10.1007/s11881-022-00273-1>
19. Carhart-Harris, R. L., & Nutt, D. J. (2017). Serotonin and brain function: A tale of two receptors. *Journal of Psychopharmacology*, *31*(9), 1091–1120. <https://doi.org/10.1177/0269881117725915>
20. Christopher P Ford. (2014). The Role of D2-Autoreceptors in Regulating Dopamine Neuron Activity and Transmission. *Neuroscience*, *282*, 13–22. <https://doi.org/10.1016/j.neuroscience.2014.01.025>
21. Chu, B., Marwaha, K., & Ayers, D. (2019). Physiology, Stress Reaction. *StatPearls*. <http://www.ncbi.nlm.nih.gov/pubmed/31082164>
22. Ciancarelli, I., Morone, G., Iosa, M., Cerasa, A., Calabrò, R. S., Iolascon, G., Gimigliano, F., Tonin, P., & Tozzi Ciancarelli, M. G. (2023). Influence of Oxidative Stress and Inflammation on Nutritional Status and Neural Plasticity: New Perspectives on Post-Stroke Neurorehabilitative Outcome. *Nutrients*, *15*(1), 1–18. <https://doi.org/10.3390/nu15010108>
23. Córdova, A., Caballero-García, A., Drobnic, F., Roche, E., & Noriega, D. C. (2023). Influence of Stress and Emotions in the Learning Process: The Example of COVID-19 on University Students: A Narrative Review. *Healthcare (Switzerland)*, *11*(12), 1–15. <https://doi.org/10.3390/healthcare11121787>
24. Den Buuse, M., & Hale, M. W. (2019). Serotonin in stress. In *Stress: Physiology, Biochemistry, and Pathology Handbook of Stress Series, Volume 3*. Elsevier Inc. <https://doi.org/10.1016/B978-0-12-813146-6.00010-2>

25. Denny, B. T., Fan, J., Liu, X., Guerreri, S., Mayson, S. J., Rimsky, L., New, A. S., Siever, L. J., & Koenigsberg, H. W. (2014). Insula-amygdala functional connectivity is correlated with habituation to repeated negative images. *Social Cognitive and Affective Neuroscience*, *9*(11), 1660–1667. <https://doi.org/10.1093/scan/nst160>

26. Di, T., Zhang, L., Meng, S., Liu, W., Guo, Y., Zheng, E., Xie, C., Xiang, S., Jia, T., Lu, L., Sun, Y., & Shi, J. (2024). The impact of REM sleep loss on human brain connectivity. *Translational Psychiatry*, *14*(1). <https://doi.org/10.1038/s41398-024-02985-x>

27. Dixon, M. L., Thiruchselvam, R., Todd, R., & Christoff, K. (2017). Emotion and the prefrontal cortex: An integrative review. *Psychological Bulletin*, *143*(10), 1033–1081. <https://doi.org/10.1037/bul0000096>

28. Dobielska, M., Bartosik, N. K., Zyzik, K. A., Kowalczyk, E., & Karbownik, M. S. (2022). Mechanisms of Cognitive Impairment in Depression. May Probiotics Help? *Frontiers in Psychiatry*, *13*(June), 1–17. <https://doi.org/10.3389/fpsyt.2022.904426>

29. Dumont, E., Syurina, E. V., Feron, F. J. M., & van Hooren, S. (2017). Music interventions and child development: A critical review and further directions. *Frontiers in Psychology*, *8*(SEP). <https://doi.org/10.3389/fpsyg.2017.01694>

30. Eiden, L. E., Hernández, V. S., Jiang, S. Z., & Zhang, L. (2022). Neuropeptides and small-molecule amine transmitters: cooperative signaling in the nervous system. *Cellular and Molecular Life Sciences*, *79*(9), 1–30. <https://doi.org/10.1007/s00018-022-04451-7>

31. Erunuma, B. M. T. (2018). *Review Diversity of structure and function of GABA B receptors : a complexity of GABA B -mediated signaling. 94*, 390–411.

32. Eslinger, P. J., Anders, S., Ballarini, T., Boutros, S., Krach, S., Mayer, A. V., Moll, J., Newton, T. L., Schroeter, M. L., de Oliveira-Souza, R., Raber, J., Sullivan, G. B., Swain, J. E., Lowe, L., & Zahn, R. (2021). The neuroscience of social feelings: mechanisms of adaptive social functioning. *Neuroscience and Biobehavioral Reviews*, *128*, 592–620. <https://doi.org/10.1016/j.neubiorev.2021.05.028>

33. Espinosa, F., & Kavalali, E. T. (2009). NMDA receptor activation by spontaneous glutamatergic neurotransmission. *Journal of*

Neurophysiology, *101*(5), 2290–2296. <https://doi.org/10.1152/jn.90754.2008>

34. Etkin, A., Büchel, C., & Gross, J. J. (2015). The neural bases of emotion regulation. *Nature Reviews Neuroscience*, *16*(11), 693–700. <https://doi.org/10.1038/nrn4044>
35. Excitatory, H., John Mihic, S., & Adron Harris, R. (1997). Neurotransmitter Review. *Journal of Neurochemistry*, *21*(2), 712–715.
36. Fletcher-Janzen, E. [Ed], & Reynolds, C. R. [Ed]. (2008). Neuropsychological perspectives on learning disabilities in the era of RTI: Recommendations for diagnosis and intervention. *Neuropsychological Perspectives on Learning Disabilities in the Era of RTI: Recommendations for Diagnosis and Intervention*. <http://ovidsp.ovid.com/ovidweb.cgi?T=JS&PAGE=reference&D=psyc6&NEWS=N&AN=2008-01999-000>
37. Galván, A. (2010). Neural plasticity of development and learning. *Human Brain Mapping*, *31*(6), 879–890. <https://doi.org/10.1002/hbm.21029>
38. Gariépy, J. L., Bailey, D. B., & Holochwost, S. J. (2020). Critical Periods. In *Encyclopedia of Infant and Early Childhood Development* (Vol. 1). Elsevier. <https://doi.org/10.1016/B978-0-12-809324-5.23626-2>
39. Gaviria, M., Celeghin, A., Michael-Titus, A. T., & Pallier, P. N. (2020). Editorial: Brain Plasticity and Contribution of the Emotional Brain to Neural Remodelling After Injury. *Frontiers in Neurology*, *11*(October), 10–11. <https://doi.org/10.3389/fneur.2020.606271>
40. Ghashghaei, H. T., Lai, C., & Anton, E. S. (2007). Neuronal migration in the adult brain: Are we there yet? *Nature Reviews Neuroscience*, *8*(2), 141–151. <https://doi.org/10.1038/nrn2074>
41. Giuliani, N. R., Drabant, E. M., Bhatnagar, R., & Gross, J. J. (2011). Emotion regulation and brain plasticity: Expressive suppression use predicts anterior insula volume. *NeuroImage*, *58*(1), 10–15. <https://doi.org/10.1016/j.neuroimage.2011.06.028>
42. Godoy, L. D., Rossignoli, M. T., Delfino-Pereira, P., Garcia-Cairasco, N., & Umeoka, E. H. de L. (2018). A comprehensive overview on stress neurobiology: Basic concepts and clinical implications. *Frontiers in Behavioral Neuroscience*, *12*(July), 1–23. <https://doi.org/10.3389/fnbeh.2018.00127>

43. Goto, A. (2022). Synaptic plasticity during systems memory consolidation. *Neuroscience Research*, *183*(June), 1–6. <https://doi.org/10.1016/j.neures.2022.05.008>

44. Gupta, V. K., You, Y., Gupta, V. B., Klistorner, A., & Graham, S. L. (2013). TrkB receptor signalling: Implications in neurodegenerative, psychiatric and proliferative disorders. *International Journal of Molecular Sciences*, *14*(5), 10122–10142. <https://doi.org/10.3390/ijms140510122>

45. Hare, J. E. (2022). NMDA Receptor-Dependent Long-Term Potentiation and Long-Term Depression (LTP/LTD). *CSH Perspectives in Biology*, *105*(3), 337–349. <https://doi.org/10.1093/monist/onac004>

46. Hayes, E. A., Warrier, C. M., Nicol, T. G., Zecker, S. G., & Kraus, N. (2003). Neural plasticity following auditory training in children with learning problems. *Clinical Neurophysiology*, *114*(4), 673–684. <https://doi.org/10.1016/S1388-2457(02)00414-5>

47. He, Z., Lin, Y., Xia, L., Liu, Z., Zhang, D., & Elliott, R. (2018). Critical role of the right VLPFC in emotional regulation of social exclusion: A tDCS study. *Social Cognitive and Affective Neuroscience*, *13*(4), 357–366. <https://doi.org/10.1093/scan/nsy026>

48. Hepsomali, P., Groeger, J. A., Nishihira, J., & Scholey, A. (2020). Effects of Oral Gamma-Aminobutyric Acid (GABA) Administration on Stress and Sleep in Humans: A Systematic Review. *Frontiers in Neuroscience*, *14*(September). <https://doi.org/10.3389/fnins.2020.00923>

49. Huang, E. J., & Reichardt, L. F. (2001). Neurotrophins: Roles in neuronal development and function. *Annual Review of Neuroscience*, *24*, 677–736. <https://doi.org/10.1146/annurev.neuro.24.1.677>

50. Ibáñez, C. F., Paratcha, G., & Ledda, F. (2020). RET-independent signaling by GDNF ligands and GFRα receptors. *Cell and Tissue Research*, *382*(1), 71–82. <https://doi.org/10.1007/s00441-020-03261-2>

51. Isaacowitz, D. M. (2022). What Do We Know About Aging and Emotion Regulation? *Perspectives on Psychological Science*, *17*(6), 1541–1555. <https://doi.org/10.1177/17456916211059819>

52. Iuculano, T., Rosenberg-Lee, M., Richardson, J., Tenison, C., Fuchs, L., Supekar, K., & Menon, V. (2015). Cognitive tutoring induces widespread neuroplasticity and remediates brain function in children

with mathematical learning disabilities. *Nature Communications*, *6*, 1–10. <https://doi.org/10.1038/ncomms9453>

53. Jan, Y. N., & Jan, L. Y. (2010). Branching out: Mechanisms of dendritic arborization. *Nature Reviews Neuroscience*, *11*(5), 316–328. https://doi.org/10.1038/nrn2836

54. Janet, R., Costes, N., Mérida, I., Derrington, E., & Dreher, J. C. (2023). Relationships between serotonin availability and frontolimbic response to fearful and threatening faces. *Scientific Reports*, *13*(1), 1–11. https://doi.org/10.1038/s41598-023-28667-0

55. Järvenoja, H., Näykki, P., & Törmänen, T. (2019). Emotional regulation in collaborative learning: when do higher education students activate group level regulation in the face of challenges? *Studies in Higher Education*, *44*(10), 1747–1757. https://doi.org/10.1080/03075079.2019.1665318

56. Jenkins, T. A., Nguyen, J. C. D., Polglaze, K. E., & Bertrand, P. P. (2016). Influence of tryptophan and serotonin on mood and cognition with a possible role of the gut-brain axis. *Nutrients*, *8*(1), 1–15. https://doi.org/10.3390/nu8010056

57. Jones-Tabah, J., Mohammad, H., Paulus, E. G., Clarke, P. B. S., & Hébert, T. E. (2022). The Signaling and Pharmacology of the Dopamine D1 Receptor. *Frontiers in Cellular Neuroscience*, *15*(January), 1–28. https://doi.org/10.3389/fncel.2021.806618

58. Juárez Olguín, H., Calderón Guzmán, D., Hernández García, E., & Barragán Mejía, G. (2016). The role of dopamine and its dysfunction as a consequence of oxidative stress. *Oxidative Medicine and Cellular Longevity*, *2016*. https://doi.org/10.1155/2016/9730467

59. Kadović, M., Mikšić, Š., & Lovrić, R. (2023). Ability of Emotional Regulation and Control as a Stress Predictor in Healthcare Professionals. *International Journal of Environmental Research and Public Health*, *20*(1). https://doi.org/10.3390/ijerph20010541

60. Kandel, E., Koester, J. D., Mack, S. H., & Siegelbaum, S. A. (2021). Principles of Neural Science. In *Principles of Neural Science*.

61. Kandlur, A., Satyamoorthy, K., & Gangadharan, G. (2020). Oxidative Stress in Cognitive and Epigenetic Aging: A Retrospective Glance. *Frontiers in Molecular Neuroscience*, *13*(March). https://doi.org/10.3389/fnmol.2020.00041

62. Kaufmann, L. (2008). Dyscalculia: Neuroscience and education. *Educational Research*, *50*(2), 163–175. https://doi.org/10.1080/00131880802082658

63. Kavanau, J. L., & Ph, D. (2000). Sleep, Memory Maintenance, and Mental Disorders. 199–208. https://doi.org/10.1176/jnp.12.2.199

64. Kershner, J. R. (2019). Neurobiological systems in dyslexia. *Trends in Neuroscience and Education*, *14*(September 2018), 11–24. https://doi.org/10.1016/j.tine.2018.12.001

65. Kiss, B., Laszlovszky, I., Krámos, B., Visegrády, A., Bobok, A., Lévay, G., Lendvai, B., & Román, V. (2021). Neuronal dopamine D3 receptors: Translational implications for preclinical research and cns disorders. *Biomolecules*, *11*(1), 1–39. https://doi.org/10.3390/biom11010104

66. Kraus, C., Castrén, E., Kasper, S., & Lanzenberger, R. (2017). Serotonin and neuroplasticity – Links between molecular, functional and structural pathophysiology in depression. *Neuroscience and Biobehavioral Reviews*, *77*, 317–326. https://doi.org/10.1016/j.neubiorev.2017.03.007

67. Laland, K. N., Odling-Smee, J., & Gilbert, S. F. (2008). EvoDevo and niche construction: Building bridges. *Journal of Experimental Zoology Part B: Molecular and Developmental Evolution*, *310*(7), 549–566. https://doi.org/10.1002/jez.b.21232

68. Lambert, J. D. C. (2007). GABA (Gamma Aminobutyric Acid). *Encyclopedia of Stress*, *2*, 97–108. https://doi.org/10.1016/B978-012373947-6.00168-9

69. Lüscher, C. (2012). Restless AMPA receptors: implications for synaptic transmission and. *October*, *4*(3), 707–731. https://doi.org/10.1016/s0166-2236(00)01959-7

70. Maha Putra, I. P. E. D., Indrasari Utami, D. K., & Sri Laksemi, D. A. A. (2023). Relationship Between Sleep Quality and Learning Achievement in Junior High School Students in Denpasar. *International Journal of Research and Review*, *10*(1), 11–18. https://doi.org/10.52403/ijrr.20230102

71. Maletic, V., Eramo, A., Gwin, K., Offord, S. J., & Duffy, R. A. (2017). The role of norepinephrine and its α-adrenergic receptors in the pathophysiology and treatment of major depressive disorder and

schizophrenia: A systematic Review. *Frontiers in Psychiatry*, *8*(MAR). https://doi.org/10.3389/fpsyt.2017.00042

72. Marshall, K. L., & Farah, M. (2021). Axonal regeneration and sprouting as a potential therapeutic target for nervous system disorders. *Neural Regeneration Research*, *16*(10), 1901–1910. https://doi.org/10.4103/1673-5374.308077

73. Martin, R. E., & Oschner, K. N. (2016). The Neuroscience of Emotion Regulation Development: Implications for Education. *Economic and Political Weekly*, *42*(14), 1273–1276. https://doi.org/10.1515/9780773598874-013

74. Marzola, P., Melzer, T., Pavesi, E., Gil-Mohapel, J., & Brocardo, P. S. (2023). Exploring the Role of Neuroplasticity in Development, Aging, and Neurodegeneration. *Brain Sciences*, *13*(12). https://doi.org/10.3390/brainsci13121610

75. Massaad, C. A., & Klann, E. (2011). Reactive oxygen species in the regulation of synaptic plasticity and memory. *Antioxidants and Redox Signaling*, *14*(10), 2013–2054. https://doi.org/10.1089/ars.2010.3208

76. Mateos-Aparicio, P., & Rodríguez-Moreno, A. (2019). The impact of studying brain plasticity. *Frontiers in Cellular Neuroscience*, *13*(February), 1–5. https://doi.org/10.3389/fncel.2019.00066

77. Mattson, M. P. (2009). *Glutamate and Neurotrophic Factors in Neuronal Plasticity and Disease*. 97–112. https://doi.org/10.1196/annals.1418.005

78. McCaskey, U., von Aster, M., O'Gorman, R., & Kucian, K. (2020). Persistent Differences in Brain Structure in Developmental Dyscalculia: A Longitudinal Morphometry Study. *Frontiers in Human Neuroscience*, *14*(July), 1–13. https://doi.org/10.3389/fnhum.2020.00272

79. McEwen, B. S., & Morrison, J. H. (2013). Brain On Stress: Vulnerability and Plasticity of the Prefrontal Cortex Over the Life Course. *NIH Public Access*, *538*(1), 151–169. https://doi.org/10.1016/j.neuron.2013.06.028

80. McEwen, B. S., Bowles, N. P., Gray, J. D., Hill, M. N., Hunter, R. G., Karatsoreos, I. N., & Nasca, C. (2015). Mechanisms of stress in the brain. *Nature Neuroscience*, *18*(10), 1353–1363. https://doi.org/10.1038/nn.4086

81. Meerlo, P., Misrlberger, R. E., Jacobs, B. L., Heller, C. H., & McGinty, D. (2009). New neurons in the adult brain: The role of

sleep and consequences of sleep loss. *Bone*, *17*(3), 211–224. https://doi.org/10.1016/j.smrv.2008.07.004

82. Meier, J. D., Aflalo, T. N., Kastner, S., & Graziano, M. S. A. (2008). Complex organization of human primary motor cortex: A high-resolution fMRI study. *Journal of Neurophysiology*, *100*(4), 1800–1812. https://doi.org/10.1152/jn.90531.2008
83. Menon, S., & Gupton, S. (2018). *Recent advances in branching mechanisms underlying neuronal morphogenesis [version 1 ; referees : 2 approved] Referee Status :* 7(0), 1–12. https://doi.org/10.12688/f1000research.16038.1
84. Mi, Y., Duan, H., Xu, Z., & Lei, X. (2023). The Impact of Sleep Deprivation on Brain Networks in Response to Social Evaluation Tasks. *Brain Sciences*, *13*(8). https://doi.org/10.3390/brainsci13081122
85. Miao, Z., Wang, Y., & Sun, Z. (2020). The relationships between stress, mental disorders, and epigenetic regulation of bdnf. *International Journal of Molecular Sciences*, *21*(4). https://doi.org/10.3390/ijms21041375
86. Mohanan, A. G., Gunasekaran, S., Jacob, R. S., & Omkumar, R. V. (2022). Role of Ca2+/Calmodulin-Dependent Protein Kinase Type II in Mediating Function and Dysfunction at Glutamatergic Synapses. *Frontiers in Molecular Neuroscience*, *15*(June). https://doi.org/10.3389/fnmol.2022.855752
87. Montgomery, R. M. (2024). *Impact of Chronic Stress on Physical and Mental Health : A Detailed Analysis Impact of Chronic Stress on Physical and Mental Health : A Detailed Analysis. August.* https://doi.org/10.13140/RG.2.2.35357.73445
88. Moore, D. R. (2007). Auditory processing disorders: Acquisition and treatment. *Journal of Communication Disorders*, *40*(4), 295–304. https://doi.org/10.1016/j.jcomdis.2007.03.005
89. Mora, F., Segovia, G., Del Arco, A., De Blas, M., & Garrido, P. (2012). Stress, neurotransmitters, corticosterone and body-brain integration. *Brain Research*, *1476*, 71–85. https://doi.org/10.1016/j.brainres.2011.12.049
90. Morilak, D. A., Barrera, G., Echevarria, D. J., Garcia, A. S., Hernandez, A., Ma, S., & Petre, C. O. (2005). Role of brain norepinephrine in the behavioral response to stress. *Progress in*

Neuro-Psychopharmacology and Biological Psychiatry, *29*(8), 1214–1224. https://doi.org/10.1016/j.pnpbp.2005.08.007

91. Mujawar, S., Patil, J., Chaudhari, B., & Saldanha, D. (2021). Memory: Neurobiological mechanisms and assessment. *Industrial Psychiatry Journal*, *30*(3), 311. https://doi.org/10.4103/0972-6748.328839
92. Ng, T. K. S., Coughlan, C., Heyn, P. C., Tagawa, A., Carollo, J. J., Kua, E. H., & Mahendran, R. (2021). Increased plasma brain-derived neurotrophic factor (BDNF) as a potential biomarker for and compensatory mechanism in mild cognitive impairment: A case-control study. *Aging*, *13*(19), 22666–22689. https://doi.org/10.18632/aging.203598
93. Norton, E. S., Beach, S. D., & Gabrieli, J. D. E. (2015). Neurobiology of Dyslexia Elizabeth. *NIH Public Access*, *23*(1), 1–7. https://doi.org/10.1016/j.conb.2014.09.007
94. Ono, B. H. V. S., Souza, J. C., & Paiva, T. (2020). Sleep and learning in school children. *Research, Society and Development*, *9*(10), e4399108636. https://doi.org/10.33448/rsd-v9i10.8636
95. Owens, M. T., & Tanner, K. D. (2017). Teaching as brain changing: Exploring connections between neuroscience and innovative teaching. *CBE Life Sciences Education*, *16*(2), 1–9. https://doi.org/10.1187/cbe.17-01-0005
96. Patel, D. R., Soares, N., & Wells, K. (2017). Neurodevelopmental readiness of children for participation in sports. *Translational Pediatrics*, *6*(3), 167–173. https://doi.org/10.21037/tp.2017.05.03
97. Perez, D. M. (2021). Current Developments on the Role of α1-Adrenergic Receptors in Cognition, Cardioprotection, and Metabolism. *Frontiers in Cell and Developmental Biology*, *9*(May), 1–23. https://doi.org/10.3389/fcell.2021.652152
98. Picciotto, M. R., Higley, M. J., & Mineur, Y. S. (2012). Acetylcholine as a neuromodulator. *Neuron*, *76*(1), 116–129. https://doi.org/10.1016/j.neuron.2012.08.036
99. Plata-Bello, J., Privato, N., Modroño, C., Pérez-Martín, Y., Borges, Á., & González-Mora, J. L. (2023). Empathy Modulates the Activity of the Sensorimotor Mirror Neuron System during Pain Observation. *Behavioral Sciences*, *13*(11). https://doi.org/10.3390/bs13110947

100. Poerio, G. L., Sormaz, M., Wang, H. T., Margulies, D., Jefferies, E., & Smallwood, J. (2017). The role of the default mode network in component processes underlying the wandering mind. *Social Cognitive and Affective Neuroscience*, *12*(7), 1047–1062. https://doi.org/10.1093/scan/nsx041

101. Popoli, M., Yan, Z., McEwen, B., & Sanacora, G. (2013). The stressed synpase_NIH Public Access. *Nat Rev Neurosci*, *23*(1), 1–7. https://doi.org/10.1038/nrn3138

102. Ramdhane, K. (2024). *Understanding Learning Difficulties : A Comprehensive Overview. September.* https://doi.org/10.53419/2259-009-003-057

103. Ramesh, S., & Arachchige, A. S. P. M. (2023). Depletion of dopamine in Parkinson's disease and relevant therapeutic options: A review of the literature. *AIMS Neuroscience*, *10*(3), 200–231. https://doi.org/10.3934/NEUROSCIENCE.2023017

104. Rapp, B., & McCloskey, M. (2018). Developmental Dysgraphia. In *World* (Vol. 3, Issue February 2004). file:///Users/alex.neumann/Documents/Mendeley Desktop/Edited by Edited by/World/[Darren_Swanson]_Creating_Adaptive_Policies_A_Gui(BookSee.org).pdf

105. Rusu, S. I., & Pennartz, C. M. A. (2020). Learning, memory and consolidation mechanisms for behavioral control in hierarchically organized cortico-basal ganglia systems. *Hippocampus*, *30*(1), 73–98. https://doi.org/10.1002/hipo.23167

106. Safron, A., Klimaj, V., & Hipólito, I. (2022). On the Importance of Being Flexible: Dynamic Brain Networks and Their Potential Functional Significances. *Frontiers in Systems Neuroscience*, *15*(January), 1–20. https://doi.org/10.3389/fnsys.2021.688424

107. Salzman, C. D., & Fusi, S. (2010). Emotion, cognition, and mental state representation in amygdala and prefrontal cortex. *Annual Review of Neuroscience*, *33*, 173–202. https://doi.org/10.1146/annurev.neuro.051508.135256

108. Schindowski, K., Belarbi, K., & Buée, L. (2008). Neurotrophic factors in Alzheimer's disease: Role of axonal transport. *Genes, Brain and Behavior*, 7(SUPPL. 1), 43–56. https://doi.org/10.1111/j.1601-183X.2007.00378.x

109. Selway, J. L., Moore, C. E., Mistry, R., Challiss, R. A. J., & Herbert, T. P. (2012). Molecular mechanisms of muscarinic acetylcholine receptor-stimulated increase in cytosolic free Ca2+ concentration and ERK1/2 activation in the MIN6 pancreatic b-cell line. *Acta Diabetologica*, *49*(4), 277–289. https://doi.org/10.1007/s00592-011-0314-9

110. Seo, D., Patrick, C. J., & Kennealy, P. J. (2008). Role of serotonin and dopamine system interactions in the neurobiology of impulsive aggression and its comorbidity with other clinical disorders. *Aggression and Violent Behavior*, *13*(5), 383–395. https://doi.org/10.1016/j.avb.2008.06.003

111. Serrano, F., & Klann, E. (2004). Reactive oxygen species and synaptic plasticity in the aging hippocampus. *Ageing Research Reviews*, *3*(4), 431–443. https://doi.org/10.1016/j.arr.2004.05.002

112. Shi, P., & Feng, X. (2022). Motor skills and cognitive benefits in children and adolescents: Relationship, mechanism and perspectives. *Frontiers in Psychology*, *13*(November), 1–14. https://doi.org/10.3389/fpsyg.2022.1017825

113. Shors, T., Anderson, L., Curlik II, D. M., & Nokia, S. (2014). Use it or lose it: How neurogenesis keeps the brain fit for learning. *Bone*, *23*(1), 1–7. https://doi.org/10.1016/j.coph.2007.10.002

114. Sibley, D. R., & Neve, K. (2007). D5 dopamine receptor. *XPharm: The Comprehensive Pharmacology Reference*, 1–9. https://doi.org/10.1016/B978-008055232-3.60155-2

115. Sippel, D., Schwabedal, J., Snyder, J. C., Oyanedel, C. N., Bernas, S. N., Garthe, A., Tröndle, A., Storch, A., Kempermann, G., & Brandt, M. D. (2020). Disruption of NREM sleep and sleep-related spatial memory consolidation in mice lacking adult hippocampal neurogenesis. *Scientific Reports*, *10*(1), 1–14. https://doi.org/10.1038/s41598-020-72362-3

116. Siu, C. R., & Murphy, K. M. (2018). The development of human visual cortex and clinical implications. *Eye and Brain*, *10*, 25–36. https://doi.org/10.2147/EB.S130893

117. Speranza, L., Di Porzio, U., Viggiano, D., de Donato, A., & Volpicelli, F. (2021). Dopamine: The neuromodulator of long-term synaptic plasticity, reward and movement control. *Cells*, *10*(4). https://doi.org/10.3390/cells10040735

118. Steber, S., & Rossi, S. (2021). The challenge of learning a new language in adulthood: Evidence from a multi-methodological neuroscientific approach. *PLoS ONE*, *16*(2 February 2021), 1–23. https://doi.org/10.1371/journal.pone.0246421

119. Sutapa, P., Pratama, K. W., Rosly, M. M., Ali, S. K. S., & Karakauki, M. (2021). Improving motor skills in early childhood through goal-oriented play activity. *Children*, *8*(11), 1–11. https://doi.org/10.3390/children8110994

120. Tang, H., Riley, M. R., Singh, B., Qi, X. L., Blake, D. T., & Constantinidis, C. (2022). Prefrontal cortical plasticity during learning of cognitive tasks. *Nature Communications*, *13*(1), 1–12. https://doi.org/10.1038/s41467-021-27695-6

121. Teleanu, D. M., Niculescu, A., Lungu, I. I., Radu, C. I., Vlad, O., Roza, E., Cost, B., Grumezescu, A. M., & Teleanu, R. I. (2022a). An Overview of Oxidative Stress, Neuroinflammation, and Neurodegenerative Diseases. https://doi.org/10.3390/ijms23115938

122. Teleanu, R. I., Niculescu, A. G., Roza, E., Vladâcenco, O., Grumezescu, A. M., & Teleanu, D. M. (2022b). Neurotransmitters—Key Factors in Neurological and Neurodegenerative Disorders of the Central Nervous System. *International Journal of Molecular Sciences*, *23*(11). https://doi.org/10.3390/ijms23115954

123. Thompson, S. M., Kallarackal, A. J., Kvarta, M. D., Van Dyke, A. M., LeGates, T. A., & Cai, X. (2015). An excitatory synapse hypothesis of depression. *Trends in Neurosciences*, *38*(5), 279–294. https://doi.org/10.1016/j.tins.2015.03.003

124. Tierney, A. L., & Nelson, C. A. (2009). Brain Development and the Role of Experience in the Early Years. *Zero to Three*, *30*(2), 9–13. PMID: 23894221; PMCID: PMC3722610.

125. Vidal-Piñeiro, D., Valls-Pedret, C., Fernández-Cabello, S., Arenaza-Urquijo, E. M., Sala-Llonch, R., Solana, E., Bargalló, N., Junqué, C., Ros, E., & Bartrés-Faz, D. (2014). Decreased Default Mode Network connectivity correlates with age-associated structural and cognitive changes. *Frontiers in Aging Neuroscience*, *6*, 256. https://doi.org/10.3389/fnagi.2014.00256

126. Vogel, S., & Schwabe, L. (2016). Learning and memory under stress: implications for the classroom. *Npj Science of Learning*, *1*(1), 1–10. https://doi.org/10.1038/npjscilearn.2016.11

127. Voglis, G., & Tavernarakis, N. (2006). The role of synaptic ion channels in synaptic plasticity. *EMBO Reports*, *7*(11), 1104–1110. https://doi.org/10.1038/sj.embor.7400830

128. Wamsley, E. J., & Stickgold, R. (2011). Memory, sleep, and dreaming: Experiencing consolidation. *Sleep Medicine Clinics*, *6*(1), 97–108. https://doi.org/10.1016/j.jsmc.2010.12.008

129. Wang, H., Xu, J., Lazarovici, P., Quirion, R., & Zheng, W. (2018). cAMP Response Element-Binding Protein (CREB): A Possible Signaling Molecule Link in the Pathophysiology of Schizophrenia. *Frontiers in Molecular Neuroscience*, *11*(August), 1–14. https://doi.org/10.3389/fnmol.2018.00255

130. Wei, F., Vadakkan, K. I., Toyoda, H., Wu, L. J., Zhao, M. G., Xu, H., Shum, F. W. F., Yong, H. J., & Zhuo, M. (2006). Calcium calmodulin-stimulated adenylyl cyclases contribute to activation of extracellular signal-regulated kinase in spinal dorsal horn neurons in adult rats and mice. *Journal of Neuroscience*, *26*(3), 851–861. https://doi.org/10.1523/JNEUROSCI.3292-05.2006

131. Woods, A. S. (2010). The dopamine D4 receptor, the ultimate disordered protein. *Journal of Receptors and Signal Transduction*, *30*(5), 331–336. https://doi.org/10.3109/10799893.2010.513842

132. Wurtz, R. H. (2009). Recounting the impact of Hubel and Wiesel. *Journal of Physiology*, *587*(12), 2817–2823. https://doi.org/10.1113/jphysiol.2009.170209

133. Youn, D. H., Gerber, G., & Sather, W. A. (2013). Ionotropic glutamate receptors and voltage-gated Ca2+ channels in long-term potentiation of spinal dorsal horn synapses and pain hypersensitivity. *Neural Plasticity*, *2013*. https://doi.org/10.1155/2013/654257

134. Zotey, V., Andhale, A., Shegekar, T., & Juganavar, A. (2023). Adaptive Neuroplasticity in Brain Injury Recovery: Strategies and Insights. *Cureus*, *15*(9). https://doi.org/10.7759/cureus.45873

Multilingualism and Neuroplasticity: Cognitive and Neural Implications

Irmak Özcan and Seda Eyilikeder Tekin

1. Introduction

Multilingualism is the ability to use more than one language fluently and effectively. Research has indicated that multilingualism enhances communication skills and causes profound brain structure and function alterations (Bialystok, 2017). These changes affect the brain's capacity for neuroplasticity and give us profound information about language processing and cognitive control mechanisms (Grundy & Timmer, 2017).

This chapter aims to investigate the effect of multilingualism on the brain's neuroplasticity capacity in light of existing literature and scientific research. Realizing the role of multilingualism across a broad spectrum, from individual development to social interactions, is very important from a cognitive and neuroscientific perspective.

2. The Impact of Language and Language Acquisition on Brain Structure

2.1. Definition and Importance of Language

Language is a system of symbols and rules used for communication from the beginning of human history. It makes it easier for people to communicate ideas, emotions, and information to one another via sounds, words, and grammatical forms (Yuan, 2023). Language plays a huge role in interpersonal communication and building social interactions and cultural identities in a social environment. It also contributes to educational and learning processes as a tool for transferring knowledge and promoting cognitive development. For instance, children's language skills facilitate the organization of thought, enhance their learning abilities, and support effective problem-solving (Smith et al., 2023).

Technological developments and cultural interactions have had a significant impact on the evolution of language and human communication.

Moreover, the global reach of digital platforms has created chances for cultural exchange, leading to the blending of language features and communication styles (Xia et al., 2024).

2.2. Critical Language Areas in the Brain

The brain is a complex language production, comprehension, and processing system. Different brain parts have active roles in language and speech skills during these processes. Broca's area, Wernicke's area, the angular gyrus, and the arcuate fasciculus are critical regions that control different language components. The cooperation within these areas enables us to comprehend, produce, and reconstruct language.

Broca's Area is located in the inferior frontal gyrus of the left hemisphere (Brodmann areas 44 and 45). This area is active in motor control, speech planning, and production (Friederici, 2019). Broca's area is responsible for coordinating the movement of speech organs, finding the right words during speech, and organizing speech according to grammatical rules. When Broca's area is damaged for different reasons (stroke, TBI, etc.), '*Broca's aphasia*' may occur. Individuals with Broca's aphasia typically have intact language comprehension abilities but have problems with fluent and grammatically correct speech (Gajardo-Vidal et al., 2021). The Broca's Area is shown in Figure 1.

Wernicke's Area is located in the superior temporal gyrus of the left hemisphere (Broadmann Area 22). It is responsible for processing the semantic aspects of language. While Broca's area is primarily related to speech production, the Wernicke's area is essential in comprehending spoken and written language. If this region is damaged, it may result in incomprehensible speech, difficulties in understanding language, inappropriate word selections, and deviant sentence structures (Swanberg et al., 2007), a condition known as Wernicke's aphasia. The Wernicke's Area is shown in Figure 1.

The Arcuate Fasciculus is a group of white matter pathways that link Broca's and Wernicke's regions. This part of the brain transfers information between language production and comprehension centers. When the arcuate fasciculus is damaged, '*conduction aphasia*' may occur. Individuals with conduction aphasia are generally able to comprehend and produce language; however, they experience difficulties with word repetition and self-correction of speech errors (Berwick & Chomsky, 2017). The Arcuate Fasciculus is shown in Figure 2.

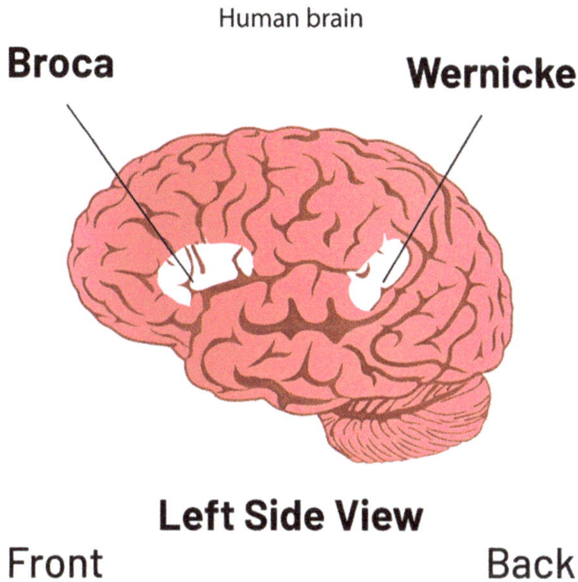

Figure 1: Broca's area and Wernicke's area
<https://commons.m.wikimedia.org/wiki/Main_Page> Date: September 1, 2024.

Significant parts of the brain are responsible for language comprehension and production: Broca's Area, Wernicke's Area

The Angular Gyrus (AG) is located in the posterior part of the inferior parietal lobule and is consistently active across different tasks. It is fundamental in word reading and reading comprehension, semantic and number processing, memory retrieval, attention, reasoning, and social and spatial cognition. More specifically, the AG gathers multisensory information to understand and interpret meaning, overcome familiar problems, and direct attention to relevant existing information (Seghier, 2013). The AG is shown in Figure 3.

3. Neural Changes During Multiple Language Acquisition

Learning multiple languages can alter brain structure and neuroplasticity. Multilingual people often show better attention and executive functions. These changes begin even before learning a language and continue during

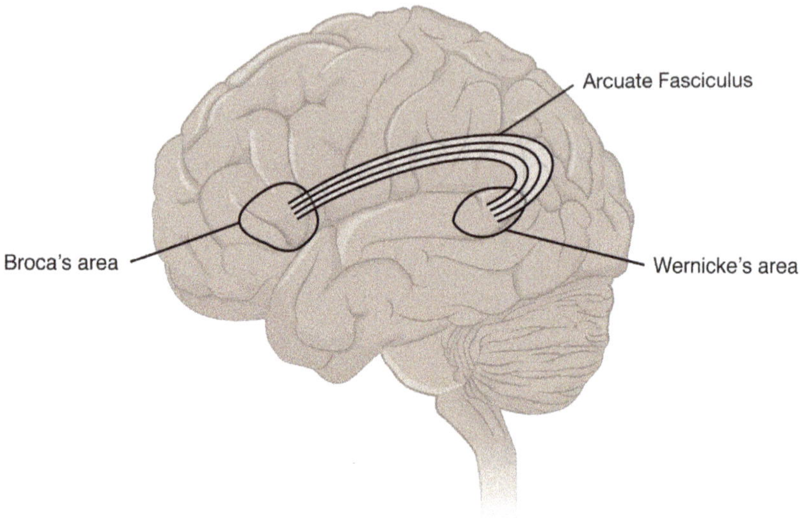

Figure 2: The Arcuate Fasciculus
<https://commons.m.wikimedia.org/wiki/Main_Page> Date: September 1, 2024.

the process, showing that multilingualism is dynamic and adaptive. This suggests that multilingualism is not just a passive state but a dynamic, adaptive process that actively shapes the brain. The continuous engagement with multiple languages helps to strengthen neural pathways and refine cognitive control, making it an effective way to promote cognitive health and mental flexibility (Kroll & Chiarello, 2016).

Research shows multilingual individuals have more gray matter in their brains. This boost aids neuroplasticity and improves language processing. It is especially evident in specific brain areas linked to language, such as the left inferior frontal gyrus, superior temporal gyrus, anterior cingulate cortex, and caudate nucleus (Grundy & Timmer, 2017; Abutalebi et al., 2012).

Multilingual people may process complex language better due to changes in the left parietal lobe. This area is critical for understanding and processing language. Such changes could make the brain work more efficiently, using

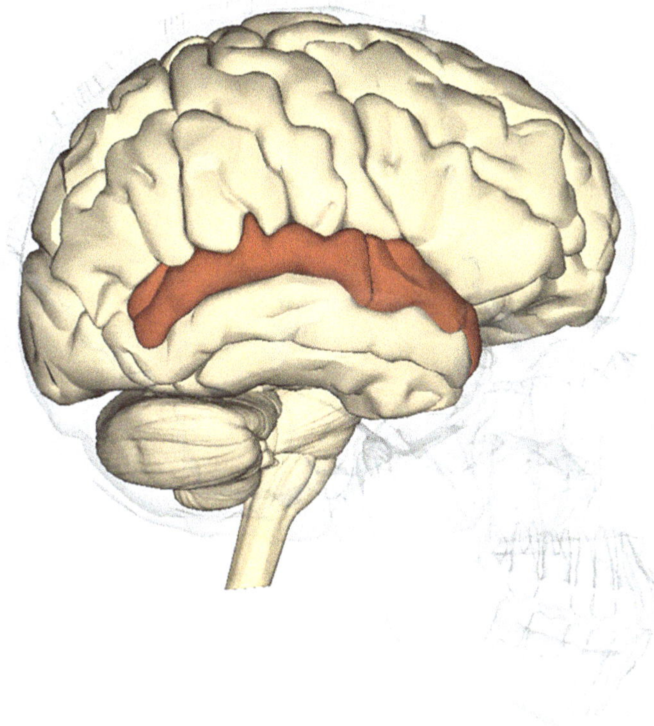

Figure 3: Angular Gyrus (AG)
<https://commons.m.wikimedia.org/wiki/Main_Page> Date: September 1, 2024.

less effort. Bilinguals, with strong connections in their brains, are more efficient. They use less energy in language tasks. Multilingual excel in complex language tasks and stay flexible in thinking. This is useful under cognitive stress (Bialystok, 2023; Abutalebi & Green, 2023).

Moreover, being multilingual boosts white matter in the brain. This suggests better information flow (Li et al., 2020). Thus, multilingualism aids brain structure and improves cognitive functions.

4. Critical Period and Age-Related Changes in Language Acquisition

4.1. Critical Period in Language Acquisition

Critical periods in language learning show how age affects our ability to learn languages. The Critical Period Hypothesis (CPH) covers both first (L1) and second languages (L2). Generally, people learn L1 better than L2, with adults and children reaching different L2 levels (Johnson & Newport, 1989).

Eric Lenneberg introduced the CPH for L1 in his 1967 book *Biological Foundations of Language*. He argued that high-level language learning happens only during this critical period. After, it gets more brutal, and achieving native-like fluency becomes unlikely. His theory came from studying children whose early experiences hindered their L1 learning. Lenneberg pinpointed puberty as the cutoff for learning a language like a native speaker. He noted that most L2 learners have accents after this period (Lenneberg, 1997), backing his theory (Johnson & Newport, 1989).

While many studies support the CPH, some contradict it. For instance, some late L2 learners speak almost like natives in everyday conversations (Bongaerts et al., 2000; Hyltenstam & Abrahamsson, 2003; Birdsong, 2007).

4.2. Age-Related Changes and Language Acquisition

Children learn languages quickly and effectively from a young age. Their brains undergo significant changes that boost language learning. This period shows high brain plasticity, indicating a strong capacity for learning languages (Hoff, 2021).

As people age, their language learning abilities change. Adults and seniors may learn languages less quickly and fluently than children. However, being multilingual and using languages regularly can maintain brain plasticity in older age (Antoniou et al., 2013). This means they can still learn new languages. Age can affect how the brain processes language, but multilingualism and active use counter these effects.

For older adults, learning languages takes more effort and strategic thinking. However, being multilingual boosts cognitive reserve and promotes brain adaptability (Bialystok & DePape, 2023). Ongoing language practice helps preserve brain function and increases flexibility in thinking.

Neuroplasticity helps us learn and remember new words (Cassilhas et al., 2016). It also affects how our brains adapt when learning a new language. The differences between languages can change brain connections. This is especially true for new language learners, as their brains adapt to new sounds and rules. Scans show multilingual people's brains react differently when learning new rules. This happens because their neural networks become stronger and reorganized. Neuroplasticity is vital for brain rehab after injuries. It also aids in improving language and cognitive issues (Antoniou et al., 2013).

Neuroplasticity is not just for learning multiple languages but also helps improve grammar skills and relearn languages after injuries. Which is crucial in rehabilitation. It shows how adaptable our brains are, enhancing our flexibility in thinking. Studies in neuroscience and genetics have shed light on how our brains change with learning and experiences.

4.3. Challenges of Learning a New Language in Elderhood

Reduced Neuroplasticity: Through age, the brain's neuroplastic capacity—its ability to form new neural connections and modify existing ones—tends to decrease. This can deteriorate the process of learning a new language and require more effort. Older adults may need more repetition and practice to learn a new language (Li & Grant, 2023).

a. Cognitive Slowing: Aging is associated with a slowdown in cognitive processes such as memory, attention, and fundamental functions compared to earlier stages. Acquiring a new language can affect capacities such as processing information quickly and short-term and long-term memory. Learning new words or grammatical structures, retaining them in short-term memory, and transferring them to long-term memory may become more challenging as one age (Bak & Ang, 2022).
b. Effects on Existing Linguistic Knowledge: Older individuals trying to learn a new language may struggle to integrate or adapt their linguistic knowledge to the new language. This can lead to code-switching and language interference (Bialystok et al., 2020).

4.4. Language Impairments and Rehabilitation in Elderhood

Older individuals may face language issues due to strokes, brain injuries, or other neurological problems. These conditions can lead to disorders like aphasia, affecting speaking, understanding, reading, and writing. Neuroplasticity is critical for recovery (Nadeau, 2015). Treatment methods include:

a. Intensive Language Therapy: This standard method helps regain and improve language skills. It targets understanding, speaking, reading, and writing. Studies show that intensive therapy boosts recovery by tapping into the brain's neuroplasticity (Kurland & Tauer, 2012; Webster et al., 2021).
b. Computer-Assisted Rehabilitation: These programs use technology to offer tailored exercises. People can practice at their own pace, promoting neuroplastic changes (Palmer et al., 2023).
c. Cognitive Rehabilitation Techniques focus on improving memory, attention, and problem-solving. They indirectly support language recovery by enhancing overall cognitive functions (Marshall & Freed, 2022).
d. Transcranial Magnetic Stimulation (TMS): TMS uses magnetic pulses to boost language therapies. It is effective for disorders like aphasia (Hamilton & Shapiro, 2022).

5. Strategies to Promote Neuroplasticity

Neuroplasticity plays a critical role in language acquisition and the treatment of language disorders in older adults. Strategies that support neuroplasticity can help older individuals acquire a new language or recover language skills following a neurological condition.

a. Complex and Rich Language Environment: Providing a complex and rich language environment during language acquisition and rehabilitation can enhance neuroplasticity. Various resources, such as social interactions, language courses, reading books, and using media sources can contribute to developing language skills (de Bruin et al., 2023).
b. Cognitive Flexibility Exercises: Exercises designed to improve cognitive flexibility in older adults can support the language learning process. Solving puzzles, memory games, and problem-solving tasks can enhance

cognitive functions and promote neuroplasticity (Woumans & Duyck, 2020).

c. Healthy Lifestyle: Regular physical activity is known to help improve neurogenesis and synaptic plasticity. In these processes, older individuals can enhance their language learning abilities by maintaining cognitive flexibility and overall brain health (Phillips, 2017).

6. Fundamental Research Areas in Bilingualism and Multilingualism

Today, being bilingual or multilingual is common and viewed positively. People have always communicated across language barriers, and international travel and digital technologies have made this easier (Grosjean, 2019).

The cognitive effects of multilingualism are critical research. It focuses on how learning multiple languages affects brain functions like executive functions and working memory (Bialystok, 2017). Speaking multiple languages boosts cognitive reserve, slowing down age-related decline. It also supports brain adaptability and enhances flexibility. This suggests that multilingualism positively alters the brain. For instance, the regular use of multiple languages improves the ability to balance and adapt (Mårtensson et al., 2022).

Another critical area is "multilingualism's effect on education." It questions whether multilingual kids need exceptional help for language growth (Core et al., 2023). Some studies also link bilingualism to mental health issues like schizophrenia and identity disorders. They aim to see if multilingualism improves mental health (Costa et al., 2019). Researchers are also looking into "multilingualism's societal effects." They ask if multilingual people face unique community challenges or benefits (Antoniou, 2013). These questions are vital for understanding multilingualism's broader effects. Ongoing research examines them in various cultural and educational settings.

6.1. Linguistic Fundamental Questions

Research on bilingualism and multilingualism is at the center of contemporary linguistic agendas. Chomsky (1986) identified three fundamental linguistic questions: What is grammar? How is grammar acquired? How is grammar used?

In studies of bilingualism and multilingualism, these questions can be reformulated to consider multiple grammatical systems:

a. What is the nature of the grammar or linguistic knowledge in the mind of a bilingual or multilingual person, and how do different grammatical systems coexist and interact?
b. How is the acquisition of multiple grammatical systems achieved, and does this acquisition occur simultaneously or sequentially? In what ways does bilingual or multilingual acquisition differ from monolingual acquisition?
c. How does a speaker use two or more languages in bilingual interactions?

6.1.1. The Nature of Multilingualism First Question: '*What is the nature of grammar or linguistic knowledge in bilingual or multilingual people, and how do their grammatical systems coexist and interact?*'

Understanding how bilingual or multilingual people manage multiple grammatical systems is critical to grasping multilingualism. Cross-linguistic influence (CLI) occurs when one language's grammar affects another. This influence is vital for multilingual grammar processing (Costa et al., 2019).

Research indicates that vocabulary and syntax interact across languages. The Bilingual Interactive Activation (BIA) Model explains bilingual language processing. It suggests that all language knowledge is active during processing (Dijkstra & van Heuven, 1998). The model also tracks how language processing and cognitive systems evolve. This is crucial for young bilinguals. As their language skills grow, so do cross-linguistic effects and cognitive control (Grainger et al., 2010).

The Revised Hierarchical Model (RHM), created by Kroll and Stewart in 1994, explains how grammatical rules are stored and how languages influence each other. This influence is more potent when languages are structurally similar (Kroll et al., 2015). Recent studies have improved our understanding of words and concepts in L1 and L2 languages. They show that early L2 learners mainly translate from L1. However, as they get better, they start linking ideas directly (Branzi et al., 2014; Kroll et al., 2015; Bailey et al., 2024). This shows how active bilinguals are with their vocabulary.

Also, recent brain studies show that as L2 skills improve, different brain areas, like the left inferior frontal gyrus, are activated. This suggests that more proficient bilinguals rely less on L1 and more on direct access to concepts in L2. This shift reduces the need for translation and makes using both languages easier.

6.1.2. Language Acquisition Processes Second Question: '*How do people learn multiple grammatical systems? Does this happen at the same time or one after the other? How is this different for bilinguals or multilinguals compared to those who speak only one language?*'

6.1.2.1. Simultaneous and sequential language acquisition

Bilinguals or multilinguals can learn languages either simultaneously or sequentially. Simultaneous learners pick up multiple languages from birth. Meanwhile, sequential learners start with one language (L1) and then learn another (L2) later.

Children learning languages simultaneously often become proficient in both at a similar pace. Studies show that children develop grammatical rules in each language, just as monolingual children do. However, if the languages share rules, one language might influence the other.

The brain adapts to learning two languages at once. This offers insights into how the brain functions during language learning (Li et al., 2023). Similar processes are used for both languages, allowing vocabulary and grammar to develop together. Children switch between languages using cognitive control (Green & Abutalebi, 2016).

In sequential learning, L1 affects how L2 is learned. This is especially true for adults. They might struggle with L2 due to L1 rules (Paradis, 2018). However, younger learners are more likely to achieve native-like proficiency (Kuhl, 2020).

Learning languages one after the other triggers unique brain changes. Picking up a second language after the first alters the brain's structure. It involves reorganizing language systems and forming new connections. Thus, this method might lead to greater brain changes (Thomas & Pliatsikas, 2023).

Cognitive control in bilingual and multilingual people differs from monolinguals. Multilingualism manages multiple grammar systems based on the situation. This constant language switching boosts attention and working memory more than in monolinguals (Bialystok, 2020).

6.1.2.2. Developmental similarities between monolinguals and multilinguals

Studies show bilingual and monolingual children reach developmental milestones at similar rates. This includes aspects like meaning, grammar, and sound. For example, both groups typically say their first words around age one. When looking at each language separately, bilingual children have vocabularies and phonetic skills similar to monolinguals. Moreover, both groups are alike in gesture use and early communication skills. Gesture use, which precedes spoken language, aids communication before verbal skills fully develop (Cankuvvet Aykut & Esersin, 2018).

6.1.2.3. Developmental differences between monolinguals and multilingual

Learning multiple languages is different from learning just one. Multilingual people often have better control over their thinking and decision-making. This is because they regularly decide which language to use, blocking one language while using another. This process changes areas in the brain linked to these functions (Bialystok, 2020). When switching languages, they show more flexibility and better attention. Research shows they also have better memory and focus (Costa et al., 2019). Multilingualism might even slow cognitive decline and prevent diseases like Alzheimer's (Perani & Abutalebi, 2015).

6.1.3. Language Use and Social Factors Third Question: *'How does a speaker use two or more languages in bilingual interactions?'*

The third question examines how a speaker uses multiple languages. It highlights behaviors like code-switching and linguistic engagement. When switching languages, speakers blend rules from both languages (Li, 2022). Language choice is affected by social and communicative factors such as the person speaker is talking to, the situation that the speaker is in, and the purpose of the conversation (Myers-Scotton, 2020; Auer, 2022).

7. Therapy Strategies Supporting Neuroplasticity

Speech and language therapists can boost the benefits of multilingualism on neuroplasticity. This support aids language learning and development. Strategies include:

a) Providing Linguistic Richness: Creating a rich language environment for multilingual enhances neuroplasticity and language skills. Active participation from family members is vital for multilingualism (Core et al., 2023).
b) Cognitive Exercises and Applications: Cognitive exercises that support cognitive control and executive functions can help strengthen the language skills of multilingual individuals. Such applications support fundamental cognitive functions like attention and memory (Pliatsikas et al., 2023).
c) Interactive and Social Language Learning: Social interactions are a significant factor in promoting neuroplasticity during the language learning process. Speech and language therapists can use social interaction-based language learning techniques to enhance individuals' language skills (Antoniou et al., 2023).

8. Speech and Language Therapy and Slts' Role in Multilingualism

Speech and language therapy aims to enhance communication and treat related disorders. Therapists can leverage the benefits of multilingualism to tailor treatments.

They develop strategies for language acquisition and use these to address language disorders. Conditions such as developmental language disorders, autism, and aphasia disrupt language learning. Therapists create personalized plans to boost language skills. They aim to strengthen or form new language pathways by tapping into neuroplasticity. For instance, encouraging aphasia patients to learn a second language can aid recovery. This approach activates different brain pathways. Studies show that learning a second language during treatment stimulates brain networks. It helps create new connections (Abutalebi & Green, 2016).

8.1. Applying Multilingualism in Therapy

Using multiple languages in therapy helps speech and language therapists (SLTs) work better with patients. First, therapists check a patient's language skills. This shows which languages the patient is good or bad at. Then, therapy can be tailored to the patient's strong language. For example, if a patient is good in a second language (L2), it can aid therapy. This approach uses the patient's language skills to improve therapy.

In treating aphasia, using the patient's L2 can be very effective. It speeds up recovery by encouraging brain changes and using existing language knowledge. Moreover, SLTs can boost cognitive health by promoting multilingualism. Bilingualism in therapy triggers brain changes and helps adapt to new structures (Marian & Shook, 2012). By doing this, SLTs can improve both language skills and cognitive functions.

- Cognitive Health: Using multiple languages in therapy can boost cognitive health. Research shows that being multilingual improves cognitive flexibility and executive functions (Bialystok, 2020).
- Developing Multilingual Strategies: Therapists can create strategies to include multilingualism in therapy. This means using language-specific methods and tailoring plans to meet the needs of multilingual patients (Marian & Shook, 2012).

9. Conclusion

Being multilingual boosts brain flexibility and health by enhancing the brain's ability to adapt and grow new connections. Multilingual people have more gray matter in language areas and better cognitive control. Studies have linked multilingualism to better executive functions, attention, and memory.

This text highlights the need to tap into the benefits of multilingualism through speech therapy. It shows how multilingualism aids brain recovery and health, going beyond communication to improve cognitive processes and brain adaptability. The chapter explores its impact on brain structure and function, proving that multilingualism has a profound effect.

The chapter has also examined the critical periods and age-related aspects of language acquisition, emphasizing how neuroplasticity supports the capacity to learn new languages at different life stages. Although language learning is more efficient in early childhood, older adults can still benefit from multilingualism through continued cognitive engagement and strategic learning. The brain's adaptability, facilitated by neuroplasticity, allows older individuals to acquire new languages and rehabilitate language skills following neurological impairments.

Furthermore, the discussion on bilingualism and multilingualism has covered the fundamental issues of language usage and social factors. These interactions highlight the brain's remarkable capacity to manage multiple languages and adjust to various linguistic contexts.

In conclusion, multilingualism provides a compelling lens through which to explore neuroplasticity. The profound effects of learning multiple languages on brain structure and function highlight the brain's inherent flexibility and capacity for lifelong learning. As our understanding of these processes advances, it is becoming clear that multilingualism enhances cognitive abilities and supports the brain's ongoing development and health throughout life.

10. Takeaways

1. Multilingualism is defined as the ability of an individual to use more than one language fluently and effectively.
2. Broca's area, Wernicke's area, the angular gyrus, and the arcuate fasciculus are critical regions that control different components of language.
3. The acquisition of multiple languages (multilingualism) can lead to permanent changes in brain structure, which are important indicators of neuroplasticity.
4. Critical periods in language acquisition are crucial in understanding how people's ability to learn languages varies with age.
5. One of the key areas of research in multilingualism is the cognitive effects of multilingualism, particularly how acquiring multiple languages impacts cognitive development, such as executive functions and working memory.
6. Numerous studies demonstrate that bilingual people develop at a similar rate to monolingual people in terms of reaching developmental milestones.

7. Speech and language therapy aims to improve individuals' communication skills and treat speech, language, voice, and swallowing disorders.
8. Studies have shown that learning a second language stimulates different brain networks that help create new synaptic connections during aphasia treatment.
9. It has been shown that bilingualism during therapy can trigger neuroplastic changes and speed up individuals' adaptation to brain structure.
10. Investigating the interactions between learning multiple languages and the brain's structure and functions has provided insights into how multilingualism is not only limited to communication but also affects cognitive processes and brain adaptation.
11. The adaptability of the brain, facilitated by neuroplasticity, allows older individuals to acquire new languages and rehabilitate language skills following neurological impairments.

Abbreviations

CPH: Critical period hypothesis
TBI: Traumatic brain injury
AG: Angular gyrus
L1: First language
L2: Second language
TMS: Transcranial magnetic stimulation
CLI: Cross-linguistic influence
BIA: Bilingual interactive activation
RHM: Revised hierarchical model
SLTs: Speech language pathologist/therapist

Bibliography

1. Abutalebi, J., Della Rosa, P. A., Green, D. W., Hernandez, M., Scifo, P., Keim, R., Cappa, S. F., & Costa, A. (2012). Bilingualism tunes the anterior cingulate cortex for conflict monitoring. Cerebral cortex (New York, N.Y. : 1991), 22(9), 2076–2086.

2. Abutalebi, J., & Green, D. W. (2016). Neuroimaging of language control in bilinguals: Neural adaptation and reserve. *Bilingualism: Language and Cognition*, 19(4), 689–698. https://doi.org/10.1017/S1366728916000225
3. Abutalebi, J., & Green, D. W. (2023). Bilingualism and Cognitive Control: A Review and a Theoretical Framework. *Journal of Cognitive Neuroscience*. https://doi.org/10.5406/19398298.137.2.04
4. Antoniou, M., Gunasekera, G. M., & Wong, P. C. (2013). Foreign language training as cognitive therapy for age-related cognitive decline: A hypothesis for future research. *Neuroscience & Biobehavioral Reviews*, 37(10), 2689–2698. *https://doi.org/10.1016/j.neubiorev.2013.09.004*
5. Antoniou, M., Veenstra, A., & Della Sala, S. (2023). Cognitive advantages of bilingualism: A critical update. *Annual Review of Applied Linguistics*, 43, 12–29.
6. Auer, P. (2022). Bilingual conversation. Cambridge University Press, 31–69.
7. Bailey, L. M., Lockary, K., & Higby, E. (2024). Cross-linguistic influence in the bilingual lexicon: Evidence for ubiquitous facilitation and context-dependent interference effects on lexical processing. Bilingualism: *Language and Cognition*, 27(3), 495–514.
8. Berwick, R. C., & Chomsky, N. (2017). Why Only Us: Language and Evolution. MIT Press. https://doi.org/10.1353/lan.2016.0085
9. Bialystok, E. (2017). Bilingual Education for Young Children: Review of the Effects and Consequences. *International Journal of Bilingualism*, 21(6), 725–739. https://doi.org/0.1080/13670050.2016.1203859
10. Bialystok, E. (2020). Bilingualism: The good, the bad, and the indifferent. *Bilingualism: Language and Cognition*, 23(1), 19–24. https://doi.org/10.1017/S1366728908003477
11. Bialystok, E. (2023). The cognitive benefits of bilingualism: A meta-analysis of recent studies. Frontiers in Psychology, 14, Article 1155158.
12. Bialystok, E., & DePape, A. M. (2023). Cognitive Benefits of Bilingualism in Aging: Insights from Recent Studies. Journal of Cognitive Neuroscience.

13. Birdsong, D. (2007). Nativelike pronunciation among late learners of French as a second language. In O.-S. Bohn & M. Munro (Eds.), Second language speech learning: The role of language experience in speech perception and production (pp. 99–116). Amsterdam: Benjamins. https://doi.org/10.1075/lllt.17.12bir

14. Bongaerts, T., Mennen, S., & van der Slik, F. (2000). Authenticity of pronunciation in naturalistic second language acquisition: The case of very advanced late learners of Dutch as a second language. Studia Linguistica 54(2), 298–308. https://doi.org/10.1111/1467-9582.00069

15. Branzi F. M., Martin, C. D., Abutalebi, J., & Costa, A. (2014). The after-effects of bilingual language production. *Neuropsychologia*, 52, 102–116.

16. Cassilhas, R. C., Tufik, S., & De Mello, M. T. (2016). Physical exercise, neuroplasticity, spatial learning and memory. *Cellular and molecular life sciences*, *73*, 975–983. https://doi.org/10.1007/s00018-015-2102-0

17. Chomsky, N. (1986). Knowledge of language: Its nature, origin, and use. New York: Praeger.

18. Core, C., Hoff, E., Rumiche, R., & Señor, M. (2023). Dual language input and young children's cognitive development. *Journal of Child Language*, 50(2), 241–259.

19. Costa, A., Hernández, M., Costa-Faidella, J., & Sebastián-Gallés, N. (2019). On the bilingual advantage in cognitive control: A summary of six years of research and a meta-analysis. *Annual Review of Linguistics*, 5, 17–35.

20. Dijkstra, T., van Heuven, W. J. B., & Grainger, J. (1998). Simulating cross-language competition with the bilingual interactive activation model. *Psychologica Belgica*, 38(3–4), 177–196. https://doi.org/10.5334/pb.933

21. Esersin, S., & Cankuvvet Aykut, N. (2018). Early Period Indicators of Language Development: Gesture Development. *27th International Conference on Educational Sciences* (Abstract Presentation) (Publication No: 4299746).

22. Friederici, A. D. (2019). Language in Our Brain: The Origins of a Uniquely Human Capacity. MIT Press.

23. Gajardo-Vidal, A., Lorca-Puls, D. L., Team, P., Warner, H., Pshdary, B., Crinion, J. T., Leff, A. P., Hope, T. M. H., Geva, S., Seghier, M. L., Green, D. W., Bowman, H., & Price, C. J. (2021). Damage to Broca's area does not contribute to long-term speech production outcome after stroke. Brain : a journal of neurology, 144(3), 817–832.

24. Grosjean, F. (2019). *The listening bilingual: Speech perception, comprehension, and bilingualism*. John Wiley & Sons.

25. Grainger, J., Midgley, K., & Holcomb, P. J. (2010). Re-thinking the bilingual interactive-activation model from a developmental perspective (BIA-d). Language acquisition across linguistic and cognitive systems, 52, 267–283. https://doi.org/10.1075/lald.52.18gra

26. Green, D. W., & Abutalebi, J. (2016). Language control in bilinguals: The adaptive control hypothesis. *Journal of Cognitive Psychology*, 25(5), 515–530. https://doi.org/10.1080/20445911.2013.796377

27. Grundy, J. G., & Timmer, K. (2017). Bilingualism and working memory capacity: A comprehensive meta-analysis. *Second Language Research*, 33(3), 325–340. https://doi.org/10.1177/026765831667828

28. Hoff, E. (2021). Early language development and brain plasticity. *Annual Review of Psychology*, 72, 61–86.

29. Hyltenstam, K., & Abrahamsson, N. (2003a). Age of onset and ultimate attainment in near-native speakers of Swedish. In K. Fraurud & K. Hyltenstam (Eds.), Multilingualism in global and local perspectives. Selected papers from the 8th Nordic conference on bilingualism, November 1–3, 2001, Stockholm Rinkeby (pp. 319–340). Stockholm: Centre for Research on Bilingualism, Stockholm University, and Rinkeby Institute of Multilingual Research. ISBN 91-631-5238-X

30. Johnson, J. S., & Newport, E. L. (1989). Critical period effects in second language learning: The age factor. *Cognitive Psychology*, 21(1), 60–99. https://doi.org/10.1016/0010-0285(89)90003-0

31. Kuhl, P. K. (2020). Critical Periods in Language Development: New Insights from Behavioral and Neuroscientific Studies. *Frontiers in* Psychology.

32. Kroll, J. F., & Chiarello, C. (2016). Language experience and the brain: variability, neuroplasticity, and bilingualism. Language, Cognition and Neuroscience, 31(3), 345–348. https://doi.org/10.1080/23273798.2015.1086009

33. Kroll, J. F., Dussias, P. E., Bice, K., & Perrotti, L. (2015). The role of language experience in the bilingual experience. *Annual Review of Linguistics*, 1, 377–397.

34. Li, P. (2022). *The Cognitive Neuroscience of Language*. Cambridge University Press.

35. Marian, V., & Shook, A. (2012). The cognitive benefits of being bilingual. *Cerebrum, 2012*, 13. PMID: 23447799

36. Mårtensson, J., Eriksson, J., Bodammer, N. C., Lindgren, M., Johansson, M., Nyberg, L., & Lövdén, M. (2022). Growth of language-related brain areas after foreign language learning. *NeuroImage*, 63(1), 240–244. https://doi.org/10.1016/j.neuroimage.2012.06.043

37. Myers-Scotton, C. (2020). *Multiple voices: An introduction to bilingualism*. John Wiley & Sons. Cambridge University Press.

38. Nadeau, S. E. (2015). Neuroplastic mechanisms of language recovery after stroke. *Cognitive plasticity in neurologic disorders*, 61–84.

39. Paradis, J. (2018). Bilingual children's acquisition of English verb morphology: Effects of language exposure, structure complexity, and input quality. *Journal of Child Language*, 45(4), 807–830. https://doi.org/10.1111/j.1467-9922.2010.00567.x

40. Perani, D., & Abutalebi, J. (2015). Bilingualism, dementia, cognitive and neural reserve. *Current opinion in neurology*, 28(6), 618–625. https://doi.org/10.1097/WCO.0000000000000267

41. Phillips, C. (2017). Lifestyle modulators of neuroplasticity: how physical activity, mental engagement, and diet promote cognitive health during aging. *Neural plasticity*, 2017(1), 3589271. https://doi.org/10.1155/2017/3589271

42. Seghier, M. L. (2013). The angular gyrus: multiple functions and multiple subdivisions. *The Neuroscientist*, 19(1), 43–61. https://doi.org/10.1177/1073858412440596

43. Smith, J., Brown, T., & Jones, P. (2023). Language in Education: The Role of Communication in Learning. *Oxford Uviversity Press*

44. Swanberg, M. M., Nasreddine, Z. S., Mendez, M. F., & Cummings, J. L. (2007). Speech and language. *Textbook of clinical neurology*, 79–98.

45. Yuan, J. (2023). The Interpretation of Deixis in Cognitive Linguistics: A Literature Review. *International Journal of English Language Studies*, *5*(2), 01–04. https://doi.org/10.32996/ijels.2023.5.2.1
46. Xia, Y., Shin, S. Y., & Kim, J. C. (2024). Cross-cultural intelligent language learning system (cils): Leveraging ai to facilitate language learning strategies in cross-cultural communication. *Applied Sciences*, 14(13), 5651. https://doi.org/10.3390/app14135651

Sleep and Synapses: Unlocking the Neuroplastic Power of Rest

Selen Gür Özmen and Yahya Dahhan

1. Introduction

Sleep is an indispensable biological function that has endured evolution and is pivotal in maintaining health and cognitive integrity. Despite its universal presence in the animal kingdom, its precise purpose remains enigmatic, especially in organisms without central nervous systems. What is evident, however, is that sleep deprivation exerts profound adverse effects on both physical health and cognitive performance (Banks & Dinges, 2007). By inducing a temporary state of unconsciousness and reducing responsiveness to external stimuli, sleep facilitates critical processes of brain reorganization and recovery (Dijk et al., 2010).

Although sleep is an energy-intensive activity, it is a widespread phenomenon across species, suggesting its evolutionary significance. It is believed to serve adaptive purposes by conserving energy, avoiding predation during vulnerable periods, and enhancing brain function (Frank et al., 2001). However, the evolutionary trade-off of reduced alertness during sleep raises an intriguing question: Why has sleep persisted across millennia? One compelling theory proposes that sleep plays a crucial role in reorganizing synaptic connections in the brain and optimizing learning and memory processes (Diekelmann & Born, 2010). This reorganization enhances behavioral adaptability and cognitive flexibility, particularly in more advanced species. Beyond its neurological impact, sleep supports immune function and facilitates tissue repair, underscoring its multifaceted benefits (Xie et al., 2013).

The developmental role of sleep is particularly significant. Sleep during early life plays a pivotal role in promoting synaptic remodeling and streamlining the integration of sensory and cognitive inputs (Frank et al., 2001). Nevertheless, the precise mechanisms underlying sleep-induced neuroplastic changes remain incompletely understood (Aton et al., 2013). Recent research explores how different sleep phases contribute to various cognitive and physiological functions.

Interestingly, individuals experience varying levels of resilience to sleep deprivation, influenced by factors such as age, environmental context, and genetic predisposition (Cirelli & Tononi, 2008). While some individuals recover swiftly from short-term disruptions, prolonged sleep disturbances during critical developmental windows can have enduring impacts on brain circuits and behavior (Blumberg et al., 2013). Notably, some individuals exhibit an innate resilience to sleep loss, potentially due to genetic factors that mitigate its detrimental effects (Dennis et al., 2017).

Emerging theories suggest that sleep may serve therapeutic purposes, particularly in modifying maladaptive memories. For instance, targeted sleep interventions could be employed to weaken harmful memories, offering promising avenues for treating conditions such as post-traumatic stress disorder (Pace-Schott et al., 2015). Sleep is indispensable in promoting learning, memory consolidation, and emotional flexibility, ensuring the brain remains adaptable to ever-changing environments (Puentes-Mestril et al., 2019).

2. Phases of Sleep

The sleep cycle is a complex process that consists of two primary types: Non-Rapid Eye Movement (NREM) and Rapid Eye Movement (REM) sleep. NREM sleep, in turn, is further divided into three distinct stages, each with unique characteristics and functions (Kryger et al., 2017) (Figure 1).

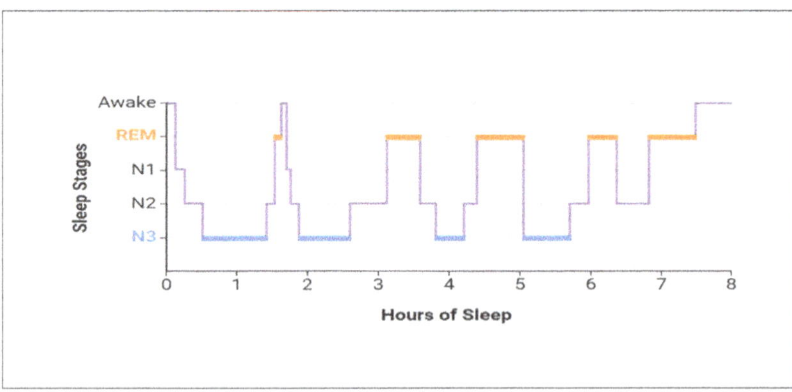

Figure 1: Sleep stages

- Stage 1 (NREM 1): This initial light sleep phase is marked by a gentle drift in and out of sleep. Imagine yourself dozing off in a comfortable armchair—you are still partially aware of your surroundings, but your eyelids are growing heavy. During this stage, your brain waves slow down, and your muscles relax, preparing your body for a deeper sleep (Kryger et al., 2017).
- Stage 2 (NREM 2): As you sink into a more profound sleep, your brain activity increases, and sleep spindles—brief bursts of rapid brain waves—and K-complexes—sudden, high-amplitude electrical waves—emerge (Ngo et al., 2013). This stage is crucial for the regulation of body temperature and heart rate. For instance, during this stage, your body conserves energy by lowering your heart rate and blood pressure. It acts as a transitional phase, bridging light and deep sleep while consolidating learned information from the day (Walker & Stickgold, 2013).
- Stage 3 (NREM 3): Often referred to as *Slow-Wave Sleep* (SWS), this phase represents the deepest sleep stage and is characterized by slow brain waves and minimal responsiveness to external stimuli. Previously divided into Stage 3 and Stage 4, modern sleep science consolidates these into one stage due to their shared characteristics (Kryger et al., 2017). During this phase, the body focuses on repairing and regenerating cells, building bone and muscle, and strengthening the immune system. It is also critical for clearing beta-amyloid via the glymphatic system (Figure 2), potentially lowering Alzheimer's risk (Xie et al., 2013). Through its role in global neural oscillations, slow-wave sleep enables synaptic homeostasis, fosters neuronal repair, and sustains neuroplastic processes (Vyazovskiy & Harris, 2013).
- REM Sleep: This stage is characterized by rapid eye movements, increased brain activity, and vivid dreaming. During REM sleep, your brain processes and consolidates memories, organizing important information and discarding what is unnecessary (Stickgold & Walker, 2013). REM sleep also plays a crucial role in emotional regulation, helping manage stress and anxiety. For instance, research has shown that REM sleep reduces the emotional intensity of traumatic experiences, aiding in resilience against stress and adversity (Walker, 2009). Furthermore, REM sleep is linked to creativity and problem-solving, enabling unique connections between disparate ideas central for innovation and learning (Walker & Stickgold, 2010).

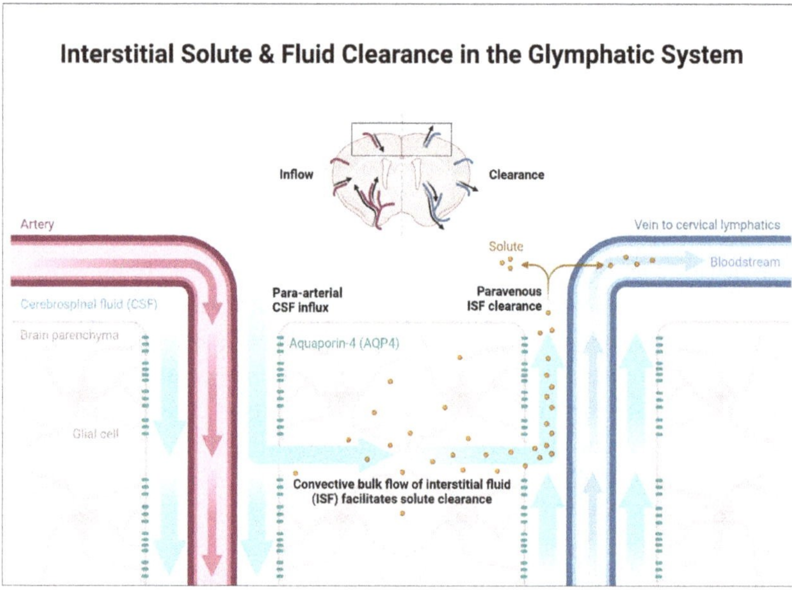

Figure 2: Illustration of the glymphatic system, depicting cerebrospinal fluid (CSF) influx and interstitial solute clearance via perivascular pathways. Aquaporin-4 channels facilitate fluid flow, contributing to beta-amyloid removal from brain tissue (created in BioRender.com).

2.1. Sleep Cycle Dynamics

Throughout the night, the complete sleep cycle typically repeats 4–6 times, each lasting around 90 minutes. Early in the night, slow-wave sleep dominates, allowing for physical restoration, while REM sleep periods become progressively longer in later cycles, emphasizing emotional and cognitive recovery. This dynamic interplay ensures that both body and brain receive the restoration and processing they need (Diekelmann & Born, 2010; Kryger et al., 2017). Recent advancements in sleep research have explored various methods to optimize these sleep cycles and improve transitions between sleep phases. One such method is closed-loop auditory stimulation, an emerging technology that delivers auditory cues in sync with an individual's physiological state, such as breathing or brainwaves, to enhance sleep quality. Unlike traditional auditory interventions, closed-loop systems continuously monitor real-time physiological signals and provide tailored auditory feedback to guide bodily rhythms. Recent studies have shown that using closed-loop auditory stimulation to guide respiration can significantly reduce the time spent initiating sleep, suggesting its potential to optimize sleep stages and support neuroplastic processes (Yoon & Yoon, 2023). This personalized approach may offer a promising avenue for improving sleep-dependent learning, memory consolidation, and emotional regulation through more efficient transitions between sleep phases.

2.2. Sleep Evolution Across Life Stages Infancy

Newborns sleep approximately 16–18 hours a day, with sleep cycles lasting about 50 minutes. REM sleep is particularly abundant, aligning with heightened synaptic formation and elimination (Jouvet-Mounier et al., 1970; Li et al., 2017).

- Childhood: Sleep duration decreases to about 10–12 hours, with more stable sleep patterns.
- Adolescence: Teenagers require about 8–10 hours of sleep but often experience shifts in circadian rhythms.
- Adulthood: Adults typically need 7–9 hours of sleep, with a more consistent sleep architecture (Mindell & Owens, 2015). In older adults, reducing REM and slow-wave sleep significantly impacts learning and memory functions (Mindell & Owens, 2015). In older adults, reductions in REM and slow-wave sleep significantly impact learning and memory and may have important consequences for neuroplasticity.

3. Impact of Sleep Phases on Cognition and Mood

REM sleep boosts creativity and problem-solving. It links unrelated ideas, necessary for creativity. Studies show insufficient REM sleep reduces innovative thinking (Walker & Stickgold, 2010). During REM, key brain areas like the prefrontal cortex and hippocampus reorganize memories. This sparks new ideas. Meanwhile, NREM sleep detoxifies the brain, especially SWS. The glymphatic system clears harmful substances, like beta-amyloid and tau proteins, during SWS. These are linked to Alzheimer's. Disrupted SWS may raise Alzheimer's risk by allowing toxic build-up (Xie et al., 2013; Ju et al., 2014). Boosting SWS could lower this risk and protect brain health. Sleep plays a foundational role in supporting cognitive processes and emotional regulation. Each phase plays a role. NREM and REM sleep have unique functions.

NREM sleep, especially SWS, is crucial for memory and learning. During SWS, the brain stabilizes and integrates new information. During this phase, hippocampal-neocortical dialogue enables memory traces to be stabilized, a mechanism critical for long-term retention (Diekelmann & Born, 2010). Research shows that boosting SWS with auditory stimulation can enhance memory (Ngo et al., 2013).

Sleep plays a crucial role in memory consolidation during infancy. It helps stabilize newly acquired information and integrate it with existing knowledge, ensuring long-term retention (Seehagen et al., 2019).

By modulating limbic system activity, REM sleep facilitates the processing of emotional experiences and the stabilization of mood. It is marked by active brain activity and vivid dreams. This phase helps manage emotional responses and integrates memories (Walker & van der Helm, 2009).

Lack of REM sleep boosts emotional reactions and disrupts regulation. For example, deprived individuals show more amygdala activity to emotional cues, indicating reduced control (Yoo et al., 2007). REM sleep also helps erase fear memories, which is necessary for conditions like PTSD (Pace-Schott et al., 2015).

Interruptions in NREM or REM sleep harm cognition and mood. Chronic lack of sleep affects attention, decision-making, and executive function (Lim & Dinges, 2010). It also raises the risk of depression and anxiety by impairing emotional processing (Goldstein & Walker, 2014). Sleep disorders in children and adolescents can significantly impact cognitive development and emotional stability, leading to long-term behavioral and academic challenges (Buckhalt et al., 2015).

Chronic deprivation can worsen neurodegenerative diseases like Parkinson's and Huntington's (Lim & Dinges, 2010). This shows the need for more research on the effects of long-term REM and SWS deprivation.

In summary, all sleep phases are crucial for cognitive and emotional health. Sustained cognitive efficiency and emotional stability rely heavily on the consistency and quality of sleep.

4. Effects of Sleep on Motor Learning Through Neuroplasticity

Neuroplasticity allows the brain to form new connections crucial for learning and adapting. It is critical in motor learning, where skills are honed through practice. Sleep, especially REM sleep, boosts neuroplasticity and aids in learning these skills.

Studies show a strong link between sleep and motor learning. For example, after piano practice, those who slept improved more than those who stayed awake. Enhanced performance in tasks like typing and finger-tapping following sufficient REM and NREM sleep underscores the dual-phase contribution of sleep to motor learning and consolidation. During REM sleep, brain circuits for motor tasks are reactivated, helping to refine skills. Neuroimaging shows increased activity in motor areas of the brain during REM sleep after learning. In NREM sleep, sleep spindles aid in moving skills from short-term to long-term memory.

Sleep improves immediate skills and long-term brain adaptability. Chronic lack of sleep harms performance and disrupts crucial brain functions. This highlights the need for good sleep to learn and maintain brain flexibility. REM and NREM sleep form the foundation of motor learning, driving both immediate performance gains and long-term plasticity. Sleep is not a luxury—but a biological necessity for cognitive refinement.

5. The Role of Sleep in Neuroplasticity

Sleep orchestrates the reorganization of neural circuits essential to learning and memory, with REM phases selectively consolidating motor competencies (Walker et al., 2002).

Studies show that REM sleep boosts the brain's motor learning circuits. After training, brain scans revealed increased activity in motor areas during REM sleep, which can indicate memory processing (Maquet et al., 2000).

5.1. Impact of Sleep Deprivation on Motor Learning and Other Cognitive Functions

Lack of sleep hampers motor learning and brain changes. Restricting sleep, especially REM, disrupts skill consolidation, affecting performance and brain plasticity (Fischer et al., 2002).

Sleep movements such as twitches are believed to play a role in motor development in infancy. Recent research suggests that early motor milestones predict the dynamics of infant sleep-related movements (DeMasi et al., 2022).

Sleep deprivation lowers the expression of crucial genes for brain adaptability. This leads to slower learning and recovery from movement skill losses (Cirelli, 2009).

Sleep deprivation significantly impacts cognitive performance and increases perceived mental workload, especially in high-demand professions such as healthcare. Studies using the NASA-TLX method have shown that inadequate rest can impair decision-making, increase error rates, and elevate stress levels among healthcare professionals (Citra et al., 2020).

By regulating synaptic remodeling and behavioral adaptability, sleep safeguards neuroplastic integrity—although its absence leads to varying degrees of dysfunction based on the nature and length of deprivation. Short-term sleep deprivation can temporarily impair memory consolidation and synaptic remodeling, but prolonged sleep loss has more profound effects, disrupting emotional regulation, cognitive performance, and long-term neural health. Interestingly, studies have shown that different brain regions and functions exhibit varying vulnerability to sleep deprivation. For instance, while some neural circuits demonstrate remarkable resilience to brief periods of sleep loss, others—particularly those involved in higher-order cognitive tasks—are more sensitive to disruptions (Weiss & Donlea, 2022). This variability highlights the complex relationship between sleep and neuroplasticity, suggesting that the brain's adaptive capacity is intricately linked to the quality and quantity of sleep. Moreover, chronic sleep deprivation can impair the balance between synaptic strengthening and weakening, potentially leading to maladaptive changes in neural circuits that contribute to mood disorders, impaired learning, and memory deficits.

5.2. Mechanisms Underlying Sleep-Dependent Motor Learning

Sleep aids motor learning through complex brain interactions. During REM sleep, acetylcholine and other substances boost synaptic plasticity. They promote long-term potentiation and strengthen connections (Rasch & Born, 2013). Reactivating neural patterns from motor tasks in sleep further enhances circuits, improving skill retention and performance (Stickgold, 2005).

Additionally, sleep spindles in NREM sleep contribute to motor learning. These bursts of brain activity help transfer information between the hippocampus and neocortex. This process integrates new motor skills into existing networks (Tamaki et al., 2013).

5.3. Synaptic Homeostasis Hypothesis

The synaptic homeostasis hypothesis suggests sleep reduces synaptic strength built up during the day. This balance prevents overload (Tononi & Cirelli, 2006). During NREM sleep, the brain consolidates useful information while discarding excess input—a neuroplastic mechanism critical for mental clarity and learning efficiency.

5.4. Circadian Rhythms and Sleep

Our body's internal clock, circadian rhythms, significantly influences sleep and brain adaptability. It determines the timing of sleep, hormone release, and brain function. Disruptions, such as shift work or jet lag, harm sleep and adaptability (Wright et al., 2013). Understanding this link can improve sleep and enhance performance.

A melatonin deficiency is a critical disruption that harms sleep and adaptability (Wright et al., 2013). It is crucial for sleep-related interventions. Circadian rhythms control sleep, hormones, temperature, and more. In teenagers, melatonin production shifts later, delaying sleep. Early school times worsen this, leading to chronic sleep deprivation and social jet lag (Roenneberg et al., 2012). Social jet lag occurs when our internal clocks do not match societal schedules, harming cognitive function, lowering grades, and increasing mental health risks (Beebe et al., 2017). It also affects neuroplasticity, disrupting memory and learning. Light exposure regulates circadian rhythms and complexly affects human metabolism, influencing energy balance, hormonal responses, and sleep quality (Ishihara et al., 2023).

In older adults, circadian rhythms weaken, disrupting sleep and reducing slow-wave sleep (SWS). This decline is linked to memory issues and a higher risk of Alzheimer's (Mander et al., 2017). Lifestyle changes, such as light therapy and regular sleep, can help. Understanding circadian rhythms across ages is critical to improving sleep and supporting brain health.

5.5. Developmental Phases and Evolution of Sleep

Synaptic plasticity is critical for brain growth, boosting learning (Cooke & Bliss, 2006). Disruptions here link to sleep disorders (Blumberg et al., 2020). Sleep is crucial in early development, aiding brain plasticity (Marks et al., 1995).

Infants often twitch in REM sleep. This may enhance coordination (Blumberg et al., 2013). Such sleep transitions infants from clumsy to precise movements (Khazipov et al., 2004).

Sleep disruptions in development can halt growth (Jones et al., 2019). Sleep guides the refinement of the visual cortex and supports healthy brain development across species (Hubel & Wiesel, 1970).

6. Conclusion

Sleep is far more than a biological necessity—it is a dynamic and active process that continuously shapes the brain's architecture. From infancy to late adulthood, sleep governs the balance between synaptic formation and elimination, preserves emotional equilibrium, and consolidates the very skills and memories that define human learning.

As research advances, sleep is no longer viewed as a passive state but as a critical period of neurobiological refinement. REM and NREM phases contribute distinctly to cognitive resilience, emotional regulation, and adaptive behavior, underscoring the complexity of sleep-dependent plasticity.

Yet, this understanding raises profound questions: What are the long-term consequences of chronic sleep restriction in developing brains? How can emerging technologies ethically modulate sleep to enhance cognition? And to what extent should educational and clinical systems integrate sleep science into policy and practice?

In the chapters ahead, we will explore the molecular pathways and behavioral outcomes through which neuroplasticity operates. As we deepen our insight into the intricate relationship between sleep and brain adaptability, one thing becomes clear: sleep is not a luxury. It is a biological imperative—essential not only for survival, but for human flourishing.

7. Takeaways

1. Synaptic Reorganization During Sleep
   Sleep actively drives synaptic remodeling essential for learning and memory. REM and NREM stages operate in tandem to stabilize emotional responses and consolidate new information across neural networks.
2. Consequences of Sleep Deprivation
   Prolonged sleep loss impairs neuroplasticity, undermines cognitive performance, and disrupts emotional regulation. It has been linked to an elevated risk of neurodegenerative disorders, including Alzheimer's disease.
3. Phase-Specific Contributions of Sleep
   Slow-wave sleep (SWS) facilitates glymphatic clearance and supports cognitive resilience, while REM sleep fosters emotional recalibration and fuels creative problem-solving.
4. Sleep-Dependent Motor Learning
   Sleep enhances procedural memory by reactivating motor circuits during REM and transferring newly acquired motor skills into long-term memory stores during NREM.
5. Developmental and Evolutionary Importance
   Across species, sleep—particularly REM—plays a foundational role in brain maturation, guiding synaptic formation, pruning, and adaptive circuit refinement in early life.
6. Synaptic Homeostasis Mechanism
   NREM sleep downscales synaptic strength to counteract daytime accumulation, preserving cognitive clarity and facilitating efficient long-term memory encoding.
7. Closed-Loop Auditory Stimulation
   This innovative neurotechnology aligns auditory cues with endogenous brain rhythms to enhance slow-wave activity, improve sleep quality, and strengthen memory consolidation.
8. Circadian Rhythms and Neuroplasticity
   Circadian timing systems orchestrate sleep-wake patterns and modulate plasticity. Disruptions impair both cognition and mood, whereas interventions like light therapy restore rhythmic balance and support brain health.

Abbreviations

NREM: Non-rapid eye movement
REM: Rapid eye movement
SWS: Slow-wave sleep

Bibliography

1. Aton, S. J., Broussard, C., Dumoulin, M., Seibt, J., Watson, A., Coleman, T., et al. (2013). Visual experience and subsequent sleep induce sequential plastic changes in putative inhibitory and excitatory cortical neurons. Proceedings of the National Academy of Sciences, 110, 3101–3106. https://doi.org/10.1073/pnas.1208093110
2. Banks, S., & Dinges, D. F. (2007). Behavioral and physiological consequences of sleep restriction. Journal of Clinical Sleep Medicine, 3, 519–528. https://doi.org/10.5664/jcsm.26918
3. Blumberg, M. S., Coleman, C. M., Gerth, A. I., & McMurray, B. (2013). Spatiotemporal structure of REM sleep twitching reveals developmental origins of motor synergies. Current Biology, 23, 2100–2109. https://doi.org/10.1016/j.cub.2013.08.055
4. Blumberg, M. S., Dooley, J. C., & Sokoloff, G. (2020). The developing brain is revealed during sleep. Current Opinion in Physiology, 15, 14–22. https://doi.org/10.1016/j.cophys.2019.11.002
5. Buckhalt, J., El-Sheikh, M., & Wolfson, A. (2015). Sleep and sleep disorders in children and adolescents. National Association of School Psychologists. Communique, 44(2), 26–27.
6. Cirelli, C. (2009). The genetic and molecular regulation of sleep: From fruit flies to humans. Nature Reviews Neuroscience, 10(8), 549–560. https://doi.org/10.1038/nrn2683
7. Cirelli, C., & Tononi, G. (2008). Is sleep essential? PLoS Biology, 6(8), e216. https://doi.org/10.1371/journal.pbio.0060216
8. Citra, R. W., Wahyudi, T., & Rahmahwati, R. (2020). Analisa beban kerja mental pada perawat di rumah sakit XYZ menggunakan metode NASA-TLX. https://core.ac.uk/download/350004197.pdf

9. Cooke, S. F., & Bliss, T. V. P. (2006). Plasticity in the human central nervous system. Brain, 129, 1659–1673. https://doi.org/10.1093/brain/awl082

10. DeMasi, A., Horger, M. N., Scher, A., & Berger, S. E. (2022). Infant motor development predicts the dynamics of movement during sleep. https://academicworks.cuny.edu/cgi/viewcontent.cgi?article=1350&context=si_pubs

11. Dennis, L. E., Wohl, R. J., Selame, L. A., & Goel, N. (2017). Healthy adults display long-term trait-like neurobehavioral resilience and vulnerability to sleep loss. Scientific Reports, 7, 14889. https://doi.org/10.1038/s41598-017-14006-7

12. Diekelmann, S., & Born, J. (2010). The memory function of sleep. Nature Reviews Neuroscience, 11, 114–126. https://doi.org/10.1038/nrn2762

13. Dijk, D.-J., Groeger, J. A., Stanley, N., & Deacon, S. (2010). Age-related reduction in daytime sleep propensity and nocturnal slow wave sleep. Sleep, 33, 211–223. https://doi.org/10.1093/sleep/33.2.211

14. Ednick, M., Cohen, A. P., McPhail, G. L., Beebe, D., Simakajornboon, N., & Amin, R. S. (2009). A review of the effects of sleep during the first year of life on cognitive, psychomotor, and temperament development. Sleep, 32, 1449–1458. https://doi.org/10.1093/sleep/32.11.1449

15. Fischer, S., Hallschmid, M., Elsner, A. L., & Born, J. (2002). Sleep forms memory for finger skills. Proceedings of the National Academy of Sciences, 99(18), 11987–11991. https://doi.org/10.1073/pnas.182178199

16. Frank, M. G., Issa, N. P., & Stryker, M. P. (2001). Sleep enhances plasticity in the developing visual cortex. Neuron, 30, 275–287. https://doi.org/10.1016/s0896-6273(01)00279-3

17. Goldstein, A. N., & Walker, M. P. (2014). The role of sleep in emotional brain function. Annual Review of Clinical Psychology, 10, 679–708. https://doi.org/10.1146/annurev-clinpsy-032813-153716

18. Hobson, J. A., & Pace-Schott, E. F. (2002). The cognitive neuroscience of sleep: Neuronal systems, consciousness and learning. Nature Reviews Neuroscience, 3(9), 679–693. https://doi.org/10.1038/nrn915

19. Hubel, D. H., & Wiesel, T. N. (1970). The period of susceptibility to the physiological effects of unilateral eye closure in kittens. Journal of Physiology, 206, 419–436. https://doi.org/10.1113/jphysiol.1970.sp009022

20. Ishihara, A., Courville, A., & Chen, K. (2023). The complex effects of light on metabolism in humans. Nutrients, 15(6), 1391. https://doi.org/10.3390/nu15061391

21. Jones, C. E., Opel, R. A., Kaiser, M. E., Chau, A. Q., Quintana, J. R., Nipper, M. A., et al. (2019). Early-life sleep disruption increases parvalbumin in the primary somatosensory cortex and impairs social bonding in prairie voles. Science Advances, 5, eaav5188. https://doi.org/10.1126/sciadv.aav5188

22. Jouvet-Mounier, D., Astic, L., & Lacote, D. (1970). Ontogenesis of sleep states in rats, cats, and guinea pigs during the first postnatal month. Developmental Psychobiology, 2, 216–239. https://doi.org/10.1002/dev.420020407

23. Khazipov, R., Sirota, A., Leinekugel, X., Holmes, G. L., Ben-Ari, Y., & Buzsáki, G. (2004). Early motor activity drives spindle bursts in the developing somatosensory cortex. Nature, 432, 758–761. https://doi.org/10.1038/nature03132

24. Kryger, M. H., Roth, T., & Dement, W. C. (2017). Principles and practice of sleep medicine. Elsevier.

25. Li, W., Ma, L., Yang, G., & Gan, W.-B. (2017). REM sleep selectively prunes and maintains new synapses in development and learning. Nature Neuroscience, 20, 427–437. https://doi.org/10.1038/nn.4479

26. Lim, J., & Dinges, D. F. (2010). A meta-analysis of the impact of short-term sleep deprivation on cognitive variables. Psychological Bulletin, 136(3), 375–389. https://doi.org/10.1037/a0018883

27. Mander, B. A., Winer, J. R., & Walker, M. P. (2017). Sleep and human aging. Neuron, 94(1), 19–36. https://doi.org/10.1016/j.neuron.2017.02.004

28. Maquet, P., Schwartz, S., Passingham, R., & Frith, C. (2000). Sleep-related consolidation of a visuomotor skill: Brain mechanisms assessed by functional magnetic resonance imaging. Journal of Neuroscience, 20(23), 8623–8630. https://doi.org/10.1523/JNEUROSCI.20-23-08623.2000

29. Marks, G. A., Shaffery, J. P., Oksenberg, A., Speciale, S. G., & Roffwarg, H. P. (1995). A functional role for REM sleep in brain maturation. Behavioural Brain Research, 69, 1–11. https://doi.org/10.1016/0166-4328(95)00018-o

30. Mindell, J. A., & Owens, J. A. (2015). A clinical guide to pediatric sleep: Diagnosis and management of sleep problems. Lippincott Williams & Wilkins.

31. Ngo, H. V. V., Martinetz, T., Born, J., & Mölle, M. (2013). Auditory closed-loop stimulation of the sleep-slow oscillation enhances memory. Neuron, 78(3), 545–553. https://doi.org/10.1016/j.neuron.2013.03.006

32. Pace-Schott, E. F., Germain, A., & Milad, M. R. (2015). Effects of sleep on memory for conditioned fear and fear extinction. Psychological Bulletin, 141, 835–857. https://doi.org/10.1037/bul0000014

33. Puentes-Mestril, C., Roach, J., Niethard, N., Zochowski, M., & Aton, S. J. (2019). How rhythms of the sleeping brain tune memory and synaptic plasticity. Sleep, 42, zsz095. https://doi.org/10.1093/sleep/zsz095

34. Rasch, B., & Born, J. (2013). About sleep's role in memory. Physiological Reviews, 93(2), 681–766. https://doi.org/10.1152/physrev.00032.2012

35. Roenneberg, T., Allebrandt, K. V., Merrow, M., & Vetter, C. (2012). Social jetlag and obesity. Current Biology, 22(10), 939–943. https://doi.org/10.016/j.cub.2012.03.038

36. Seehagen, S., Zmyj, N., & Herbert, J. S. (2019). Remembering in the context of internal states: The role of sleep for infant memory. https://core.ac.uk/download/212722645.pdf

37. Stickgold, R. (2005). Sleep-dependent memory consolidation. Nature, 437(7063), 1272–1278. https://doi.org/10.1038/nature04286

38. Stickgold, R., & Walker, M. P. (2013). Sleep-dependent memory triage: Evolving generalization through selective processing. Nature Neuroscience, 16(2), 139–145. https://doi.org/10.1038/nn.3303

39. Tamaki, M., Matsuoka, T., Nittono, H., & Hori, T. (2013). Fast sleep spindle (13–15 Hz) activity correlates with sleep-dependent

improvement in visuomotor performance. Sleep, 36(12), 1947–1955. https://doi.org/10.5665/sleep.3230

40. Tononi, G., & Cirelli, C. (2006). Sleep function and synaptic homeostasis. Sleep Medicine Reviews, 10(1), 49–62. https://doi.org/10.1016/j.smrv.2005.05.002
41. Vyazovskiy, V. V., & Harris, K. D. (2013). Sleep and the single neuron: The role of global slow oscillations in individual cell rest. Nature Reviews Neuroscience, 14(6), 443–451. https://doi.org/10.1038/nrn3494
42. Walker, M. P. (2009). The role of sleep in cognition and emotion. Nature Reviews Neuroscience, 10(11), 741–752. https://doi.org/10.1111/j.1749-6632.2009.04416.x
43. Walker, M. P., & Stickgold, R. (2010). Overnight alchemy: Sleep-dependent memory evolution. Nature Reviews Neuroscience, 11(3), 218–219. https://doi.org/10.1038/nrn2762-c1
44. Walker, M. P., & van der Helm, E. (2009). Overnight therapy? The role of sleep in emotional brain processing. Psychological Bulletin, 135(5), 731–748. https://doi.org/10.1037/a0016570
45. Walker, M. P., Brakefield, T., Morgan, A., Hobson, J. A., & Stickgold, R. (2002). Practice with sleep makes perfect: Sleep-dependent motor skill learning. Neuron, 35(1), 205–211. https://doi.org/10.1016/S0896-6273(02)00746-8
46. Weiss, J., & Donlea, J. (2022). Roles for sleep in neural and behavioral plasticity: Reviewing variation in the consequences of sleep loss—Frontiers in Behavioral Neuroscience. https://doi.org/10.3389/fnbeh.2021.777799
47. Wright, K. P., Jr., Bogan, R. K., & Wyatt, J. K. (2013). Shift work and the assessment and management of shift work disorder (SWD). Sleep Medicine Reviews, 17(1), 41–54. https://doi.org/10.1016/j.smrv.2012.02.002
48. Xie, L., Kang, H., Xu, Q., Chen, M. J., Liao, Y., Thiyagarajan, M., … & Nedergaard, M. (2013). Sleep drives metabolite clearance from the adult brain. Science, 342(6156), 373–377. https://doi.org/10.1126/science.1241224

49. Yoo, S. S., Gujar, N., Hu, P., Jolesz, F. A., & Walker, M. P. (2007). The human emotional brain without sleep—a prefrontal amygdala disconnect. Current Biology, 17(20), R877–R878. https://doi.org/10.1016/j.cub.2007.08.007

50. Yoon, H., & Yoon, H. (2023). Closed-loop auditory stimulation to guide respiration: Preliminary study to evaluate the effect on time spent in sleep initiation during a nap. Sensors, 23(14), 6468. https://doi.org/10.3390/s23146468

Neural Weavers: The Art of Social Connections in Brain Shaping

Selin Yilmaz and Selen Gür Özmen

1. Introduction

Social relationships play a critical role in shaping the development and functioning of the human brain by influencing our thoughts, emotions, and behaviors in profound ways (Davidson & McEwan, 2012). Emotional support from close partners, for instance, has been shown to increase oxytocin levels, a neuropeptide associated with enhanced trust and social bonding (Kosfeld et al., 2005). This biochemical response does not only foster intimacy but also contributes to the strengthening of neural pathways involved in emotional regulation (Kosfeld et al., 2005). Furthermore, evidence from developmental research indicates that children in institutional care who receive additional emotional and social support exhibit improved functioning in brain regions associated with emotional and social processing (McLaughlin et al., 2011). These findings underscore the essential role of supportive social environments in fostering healthy brain development across the lifespan (McLaughlin et al., 2011; Kosfeld et al., 2005).

Beyond immediate emotional and behavioral outcomes, social interactions actively contribute to long-term changes in brain structure and function. These changes occur through the formation and reinforcement of neural connections, a phenomenon known as neuroplasticity (Bourassa et al., 2015). Neuroplasticity is fundamental to processes such as learning and memory, and it is highly responsive to environmental input, particularly from social contexts. For example, research involving London taxi drivers demonstrated that individuals with more extensive experience navigating the city exhibited increased volume in the hippocampus, a brain region essential for spatial memory (Maguire et al., 2000). This finding illustrate how socially embedded experiences can shape and reorganize neural architecture over time.

Building on this foundation, the present paper explores the intricate relationship between social relationships and neuroplasticity. Specifically, it investigates how various dimensions of our social world, ranging from supportive bonds to complex social interactions, affect the structure and function of neural networks. Gaining a deeper understanding of this dynamic can shed light on the pivotal role of social connections in promoting cognitive flexibility, emotional regulation, and overall psychological well-being.

2. Shaping Neural Pathways Through Social Interactions

2.1. Mechanisms of Influence

Social interactions play a fundamental role in shaping the brain's neural architecture. Rather than being static, the brain is a dynamic network of pathways that are continually influenced by interpersonal engagement. Research has demonstrated that engaging in conversations, sharing experiences, and spending time with close others can significantly promote synaptic growth, particularly in regions associated with social cognition and emotional processing (Hari & Kujala, 2009). Activities that involve empathy, cooperation, and positive affect not only maintain existing neural connections but also enhance the development of new ones (Seeman et al., 2001; Hari & Kujala, 2009).

Importantly, these interactions contribute to more than just cognitive maintenance, they actively reorganize the brain. Hari and Kujala (2009) emphasized that repeated exposure to socially rich environments enhances circuits responsible for empathy and social reasoning. Such repeated stimulation results in enduring neurobiological changes, contributing to improved social understanding and emotional attunement.

Maintaining an active social life is also associated with greater cognitive flexibility, the capacity to adjust thoughts and behaviors in response to shifting social demands (Wang et al., 2021). This adaptability is particularly critical in navigating the fluid and often ambiguous nature of interpersonal interactions. Studies on early childhood development provide compelling evidence for the foundational role of cognitive flexibility. For instance, Wong et al. (2008) and Silkenbeumer et al. (2016) demonstrated that children's ability to adapt to new emotional and contextual demands is crucial for healthy emotional development. Between the ages of three and six, children begin to comprehend the multidimensional nature of emotions, which requires both verbal ability and flexible thinking (Baron-Cohen, 1998; Pons et al., 2000, 2004).

Verbal ability, in particular, serves as a key mechanism facilitating cognitive flexibility. It enables more precise communication, allowing children to interpret and respond to social cues more effectively (Pons et al., 2003; Lindquist et al., 2015). Also enhanced verbal skills contribute to an improved capacity to navigate social norms and understand abstract emotional states. In this regard, neuroplasticity supports the strengthening of connections between language centers and higher-order cognitive processes (Harris et al., 2005), illustrating the interplay between linguistic and social-cognitive development.

Moreover, cognitive flexibility underpins the ability to manage uncertainty and respond adaptively in complex social situations. Individuals with high cognitive flexibility are better equipped to interpret emotions, adjust their behavior, and communicate effectively, skills essential for building and sustaining meaningful relationships (Brennan & Clark, 1996).

These findings underscore the centrality of cognitive flexibility, verbal competence, and emotional understanding in shaping social functioning. Together, they form a triadic framework that influences neural plasticity and promotes the development of robust, adaptable social brains.

This broader capacity for change, referred to as social plasticity, involves the reorganization of behavior, cognition, and emotional responses based on social experience (Teles et al., 2016). It enables individuals to fine-tune their mental states and behavioral strategies in response to the social environment, further reinforcing the adaptability and resilience of the human brain.

3. Early Life Stress and Brain Development

3.1. Adverse Childhood Experiences

Adverse childhood experiences have been shown to exert profound and lasting effects on brain development (Felitti et al., 1998). Chronic exposure to stress in early life can disrupt the maturation of key brain regions, such as the hippocampus and prefrontal cortex (PFC), both of which are essential for memory, learning, and executive functions (Felitti et al., 1998). Elevated levels of stress hormones during these critical periods can impair neurodevelopment, increasing the vulnerability to mental health disorders including anxiety, depression, and post-traumatic stress disorder (Felitti et al., 1998).

In contrast, positive early social experiences, such as cooperative play and successful conflict resolution, contribute to the development of adaptive coping mechanisms (Dettling et al., 2002). These interactions reinforce neural pathways responsible for emotional regulation and social behavior (Larson & Woolley, 2018; Dettling et al., 2002). Conversely, negative experiences or social deprivation can weaken these circuits, reducing an individual's capacity to manage stress and emotions in later life. These findings underscore the importance of nurturing and supportive environments during early developmental stages (Larson & Woolley, 2018; Dettling et al., 2002).

3.2. Neural Basis of Social Bonding and Attachment

Social bonding and attachment behaviors are fundamentally rooted in neurobiological mechanisms. Core brain structures involved in these processes include the hypothalamus, amygdala, and PFC (Carter, 1998). The hypothalamus plays a key role by releasing oxytocin and vasopressin, neuropeptides that facilitate social bonding and attachment (Carter, 1998). Oxytocin, often referred to as the love hormone, enhances social recognition and promotes trust and empathy, thereby facilitating the formation of close interpersonal connections (Heinrichs et al., 2009).

The amygdala is critical for interpreting emotional stimuli and evaluating the salience of social cues, particularly in relation to fear and reward processing, functions essential for forming attachments (LeDoux, 2000). Meanwhile, the medial PFC is implicated in higher-order social cognition, including empathy, theory of mind, and moral reasoning. It integrates input from the amygdala and other regions to guide contextually appropriate social behavior (LeDoux, 2000).

Additionally, the dopaminergic reward system, particularly pathways involving the ventral tegmental are and nucleus accumbens, plays a central role in reinforcing social bonds. Dopamine release during positive social interactions strengthens interpersonal connections through reward-based learning (Insel, 2003).

3.3. Neuroplasticity and Social Bonding

Neuroplasticity, the brain's capacity to reorganize and form new synaptic connections throughout life, is intricately linked to social bonding and attachment (Davidson & McEwen, 2012). Experiences within close relationships can shape the neural architecture, particularly in regions associated with emotional processing and social cognition. Positive social interactions enhance neuroplasticity by increasing the formation and strengthening of synapses, thereby reinforcing the brain's ability to adapt and respond to social and emotional demands across the lifespan (Davidson & McEwen, 2012).

Supportive social environments have been associated with elevated levels of brain-derived neurotrophic factor (BDNF), a protein crucial for neuronal survival, synaptic growth, and cognitive functions such as learning and memory (Park & Poo, 2013). These mechanisms are especially relevant for the maintenance of social bonds. Conversely, social isolation has been found to diminish synaptic density, especially in the hippocampus and PFC, impairing both emotional regulation and cognitive flexibility (Cacioppo & Hawkley, 2009).

The impact of social experiences on neuroplasticity is bidirectional. While positive interactions support neural development and resilience, chronic stress and social deprivation contribute to structural and functional impairments in the brain (McEwen, 2007). These effects highlight the critical role of environmental and psychosocial factors in shaping brain health.

Bond formation, particularly in early life, exemplifies this dynamic. For instance, oxytocin and vasopressin are neurochemicals centrally involved in mother-infant attachment, with oxytocin facilitating trust and closeness during nurturing behaviors and vasopressin contributing to the stability of long-term social bonds (Carter, 1998; Heinrichs et al., 2009). These interactions stimulate neuroplastic changes by reinforcing synaptic connectivity in emotion-related neural circuits (Davidson & McEwen, 2012).

However, adverse environmental conditions such as high levels of psychosocial stress can disrupt these processes. Elevated cortisol levels resulting from chronic stress have been shown to inhibit synapse formation and damage brain circuits involved in attachment and emotional regulation (McEwen, 2007). Thus, the interplay between neurobiology and environment is critical for understanding the developmental trajectory of social and emotional functioning.

4. Cultural Influences on Brain Function

4.1. Cultural Practices and Beliefs

Cultural practices and belief systems vary greatly across societies and exert a significant influence on brain function. Specific cultural habits, such as meditation, communal rituals, and dietary patterns, can shape stress regulation, social connectivity, and cognitive processes. For example, regular meditation has been linked to enhanced neuroplasticity and improved mental well-being through its effects on attention regulation, emotional control, and structural brain changes (Tang et al., 2015). Communal activities, such as group rituals and collective celebrations, foster social bonding and empathy by stimulating neural circuits involved in social cognition (Whitehouse et al., 2014).

Moreover, dietary customs that emphasize nutrients such as omega-3 fatty acids and antioxidants contribute positively to cognitive performance by supporting neuronal function and reducing inflammation (Gómez-Pinilla, 2008).

Culturally specific behaviors also shape brain development through language use, social norms, and ritualistic practices. For instance, bilingual individuals exhibit increased connectivity and efficiency in language-related brain regions, such as the left inferior frontal gyrus and superior temporal sulcus, leading to improved cognitive flexibility, working memory, and task-switching abilities (Abutalebi & Green, 2016). Furthermore, adherence to social norms and participation in rituals engage neural systems associated with social evaluation, self-regulation, and emotional processing. These shared experiences may promote similar neural activation patterns within cultural groups, reinforcing collective identity and interpersonal cohesion (Immordino-Yang et al., 2014).

4.2. Language and Cultural Experiences

Language acquisition and cultural immersion during critical developmental periods significantly influence neural architecture and cognitive flexibility. Early multilingual exposure enhances the efficiency of neural networks involved in auditory discrimination and language processing (Kuhl, 2010). Bilingualism has been associated with heightened executive functioning, improved attentional control, and greater cognitive resilience in aging populations (Bialystok et al., 2012).

Beyond language, diverse cultural experiences contribute to cognitive enrichment by encouraging perspective-taking, adaptive reasoning, and creativity. Individuals exposed to multiple cultures often demonstrate superior problem-solving skills, increased mental flexibility, and innovative thinking (Leung et al., 2008). These effects are attributed to the brain's capacity for adaptive change in response to complex and novel environments, highlighting the importance of sociocultural engagement in cognitive development.

Overall, culture plays a critical role in shaping brain function by influencing neural pathways through unique practices, beliefs, and experiences. Understanding these cultural contributions is essential for promoting cognitive health, supporting neurodevelopment, and addressing disparities in mental health outcomes across populations (Kuhl, 2010).

5. Role of Social Support in Recovery From Neurological Diseases or Brain Injuries

5.1. Recovery and Rehabilitation

The recovery process following neurological diseases or brain injuries, such as stroke or traumatic brain injury, is significantly influenced by the availability of social support. Research has demonstrated that stroke survivors experience improved recovery outcomes when supported by family, friends, and close social networks. This support not only provides emotional comfort but also offers practical assistance, which collectively fosters motivation and enhances rehabilitation efforts. Johansson (2011) notes that strong social support can accelerate recovery by increasing the brain's ability to adapt to damage. In a similar vein, individuals recovering from traumatic brain injuries also benefit from social interactions. Participation in group therapy or community-based activities can facilitate recovery by stimulating the production of proteins that support neuronal growth and repair (Pajevic et al., 2012). These findings underscore the critical role that social engagement plays in the rehabilitation process.

5.2. Buffer Against Stress

Social support serves as a crucial buffer against the detrimental effects of chronic stress, which can impede recovery from neurological conditions (Johansson, 2011). Chronic stress is associated with elevated cortisol levels, a hormone that, in excessive amounts, hampers the brain's healing processes. Conversely, the presence of supportive relationships has been shown to reduce cortisol levels, creating a more favorable environment for recovery (McEwan & Gianaros, 2010).

Attachment theory provides further insight into the mechanisms behind this process. According to Bowlby (1982), early caregiver relationships are foundational in shaping brain development and emotional regulation. Secure attachment bonds formed in childhood promote healthier emotional responses and social behaviors in adulthood, contributing to improved stress management and overall mental well-being. Thus, supportive relationships, both early in life and during recovery, are essential for optimizing brain health and emotional stability.

5.3. Resilience Mechanisms

The resilience exhibited by certain children, even in the face of adversity, can be attributed to a combination of supportive interventions and positive environmental factors. These include therapeutic support, stable relationships, and participation in community programs, all of which can help mitigate the negative effects of early stress and foster better brain development (Luthar, 2015).

Research further highlights the impact of socioeconomic status (SES) on cognitive development. Children from low SES backgrounds face numerous challenges, such as limited access to education, inadequate nutrition, and heightened exposure to stress, all of which can hinder neural development (Hackman et al., 2010). The effects of poverty on brain development are particularly pronounced in areas related to decision-making, learning, and emotional regulation. Blair and Raver (2016) emphasize that poverty can lead to changes in brain structure that impair impulse control, memory, and cognitive flexibility. These neural changes have lasting consequences, potentially affecting a child's learning abilities and overall cognitive functioning in the long term.

In conclusion, the role of social support in recovery from neurological diseases or brain injuries cannot be overstated. It not only aids in the rehabilitation process but also protects the brain from the harmful effects of stress, enhances emotional regulation, and promotes resilience. Furthermore, the influence of socioeconomic factors highlights the need for targeted interventions that address both social and environmental determinants to optimize brain health and recovery outcomes.

6. Stress, Coping Mechanisms, and Neuroendocrine Systems

6.1. Chronic Stress Impact

Chronic stress has a profound and detrimental effect on brain function, primarily by disrupting the Hypothalamic–Pituitary–Adrenal axis (HPA), which regulates cortisol release in response to stress. When stress becomes prolonged, the HPA axis remains overly active, resulting in sustained high levels of cortisol. McEwen (2007) highlights that this prolonged exposure to elevated cortisol levels is harmful not only to the brain but also to the body, as it impedes various physiological processes.

One of the primary consequences of chronic stress is its negative impact on neuroplasticity, the brain's ability to form new neural connections. The hippocampus, the brain region essential for learning and memory, is particularly vulnerable to the effects of elevated cortisol. Chronic stress causes shrinkage of the hippocampus, which leads to a decline in cognitive function (McEwen, 2007). In addition, the PFC, which plays a crucial role in decision-making, impulse control, and emotional regulation, is also adversely affected. High cortisol levels reduce the complexity of neurons within the PFC, impairing these vital cognitive functions (Arnsten, 2009).

Moreover, chronic stress impacts the amygdala, the brain region involved in processing emotions and memories (Roozendaal et al., 2009). Stress-induced enlargement and overactivation of the amygdala occur in tandem with the shrinkage of the hippocampus and PFC. This imbalance increases sensitivity to stress and heightens anxiety levels, creating a feedback loop that further exacerbates the effects of chronic stress (Roozendaal et al., 2009). Thus, chronic stress not only diminishes cognitive capabilities but also amplifies emotional responses, contributing to a vicious cycle of stress and anxiety.

6.2. Coping Mechanisms

While chronic stress can undermine brain health, effective coping mechanisms can mitigate its harmful effects and promote overall well-being. Mindfulness practices, including meditation and deep breathing, have been shown to reduce cortisol levels and enhance neural connectivity, thereby supporting emotional regulation (Davidson & McEwen, 2012). These practices help reduce stress-related physiological arousal, which, in turn protects the brain from the damaging effects of chronic stress.

Exercise also plays a significant role in counteracting stress. Regular physical activity reduces cortisol levels and increases the production of BDNF (Cotman & Berchtold, 2002). In addition to managing stress, exercise improves mood and cognitive function, thereby fostering resilience against stress.

Cognitive behavioral therapy (CBT) is another effective strategy for managing stress which helps individuals identify and modify negative thought patterns and behaviors, thereby reducing the psychological impact of stress (Hofmann et al., 2012). By reframing harmful thoughts and developing problem-solving skills, individuals can better cope with stress and its effects on the brain (Hofmann et al., 2012). These coping strategies are essential for enhancing mental resilience and protecting brain function from chronic stress.

6.3. Stress Hormones

The role of stress hormones, particularly cortisol, in influencing brain health is well-documented. Prolonged exposure to high levels of cortisol can damage neurons and inhibit the growth of new neurons in the hippocampus, a critical area for cognitive and emotional health. Lucassen et al. (2014) emphasize that chronic stress leads to reduced neurogenesis in the hippocampus, impairing memory function and increasing susceptibility to mood disorders such as depression.

Effective stress management is crucial for maintaining low cortisol levels and supporting brain flexibility. Social support plays an important role in this process. Positive social interactions have been shown to reduce cortisol levels and improve overall well-being, which in turn supports brain health (Heinrichs et al., 2009). Building strong social networks and seeking help during stressful times can therefore protect the brain from the adverse effects of stress.

Sleep is another critical factor in managing stress and promoting brain flexibility. Chronic stress often disrupts sleep patterns, exacerbating cognitive and emotional difficulties. Adequate sleep is essential for regulating cortisol levels and facilitating brain repair processes. Walker and Stickgold (2006) note that quality sleep is vital for restoring brain function and maintaining mental resilience, as it allows the brain to recover from the effects of stress.

In conclusion, the relationship between stress, coping mechanisms, and neuroendocrine systems is complex and interconnected. Chronic stress can significantly impair brain function and neuroplasticity, leading to diminished cognitive abilities and heightened emotional responses. However, coping strategies such as mindfulness, exercise, therapy, social support, and adequate sleep can help mitigate these effects. These strategies promote brain health, enhance resilience, and contribute to the overall well-being of individuals facing chronic stress (Walker & Stickgold, 2006).

7. The Role of Emotions in Neurodevelopment

7.1. Influence of Emotional Experiences

Emotional experiences play a critical role in shaping brain development, particularly during sensitive periods in childhood and adolescence. Positive emotions, such as joy, love, and gratitude, support healthy neurodevelopment by enhancing neuroplasticity, the brain's capacity to adapt and reorganize. These emotional states are associated with increased neurogenesis and synaptic plasticity in the hippocampus, a brain region crucial for learning and memory (Kumar et al., 2014). In contrast, persistent exposure to negative emotions such as fear, anger, and sadness can hinder brain development and elevate the risk of psychological disorders. Chronic negative emotional states may impair hippocampal growth and disrupt emotional regulation systems, particularly during critical developmental windows. Therefore, creating emotionally supportive environments that foster positive experiences is essential for optimal brain development and long-term mental health.

7.2. Emotional Regulation

Emotional regulation, the ability to manage and respond to emotional experiences in a balanced way, is a key factor in both neurodevelopment and mental health. This process primarily involves the PFC, which exerts top-down control over subcortical regions such as the amygdala, a center for emotional processing (Ochsner et al., 2012). The development of emotional regulation skills enhances neural plasticity by strengthening connectivity between the PFC and the amygdala (Ochsner et al., 2012). Improved communication between these regions allows for more adaptive emotional responses, reduced reactivity, and greater psychological resilience. Effective emotional regulation not only supports cognitive and emotional growth but also serves as a protective factor against mood and anxiety disorders. Thus, nurturing these skills through supportive relationships, education, and targeted interventions is crucial for healthy emotional and neural development.

7.3. Emotional Intelligence

Positive emotions and emotional intelligence play a synergistic role in enhancing cognitive function and emotional resilience. Emotional intelligence is defined as the ability to recognize, understand, manage, and use emotions effectively (Mayer et al., 2016). Individuals with high emotional intelligence tend to exhibit better social functioning, greater psychological well-being, and enhanced brain efficiency in areas related to emotional and social cognition (Mayer et al., 2016). Neuroimaging studies have shown that people with high emotional intelligence display greater activity and connectivity in brain regions such as the ventromedial prefrontal cortex and the anterior cingulate cortex, areas involved in emotional awareness, empathy, and regulation (Mayer et al., 2016). These neural adaptations contribute to more effective emotional processing and improved interpersonal interactions.

8. Conclusion

The relationship between social connections and brain development reveals that our minds are not only shaped by our biology but also deeply sculpted by the quality of our relationships. From the formation of emotional bonds in early childhood to the healing power of supportive communities in recovery from brain injury, social environments profoundly influence neural pathways, cognitive flexibility, and emotional well-being. Social interactions serve not just as enriching experiences but as biological catalysts for neuroplastic change promoting resilience, shaping behavior, and sustaining mental health across the lifespan. Understanding these dynamics emphasizes the need to foster compassionate, connected societies where nurturing relationships are seen not only as emotional assets but as essential contributors to cognitive and neurological health. As neuroscience continues to uncover the social fabric of the brain, it becomes increasingly clear that connection is not a luxury, but a fundamental element of human thriving.

9. Takeaways

1. Social relationships significantly influence brain development, shaping thoughts, emotions, and behaviors through mechanisms like oxytocin release, which enhances trust and bonding.
2. Neuroplasticity, the brain's ability to form and reorganize connections, is deeply influenced by social interactions, driving learning, memory, and emotional understanding.
3. Social experiences, such as conversations and cooperation, strengthen neural pathways and improve cognitive flexibility, essential for interpreting emotions and adapting to social norms.
4. Early life stress, including adverse childhood experiences, can harm brain regions like the hippocampus and prefrontal cortex, increasing vulnerability to mental health issues.
5. Positive social environments in childhood, including verbal and emotional support, enhance cognitive flexibility, emotional regulation, and neural development.
6. Social bonding is rooted in neurobiology, with oxytocin, dopamine, and key brain regions like the amygdala and prefrontal cortex playing pivotal roles in attachment and empathy.
7. Neuroplasticity supports social bonding by reinforcing synaptic connections, while chronic stress and isolation weaken these pathways, highlighting the importance of supportive environments.
8. Cultural practices and beliefs, such as meditation and communal rituals, influence brain function by fostering social bonds, improving empathy, and enhancing cognitive performance.
9. Bilingualism and diverse cultural experiences improve neural connectivity, cognitive flexibility, and problem-solving abilities, showcasing the role of culture in brain adaptation.
10. Social support is crucial for recovery from neurological diseases or brain injuries, enhancing motivation, promoting neuron repair, and buffering against stress-induced damage.
11. Early secure attachments with caregivers foster emotional regulation and resilience, leading to stronger neural pathways for managing emotions and stress.

12. Resilience in children facing adversity is bolstered by supportive relationships, therapy, and community programs, which counteract the effects of early life stress.
13. Chronic stress elevates cortisol levels, harming brain plasticity and recovery, while social support lowers cortisol, creating an environment conducive to healing.
14. Understanding the interplay between social interactions, neuroplasticity, and cultural influences can inform strategies to improve mental health and cognitive resilience.
15. Positive social environments, including nurturing relationships and cultural engagement, are vital for lifelong brain health and emotional well-being.

Abbreviations

BDNF: Brain-derived neurotrophic gactor
CBT: Cognitive Behavioral Therapy
PFC: Prefrontal cortex
SES: Socioeconomic status
HPA: Hypothalamic–Pituitary–Adrenal axis

Bibliography

1. Abutalebi, J., & Green, D. W. (2016). Neuroimaging of language control in bilinguals: Neural adaptation and reserve. Bilingualism: Language and Cognition, 19(4), 689–698. https://doi.org/10.1017/S1366728916000225
2. Arnsten, A. F. T. (2009). Stress signaling pathways that impair prefrontal cortex structure and function. Nature Reviews Neuroscience, 10 (6), 410–422. https://doi.org/10.1038/nrn2648
3. Baron-Cohen, S. (1998). Development of the theory of mind: Representation, self-awareness, and social relationships. In M. D. Ferrari & R. J. Sternberg (Eds.), Self-awareness: Its nature and development (pp. 221–243). The Guilford Press.

4. Bialystok, E., Craik, F. I. M., & Luk, G. (2012). Bilingualism: Consequences for mind and brain. Trends in Cognitive Sciences, 16(4), 240–250. https://doi.org/10.1016/j.tics.2012.03.001
5. Blair, C., & Raver, C. C. (2016). Poverty, stress, and brain development: New directions for prevention and intervention. *Academic Pediatrics, 16*(3), S30-S36. https://doi.org/10.1016/j.acap.2016.01.010
6. Bourassa, K. J., Memel, M., Woolverton, C., & Sbarra, D. A. (2015). Social participation predicts cognitive functioning in aging adults over time: Comparisons with physical health, depression, and physical activity. Aging & Mental Health, 19(12), 747–758. https://doi.org/10.1080/13607863.2014.962007
7. Bowlby, J. (1982). Attachment and loss: Vol. 1. Attachment (2nd ed.). Basic Books.
8. Brennan, S. E., & Clark, H. H. (1996). Conceptual pacts and lexical choice in conversation. Journal of Experimental Psychology: Learning, Memory, and Cognition, 22(6), 1482–1493. https://doi.org/10.1037/0278-7393.22.6.1482
9. Cacioppo, J. T., & Hawkley, L. C. (2009). Perceived social isolation and cognition. Trends in Cognitive Sciences, 13(10), 447–454. https://doi.org/10.1016/j.tics.2009.06.005
10. Carter, C. S. (1998). Neuroendocrine perspectives on social attachment and love. Psychoneuroendocrinology, 23(8), 779–818. https://doi.org/10.1016/S0306-4530(98)00055-9
11. Cotman, C. W., & Berchtold, N. C. (2002). Exercise: A behavioral intervention to enhance brain health and plasticity. Trends in Neurosciences, 25 (6), 295–301. https://doi.org/10.1016/S0166-2236(02)02143-4
12. Davidson, R. J., & McEwen, B. S. (2012). Social influences on neuroplasticity: stress and interventions to promote well-being. *Nature Neuroscience, 15*(5), 689–695. https://doi.org/10.1038/nn.3093
13. Dettling, H. C., Rüedi-Bettschen, A., Feldon, M., & Pryce, C. L. (2002). Early social experience affects the neuronal structure and emotional behavior in adult mice. Physiology & Behavior, 77(2–3), 365–372. https://doi.org/10.1016/S0031-9384(02)00904-2

14. Felitti, V. J., Anda, R. F., Nordenberg, D., Williamson, D. F., Spitz, A. M., Edwards, V., … & Marks, J. S. (1998). Relationship of childhood abuse and household dysfunction to many of the leading causes of death in adults: The Adverse Childhood Experiences (ACE) Study. American Journal of Preventive Medicine, 14(4), 245–258. https://doi.org/10.1016/S0749-3797(98)00017-8
15. Gómez-Pinilla, F. (2008). Brain foods: The effects of nutrients on brain function. Nature Reviews Neuroscience, 9(7), 568–578. https://doi.org/10.1038/nrn2421
16. Hackman, D. A., Farah, M. J., & Meaney, M. J. (2010). Socioeconomic status and the brain: Mechanistic insights from human and animal research. Nature Reviews Neuroscience, 11(9), 651–659. https://doi.org/10.1038/nrn2897
17. Hari R, Kujala MV. Brain basis of human social interaction: from concepts to brain imaging. Physiol Rev. 2009 Apr;89(2):453–79. https://doi.org/10.1152/physrev.00041.2007
18. Harris, P. L., de Rosnay, M., & Pons, F. (2005). Understanding emotion. In M. E. Lamb & R. M. Lerner (Eds.), Handbook of child psychology (6th ed., pp. 18–30). Wiley.
19. Heinrichs, M., von Dawans, B., & Domes, G. (2009). Oxytocin, vasopressin, and human social behavior. Frontiers in Neuroendocrinology, 30(4), 548–557. https://doi.org/10.1016/j.yfrne.2009.05.005
20. Hofmann, S. G., Asnaani, A., Vonk, I. J., Sawyer, A. T., & Fang, A. (2012). The efficacy of cognitive behavioral therapy: A review of meta-analyses. Cognitive Therapy and Research, 36(5), 427–440. https://doi.org/10.1007/s10608-012-9476-1
21. Immordino-Yang, M. H., Darling-Hammond, L., & Krone, C. R. (2014). Neuroscience and the classroom: Making connections. Harvard Education Press.
22. Insel, T. R. (2003). Is social attachment an addictive disorder? Physiology & Behavior, 79(3), 351–357. https://doi.org/10.1016/S0031-9384(03)00148-3
23. Johansson, B. B. (2011). Current trends in stroke rehabilitation: A review focusing on brain plasticity. Acta Neurologica Scandinavica, 123(3), 147–159. https://doi.org/10.1111/j.1600-0404.2010.01417.x

24. Kosfeld, M., Heinrichs, M., Zak, P. J., Fischbacher, U., & Fehr, E. (2005). Oxytocin increases trust in humans. Nature, 435(7042), 673–676. https://doi.org/10.1038/nature03701

25. Kuhl, P. K. (2010). Brain mechanisms in early language acquisition. Neuron, 67 (5), 713–727. https://doi.org/10.1016/j.neuron.2010.08.038

26. Kumar, A., Malabika, P., Sharma, R., & Dhiman, V. (2014). Positive emotions boost brain health: Neurogenesis, cognitive function, and resilience. Nature Reviews Neuroscience, 10 (4), 260–270. https://doi.org/10.1002/ab.21682

27. Larson, T. A., & Woolley, S. C. (2018). Social interactions shape the development of neural circuits in the juvenile songbird. Developmental Neurobiology, 78(7), 796–807. https://doi.org/10.1002/dneu.22599

28. LeDoux, J. E. (2000). Emotion circuits in the brain. Annual Review of Neuroscience, 23, 155–184. https://doi.org/10.1146/annurev.neuro.23.1.155

29. Leung, A. K. Y., Maddux, W. W., Galinsky, A. D., & Chiu, C. Y. (2008). Multicultural experience enhances creativity: The when and how. American Psychologist, 63 (3), 169–181. https://doi.org/10.1037/0003-066X.63.3.169

30. Lindquist, K. A., MacCormack, J. K., & Shablack, H. (2015). The role of language in emotion: Predictions from psychological constructionism. Frontiers in Psychology, 6, 444. https://doi.org/10.3389/fpsyg.2015.00444

31. Lucassen, P. J., Stumpel, M. W., Wang, Q., & Aronica, E. (2014). Decreased neurogenesis in depression: Role of serotonin receptor signaling in the adult hippocampus. Frontiers in Pharmacology, 5, 196. https://doi.org/10.3389/fphar.2014.00196

32. Luthar, S. S. (2015). Resilience in development: A synthesis of research across five decades. In D. Cicchetti (Ed.), *Developmental Psychopathology* (3rd ed., Vol. 4, pp. 739–795). John Wiley & Sons.

33. Maguire, E. A., Gadian, D. G., Johnsrude, I. S., Good, C. D., Ashburner, J., Frackowiak, R. S. J., & Frith, C. D. (2000). Navigation-related structural change in the hippocampi of taxi drivers. Proceedings of the National Academy of Sciences, 97(8), 4398–4403. https://doi.org/10.1073/pnas.070039597

34. Mayer, J. D., Roberts, R. D., & Barsade, S. G. (2016). Human abilities: Emotional intelligence. Annual Review of Psychology, 59 (1), 507–536. https://doi.org/10.1146/annurev.psych.59.103006.093646

35. McEwen, B. S. (2007). Physiology and neurobiology of stress and adaptation: Central role of the brain. Physiological Reviews, 87(3), 873–904. https://doi.org/10.1152/physrev.00041.2006

36. McLaughlin, K. A., Sheridan, M. A., & Lambert, H. K. (2014). Childhood adversity and neural development: Deprivation and threat as distinct dimensions of early experience. Neuroscience & Biobehavioral Reviews, 47, 578–591. https://doi.org/10.1016/j.neubiorev.2014.10.012

37. Ochsner, K. N., Silvers, J. A., & Buhle, J. T. (2012). Functional imaging studies of emotion regulation: A synthetic review and evolving model of the cognitive control of emotion. Annals of the New York Academy of Sciences, 1251 (1), E1-E24. https://doi.org/10.1111/j.1749-6632.2012.06751.x

38. Pajevic, S., Smajlovic, D., Dostovic, Z., Sinanovic, O., & Vidovic, M. (2012). The effect of rehabilitation treatment on functional recovery after ischemic stroke. Medical Archives, 66(3), 150–153. https://doi.org/10.5455/medarh.2012.66.150-153

39. Park, H., & Poo, M. M. (2013). Neurotrophin regulation of neural circuit development and function. Nature Reviews Neuroscience, 14(1), 7–23. https://doi.org/10.1038/nrn3379

40. Roozendaal, B., McEwen, B. S., & Chattarji, S. (2009). Stress, memory, and the amygdala. Nature Reviews Neuroscience, 10(6), 423–433. https://doi.org/10.1038/nrn2651

41. Silkenbeumer, M., Schiller, E., Knopf, M., & Kamp-Becker, I. (2016). Cognitive flexibility and emotional understanding influence social behavior in children with autism spectrum disorder. Journal of Autism and Developmental Disorders, 46(1), 287–299. https://doi.org/10.1007/s10803-015-2560-5

42. Tang, Y.-Y., Hölzel, B. K., & Posner, M. I. (2015). The neuroscience of mindfulness meditation. Nature Reviews Neuroscience, 16(4), 213–225. https://doi.org/10.1038/nrn3916

43. Teles, M. C., Cardoso, S. D., & Oliveira, R. F. (2016). *Social Plasticity Relies on Different Neuroplasticity Mechanisms across the Brain*

Social Decision-Making Network in Zebrafish. Frontiers in Behavioral Neuroscience, 10. https://doi.org/10.3389/fnbeh.2016.00016

44. Walker, M. P., & Stickgold, R. (2006). Sleep, memory, and plasticity. Annual Review of Psychology, 57, 139–166. https://doi.org/10.1146/annurev.psych.56.091103.070307
45. Wang, X., Liu, X., & Feng, T. (2021). The continuous impact of cognitive flexibility on developing emotion understanding in children aged 4 and 5 years: A longitudinal study. *Journal of Experimental Child Psychology, 203*, 105018. https://doi.org/10.1016/j.jecp.2020.105018
46. Whitehouse, H., McQuinn, B., Buhrmester, M., & Swann Jr., W. B. (2014). Brothers in arms: Libyan revolutionaries bond like family. Proceedings of the National Academy of Sciences, 111(50), 17783–17785. https://doi.org/10.1073/pnas.1416284111
47. Wong, S. S., Jacques, S., & Zelazo, P. D. (2008). A preliminary investigation of the effects of emotional stimuli on 4-year-old children? s abstraction and cognitive flexibility on the flexible item selection task (FIST). *University of Toronto Journal of Undergraduate Life Sciences*, *2*(1).

Music-Induced Neuroplasticity: Cognitive and Emotional Dimensions

Selen Gür Özmen

1. Introduction

Neuroplasticity, the brain's remarkable capacity to reorganize itself by forming new neural connections, plays a crucial role in human development, learning, and adaptation (Kolb & Whishaw, 1998). This adaptability is influenced by various activities and experiences, among which music holds a significant position. Music, a ubiquitous and integral part of human culture, profoundly affects the brain, influencing emotional and cognitive processes, physical health, and overall well-being (Levitin, 2006). This chapter delves into the intricate relationship between neuroplasticity and music, examining how musical experiences shape the brain and contribute to its adaptability.

2. Neurophysiology of Listening to Music

Listening to music is a multifaceted experience that engages numerous brain regions, involving intricate neural processes that interpret rhythm, timbre, pitch, and lyrics (Zatorre et al., 2007). Studies have shown that music training and musical experience significantly enhance brain plasticity, promoting both cognitive and emotional adaptability (Habib & Besson, 2009).Recent neuroimaging studies have revealed a complex network of brain areas that combine to create rich musical experiences. These elements are received and processed in different brain areas, each contributing to our overall perception and enjoyment of music:

1. *Auditory Cortex:* The primary and secondary auditory cortices are crucial for processing the essential elements of sound, such as pitch and volume (Warren, 2008). Recent research has shown that the auditory cortex exhibits remarkable plasticity, adapting to musical training and experience (Herholz & Zatorre, 2012). For instance, professional musicians often display enlarged auditory cortices compared to non-musicians,

reflecting their enhanced ability to process complex musical stimuli (Schneider et al., 2002).

2. *Prefrontal Cortex:* This area involves cognitive appraisal and emotional response to music (Blood & Zatorre, 2001). The prefrontal cortex plays a crucial role in our aesthetic judgments of music and contributes to the experience of musical pleasure. Recent studies have shown that the dorsolateral prefrontal cortex is particularly active when listeners engage in musical analysis or experience chills in music response (Sachs et al., 2016).
3. *Motor Cortex:* The motor cortex (Figure 1) engages when we physically move to the rhythm of music, even if we are unaware of it (Chen et al., 2008). This connection between auditory and motor systems is so strong that simply listening to music can activate motor planning areas in the brain. This phenomenon, known as "motor resonance," may explain why music is such a powerful tool for movement rehabilitation in conditions like Parkinson's disease (Thaut et al., 2015).
4. *Limbic System:* This includes the amygdala and hippocampus, which are involved in music's emotional and memory-related aspects (Koelsch, 2010). The amygdala, in particular, plays a crucial role in processing the emotional content of music, responding differently to pleasant and unpleasant musical stimuli. On the other hand, the hippocampus is essential for forming and retrieving musical memories, contributing to the powerful nostalgic experiences often associated with familiar songs (Janata, 2009).
5. *Corpus Callosum:* This structure connects the brain's two hemispheres and is more significant in musicians, particularly those who started training at a young age. This enlargement facilitates the complex inter-hemispheric communication required for musical performance and appreciation (Schlaug et al., 1995).
6. *Cerebellum:* While traditionally associated with motor control, the cerebellum has been shown to play a significant role in the perception of musical timing and rhythm. It is particularly active during tasks that involve tracking or predicting musical beats (Penhune et al., 1998).
7. *Nucleus Accumbens:* This part of the brain's reward system is activated when we listen to music we enjoy. It releases dopamine, a neurotransmitter associated with pleasure and reward, explaining the euphoric feelings often experienced during music listening (Salimpoor et al., 2011).

The intricate interplay between these brain regions underlies our ability to perceive, enjoy, and respond to music. Moreover, the brain's engagement with music is not static but highly dynamic, changing with musical training, cultural exposure, and individual preferences. This neuroplasticity in response to music has significant implications for cognitive development, emotional regulation, and therapeutic applications (Herholz & Zatorre, 2012).

3. Processing Musical Elements

The human brain's ability to process and interpret music is a testament to its remarkable plasticity and complexity. Different elements of music engage various neural networks, showcasing the brain's capacity to adapt and respond to diverse auditory stimuli.

3.1. Rhythm

Rhythm, one of the fundamental aspects of music, involves the timing of sounds and silences. The perception of rhythm engages the auditory cortex, the motor cortex (Figure 1), and the basal ganglia, indicating a close link between auditory and motor systems (Grahn & Brett, 2007). This connection suggests that rhythm perception may be rooted in our ability to move and synchronize with beats, highlighting the role of neuroplasticity in adapting to rhythmic patterns.

Recent studies have further elucidated the neural mechanisms underlying rhythm processing:

- Predictive Timing: The cerebellum and basal ganglia play crucial roles in predictive timing, allowing listeners to anticipate beats and rhythmic patterns (Patel & Iversen, 2014).
- Beat Perception: The supplementary motor area (SMA) and premotor cortex are particularly active during beat perception, even without overt movement (Merchant et al., 2015).
- Rhythmic Entrainment: The ability to synchronize movements to a beat involves a network including the putamen, SMA, and premotor cortex, demonstrating the brain's capacity for sensorimotor integration (Kung et al., 2013).
- Cross-modal Plasticity: Training in rhythm perception can improve performance in other temporal processing tasks, suggesting cross-modal

plasticity in timing networks (Slater et al., 2018). These temporal processing tasks can be explained as follows:

3.1.1. Speech Perception

Phoneme Discrimination: The ability to distinguish between different phonemes in speech, which requires precise timing to differentiate sounds like "ba" and "pa" (Patel & Iversen, 2014).

Prosody Recognition: Understanding the rhythm, stress, and intonation patterns in spoken language, which is crucial for comprehending emotions and intentions in speech (Merchant et al., 2015).

3.1.2 Motor Coordination

Finger Tapping Tasks: Participants are asked to tap their fingers in time with a metronome or a rhythmic sequence, assessing their ability to maintain consistent timing (Kung et al., 2013).

Gait Synchronization involves walking in sync with a rhythmic auditory cue, which can be used in rehabilitation for motor-impaired patients (Slater et al., 2018).

3.1.3. Music Performance

Instrumental Timing: Musicians must coordinate their movements precisely to produce the correct notes at the right time, whether solo or ensemble (Patel & Iversen, 2014).

Sight-Reading: The ability to read and perform music notation in real-time, which involves rapid temporal processing to translate visual information into motor actions (Merchant et al., 2015).

3.1.4. Cognitive Tasks

Temporal Order Judgment: This involves determining the sequence of events, such as which of two stimuli appeared first, and relies on accurate temporal processing (Kung et al., 2013).

Interval Timing: Estimating the duration of time intervals, such as judging how long a light was on, is essential for various cognitive and perceptual tasks (Slater et al., 2018).

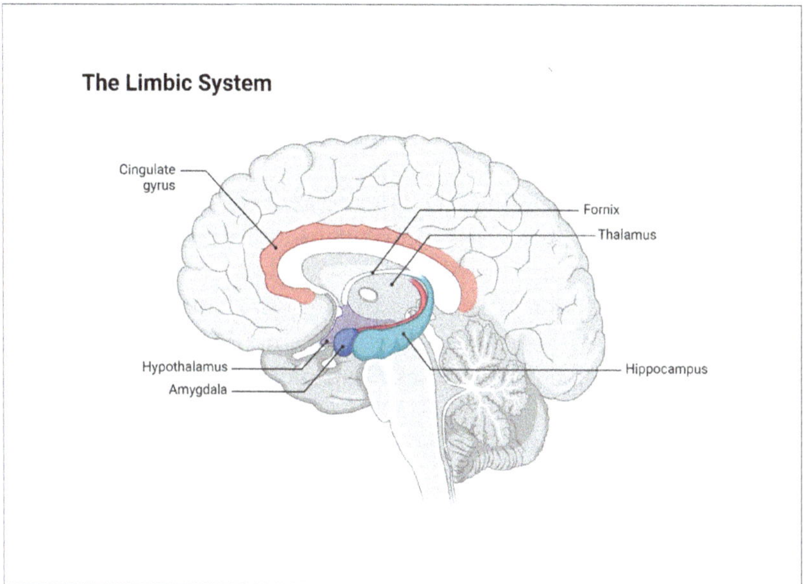

Figure 1: Topographic map of the motor cortex, illustrating its organization and functional regions. The lateral and mid-sagittal views highlight the location of the primary motor cortex, while the coronal view shows the somatotopic arrangement, including regions responsible for controlling the face, trunk, upper extremity, and lower extremity. This representation demonstrates the motor cortex's role in voluntary movement control (Created in https://BioRender.com).

3.1.5. Auditory Processing

Gap Detection: Identifying brief silent gaps in a continuous sound, a measure of auditory temporal resolution (Patel & Iversen, 2014).

Rhythm Discrimination: Differentiating between rhythmic patterns is essential for music perception and language processing (Merchant et al., 2015).

These examples illustrate the broad impact of rhythm training on various temporal processing tasks, highlighting the interconnectedness of timing networks in the brain.

3.2. Timbre

Timbre, often described as the color or quality of a sound, allows us to distinguish between different musical instruments or voices even when they play the same pitch. The auditory cortex plays a significant role in processing timbre, with different neural pathways dedicated to identifying and differentiating these complex sound qualities (Caclin et al., 2006).

Recent advancements in our understanding of timbre processing include:

· Spectrotemporal Processing: The superior temporal gyrus processes the spectrotemporal features that contribute to timbre perception (Alluri et al., 2012).
· Categorical Perception: The brain can categorize timbres into distinct groups, similar to how it processes phonemes in speech, involving bottom-up and top-down processes (Giordano et al., 2014).
· Expertise Effects: Musicians show enhanced neural responses to timbral differences, particularly in the right superior temporal gyrus, demonstrating experience-dependent plasticity (Pantev et al., 2001).

3.3. Pitch

Pitch perception involves the ability to discern different sound frequencies, which is essential for recognizing melodies and harmonies. The primary auditory cortex is heavily involved in pitch processing, with specific neurons responding to particular frequencies (Bendor & Wang, 2005).

Recent research has revealed more about the intricacies of pitch processing:

· Tonotopic Organization: The auditory cortex maintains a tonotopic organization, with different areas responding preferentially to specific frequency ranges. For example, the primary auditory cortex (A1) is organized such that neurons in the rostral (front) part respond best to low frequencies, while neurons in the caudal (rear) part respond best to high frequencies. Additionally, secondary auditory areas, such as the belt and parabelt regions, exhibit frequency-specific responses, though they may integrate more complex auditory information (Moerel et al., 2014).
· Relative Pitch: The planum temporale, particularly in the right hemisphere, is crucial for processing relative pitch relationships, which are fundamental to melody perception (Foster & Zatorre, 2010).
· Pitch Constancy: The brain's ability to maintain pitch constancy across different timbres and intensities involves complex processing in the auditory cortex and beyond (McLachlan & Wilson, 2017).

- Plasticity in Pitch Processing: Musical training can enhance pitch discrimination abilities and corresponding changes in auditory cortex structure and function (Herholz & Zatorre, 2012).

3.4. Lyrics and Vocals

Processing lyrics and vocals combine musical and linguistic elements, engaging the auditory cortex and language-processing areas such as Broca's and Wernicke's. This integration illustrates the brain's ability to process and integrate different types of information simultaneously, showcasing the plasticity of neural networks involved in music and language (Patel, 2010).

Further insights into lyrics and vocal processing include:

- Dual-Stream Processing: Lyrics are processed through dorsal (articulation and phonology) and ventral (semantics) streams, similar to speech processing. The dorsal stream, often called the "where" pathway, maps sounds to articulatory representations, crucial for speech production and phonological processing. This stream includes the posterior superior temporal gyrus and the inferior parietal lobule. The ventral stream, known as the "what" pathway, maps sounds to their meanings, involving regions such as the anterior superior temporal gyrus and the middle temporal gyrus (Sammler et al., 2015).
- Integration of Melody and Lyrics: The superior temporal sulcus is crucial in integrating melodic and linguistic information in songs (Schön et al., 2010).
- Emotional Processing: The limbic system, particularly the amygdala, processes the emotional content of vocals, contributing to the affective impact of songs (Frühholz et al., 2016).
- Cross-linguistic Plasticity: Exposure to lyrics in foreign languages can enhance phonological processing and linguistic plasticity (Ludke et al., 2014).

This intricate interplay of neural processes in music perception demonstrates the brain's remarkable plasticity and capacity for complex auditory processing. Understanding these mechanisms provides insights into cognitive neuroscience and has implications for music education, therapy, and the development of auditory processing technologies.

4. Liking and Disliking Music: Neural Correlates of Musical Preferences

The brain's response to music we like versus music we dislike involves complex interactions between reward, emotional, and cognitive systems. This intricate interplay of neural processes reflects our immediate reactions to music, shapes our long-term musical preferences, and contributes to music's profound impact on our emotional states.

4.1. Pleasurable Musical Experiences

When we listen to music we enjoy, several key brain regions are activated:

- Reward System: The nucleus accumbens, a central component of the brain's reward system, is activated when we listen to music we like, releasing dopamine, a neurotransmitter associated with pleasure and reward (Salimpoor et al., 2013). This dopamine release can occur not only during the peak emotional moments of a piece but also in anticipation of these moments, suggesting a role for musical expectancy in pleasure (Salimpoor et al., 2011).
- Prefrontal Cortex: The orbitofrontal cortex, involved in processing emotional stimuli and decision-making, shows increased activity during pleasurable music listening. This region may play a role in the conscious evaluation of music and the formation of preferences (Sachs et al., 2016).
- Limbic System: The hippocampus and amygdala, critical structures in emotional processing and memory, are engaged during enjoyable music listening. These areas contribute to forming musical memories and associating positive emotions with specific pieces of music (Koelsch, 2014) (Figure 2).
- Default Mode Network: Recent research has shown that listening to preferred music activates the default mode network, a set of brain regions associated with self-referential thinking and mind-wandering. This suggests that enjoying music may involve a form of self-reflection or autobiographical memory retrieval (Wilkins et al., 2014).

4.2. Unpleasant Musical Experiences

Conversely, when we encounter music we dislike, different neural patterns emerge:

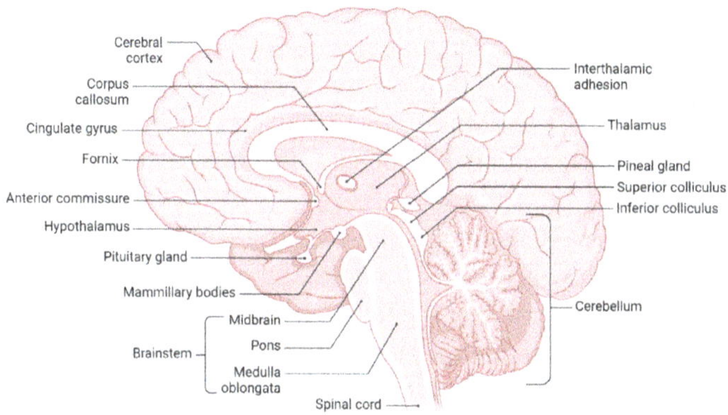

Figure 2: Limbic system (Created in <https://BioRender.com>).

- Amygdala Activation: Music that we find unpleasant can activate the amygdala and be associated with negative emotions such as fear and discomfort (Blood et al., 1999). This activation may be part of a broader aversive response to disliked stimuli.
- Parahippocampal Gyrus: Disliked music has been shown to activate the parahippocampal gyrus, a region that processes negative emotions and unpleasant auditory stimuli (Koelsch et al., 2006).
- Decreased Reward System Activity: Unlike pleasurable music, disliked music typically fails to activate the reward system and may even lead to deactivation in these areas (Mas-Herrero et al., 2018).
- Right Hemisphere Dominance: Some studies suggest that the processing of unpleasant music may be more lateralized to the right hemisphere, particularly in frontal and temporal regions (Altenmüller et al., 2002).

4.3. Factors Influencing Musical Preferences

Several factors contribute to the formation and modulation of our musical likes and dislikes:

- Familiarity: The mere exposure effect suggests that repeated exposure to a piece of music can increase liking, possibly due to the ease of processing familiar stimuli (Peretz et al., 1998).
- Cultural Context: Our cultural background significantly influences our musical preferences, shaping our expectations and emotional responses to different musical styles (Morrison & Demorest, 2009).
- Personal Associations: Music often links memories and experiences, influencing our emotional responses and preferences (Janata et al., 2007).
- Personality Traits: Research has shown correlations between personality traits and musical preferences, suggesting a link between individual differences in brain structure and function and musical tastes. For example, individuals who are highly open to experience tend to prefer more complex and novel genres such as classical, jazz, and avant-garde music. Those high in extraversion often prefer energetic and rhythmic music such as pop, dance, and hip-hop. Conversely, individuals high in neuroticism may prefer music that reflects their emotional states, such as sad or melancholic genres (Rentfrow & Gosling, 2003).
- Cognitive Factors: Music's complexity and predictability can influence preferences, with individuals often preferring music that balances novelty and familiarity (Huron, 2006).
- Neuroplasticity in Musical Preferences: Our musical preferences are not static but can change over time, demonstrating the brain's plasticity in response to musical experiences:
  - Expertise Effects: Musical training can alter brain responses to music, potentially influencing preferences. Musicians often show greater activation in auditory and motor regions when listening to music, which may enhance appreciation (Alluri et al., 2017).
  - Age-Related Changes: Musical preferences can shift throughout the lifespan, with cognitive and emotional processing changes potentially influencing these shifts (Cohrdes et al., 2017).
  - Mood-Dependent Preferences: The brain's state-dependent music processing means that our preferences can vary based on our current mood, showcasing the dynamic nature of musical liking (Eerola & Vuoskoski, 2013).

Understanding the neural basis of musical preferences provides insights into the nature of aesthetic experiences. It has implications for music therapy, marketing, and the development of personalized music recommendation systems. The complex interplay between cognitive, emotional, and reward systems in shaping our musical likes and dislikes underscores music's profound impact on the human brain and its remarkable plasticity in response to auditory experiences.

5. Ability to Appreciate Music

The ability to appreciate music is a complex cognitive and emotional process that varies widely among individuals. Genetic predispositions and environmental factors, including cultural exposure and personal experiences, influence this capacity.

5.1. Genetic Factors

Recent research has identified specific genes that may contribute to musical ability and appreciation. For instance, the AVPR1A gene, associated with social bonding and communication, has been linked to musical aptitude (Ukkola-Vuoti et al., 2011). Additionally, gene variations related to dopamine function, such as DRD2 and SLC6A3, have been associated with individual differences in music reward experiences (Mas-Herrero et al., 2018).

5.2. Environmental Influences

Cultural context plays a crucial role in shaping musical preferences and appreciation. Exposure to diverse musical styles during childhood can enhance one's ability to appreciate a broader range of music later in life (Hannon & Trainor, 2007).

5.3. Impact of Musical Training

Musical training can significantly enhance the ability to appreciate music, leading to structural and functional changes in the brain. Studies have shown that musicians have increased gray matter volume in areas related to auditory processing, motor control, and multimodal integration (Gaser & Schlaug, 2003). For example:

- Auditory Cortex: Musicians show enhanced responses in the auditory cortex, particularly for complex sounds and musical tones (Schneider et al., 2002).
- Corpus Callosum: The anterior portion of the corpus callosum, which connects the two hemispheres, is more significant in musicians, facilitating interhemispheric communication (Schlaug et al., 1995) (Figure 3).
- Motor Areas: Increased gray matter density is observed in motor regions, reflecting the fine motor control required for playing instruments (Amunts et al., 1997).
- Prefrontal Cortex: Enhanced activation in the prefrontal cortex during music listening suggests improved cognitive processing of musical structures (Zatorre et al., 2007).

These neuroplastic changes underscore the brain's remarkable ability to adapt in response to musical experiences and training.

6. Differences Between Musically Talented and Non-Talented Individuals

Research indicates that musically talented individuals exhibit distinct neural patterns compared to those without musical abilities. These differences in brain structure and function reflect the complex interplay between innate predispositions and experience-dependent plasticity.

6.1. Structural Differences

- Auditory Cortex: Musically talented individuals often have larger Heschl's gyri, the site of the primary auditory cortex, which may contribute to enhanced pitch perception (Schneider et al., 2002).
- Planum Temporale: A larger left planum temporale has been observed in individuals with absolute pitch, a rare ability often associated with musical talent (Keenan et al., 2001).
- Motor Areas: Enhanced structural connectivity in motor regions, particularly in white matter tracts, has been found in musicians, facilitating the complex motor skills required for musical performance (Bengtsson et al., 2005).

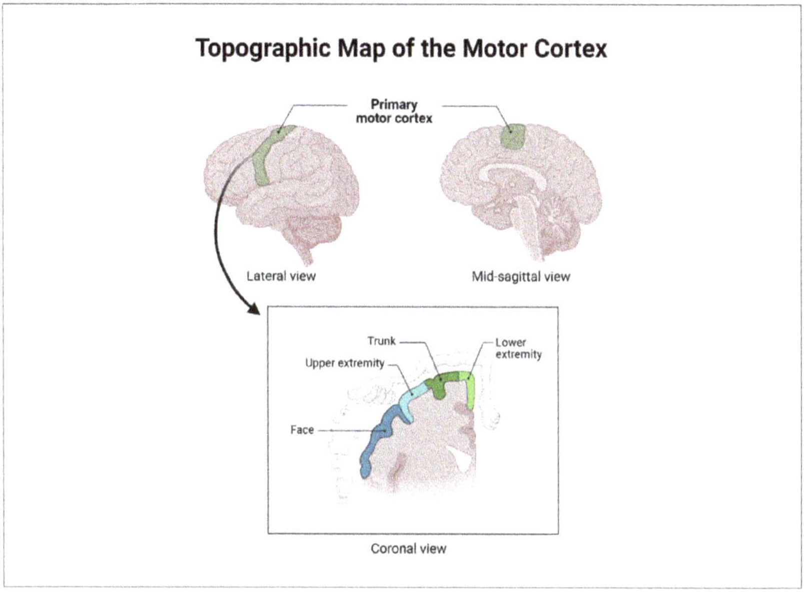

Figure 3: Illustration of the sagittal view of the human brain, highlighting key regions such as the corpus callosum, cingulate gyrus, hypothalamus, pituitary gland, and cerebellum. This diagram also includes structures involved in sensory processing, motor control, and regulatory functions, such as the thalamus, brainstem (midbrain, pons, medulla oblongata), and pineal gland (Created in https://BioRender.com).

6.2. Functional Differences

- Auditory-Motor Coupling: Talented musicians show enhanced connectivity between auditory and motor regions, allowing for more efficient translation of auditory information into motor commands (Zatorre et al., 2007).
- Prefrontal Cortex Activity: Increased activity in the prefrontal cortex during musical tasks has been observed in musicians, suggesting enhanced cognitive control and working memory for musical information (Pallesen et al., 2010).
- Emotional Processing: Musically talented individuals often show heightened responses in limbic and paralimbic regions during music listening, indicating a more intense emotional engagement with music (Brattico et al., 2016).

- Multisensory Integration: Musicians have been observed to integrate auditory, visual, and somatosensory information more effectively, reflecting their ability to coordinate multiple sensory inputs during performance (Paraskevopoulos et al., 2012).

These differences underscore the role of neuroplasticity in developing and honing musical skills through practice and training. However, it is essential to note that the relationship between brain structure/function and musical ability is complex and bidirectional—while some differences may predispose individuals to musical talent, others likely result from extensive musical training and practice.

7. Active Engagement: Playing Instruments and Singing

Active engagement with music, such as playing instruments or singing, induces even more pronounced neuroplastic changes compared to passive listening:

- Multimodal Integration: Playing an instrument requires the integration of auditory, motor, and cognitive processes, leading to extensive neural adaptation. Musicians often exhibit increased cortical representation in the areas of the brain responsible for finger movements and auditory processing (Pantev et al., 2003).
- Auditory-Motor Coupling: Musical training enhances the coupling between auditory and motor systems, improving the ability to translate heard sounds into motor actions (Zatorre et al., 2007).
- Executive Functions: Instrumental training has been associated with improvements in executive functions, including working memory, inhibitory control, and cognitive flexibility (Moreno et al., 2011).
- Vocal-Motor Integration: Singing engages the auditory and vocal-motor systems, enhancing connectivity between the auditory cortex and the motor regions responsible for vocal production (Özdemir et al., 2006).
- Respiratory Control: Singing can improve respiratory and phonatory control, which benefits speech and singing abilities (Kleber et al., 2010).

8. Latest Research on Music and the Brain

Recent advances in neuroimaging techniques, such as Electroencephalography (EEG), functional magnetic resonance imaging (fMRI), and MEG, have provided deeper insights into how music affects the brain. These studies have revealed the complex and widespread neural networks involved in music processing and shed light on music's potential therapeutic applications.

8.1. EEG Studies

- Brainwave Synchronization: EEG studies have shown that music can enhance brainwave synchronization, improving cognitive functions such as attention and memory (Lesser et al., 2020). For instance, listening to music with a strong beat can synchronize brainwaves to the beat frequency, potentially enhancing cognitive performance (Nozaradan et al., 2011). The 'Mozart effect' has been linked to enhanced brain activity, particularly in EEG studies showing increased event-related desynchronization/synchronization patterns during music listening (Jäusovec & Habe, 2003).
- Emotional Processing: EEG research has revealed distinct patterns of frontal alpha asymmetry associated with emotional responses to music, providing insights into the neural correlates of music-induced emotions. Frontal alpha asymmetry refers to the difference in alpha wave activity (8–12 Hz) between the brain's left and right frontal lobes. More excellent alpha activity in the left frontal lobe compared to the right is often associated with positive emotional states. In contrast, more excellent alpha activity in the right frontal lobe is linked to negative emotional states. This asymmetry can indicate an individual's emotional response to music, with different patterns emerging based on whether the music elicits positive or negative emotions (Schmidt & Trainor, 2001).

8.2. fMRI Studies

- Network Activation: Functional MRI studies have revealed that listening to and playing music activates extensive neural networks, including auditory, motor, limbic, and prefrontal regions, highlighting the brain's plastic response to musical stimuli (Schaefer et al., 2013).

- Reward Processing: fMRI research has shown that pleasurable music activates the brain's reward system, including the nucleus accumbens and ventral tegmental area, similar to other rewarding stimuli (Salimpoor et al., 2013).
- Structural Connectivity: Diffusion tensor imaging (DTI) studies have revealed enhanced structural connectivity in musicians' brains, particularly in auditory-motor pathways (Halwani et al., 2011).

8.3. Emerging Research Areas

- Music and Neuroplasticity: Studies explore how musical training can induce neuroplastic changes across the lifespan, with potential applications in cognitive rehabilitation and healthy aging (Herholz & Zatorre, 2012).
- Music and Social Cognition: Recent research investigates how music influences social bonding and empathy, revealing shared neural networks between music processing and social cognition (Koelsch, 2014).
- Music and Predictive Processing: New theoretical frameworks propose that musical expectancy and prediction play a crucial role in the pleasurable and emotional aspects of music, involving complex interactions between sensory, motor, and reward systems (Koelsch et al., 2019).

9. Music Therapy

Music therapy has gained recognition as an effective intervention for various health conditions, utilizing the therapeutic elements of music to promote physical, emotional, and cognitive well-being. Recent research has expanded our understanding of its mechanisms and applications:

9.1. Neurological Disorders

- Stroke Rehabilitation: Music therapy has been shown to improve motor function, gait, and speech in stroke patients. Rhythmic auditory stimulation can enhance motor coordination by engaging sensorimotor networks (Thaut et al., 2015; Rouijel, I. 2023).
- Parkinson's Disease: Rhythmic auditory cues in music therapy can improve gait parameters and reduce freezing episodes in Parkinson's patients (de Dreu et al., 2012).

- Alzheimer's Disease: Music therapy can enhance mood, reduce agitation, and improve cognitive function in individuals with Alzheimer's disease, possibly by engaging preserved musical memory networks (Särkämö et al., 2014).
- Acquired Brain Injury: Music therapy has shown to be effective in rehabilitation for individuals with acquired brain injuries, aiding in the recovery of motor functions, emotional regulation, and cognitive abilities (Bradt et al., 2010).

9.2. Mental Health

- Depression and Anxiety: Music therapy has been shown to reduce symptoms of depression and anxiety, potentially by modulating activity in the limbic system and prefrontal cortex (Fachner et al., 2013).
- PTSD: Recent studies suggest that music therapy can help reduce symptoms of PTSD by facilitating emotional processing and regulation (Landis-Shack et al., 2017).
- Autism Spectrum Disorders: Music therapy can improve social communication and emotional reciprocity in individuals with autism, possibly by engaging mirror neuron systems (Geretsegger et al., 2014).

9.3. Pain Management

Music therapy has been found effective in reducing perceived pain intensity and increasing pain tolerance, potentially by modulating activity in the pain matrix and enhancing descending pain inhibitory systems (Garza-Villarreal et al., 2017).

10. Impact of Music on Different Age Groups

10.1. Babies and Children

Music has a profound impact on the developing brains of babies and children, influencing various aspects of cognitive, emotional, and social development:

- Language Development: Early exposure to music can enhance phonological awareness and vocabulary acquisition. Singing to infants has been shown to facilitate language learning (Gerry et al., 2012).

- Cognitive Skills: Musical training in childhood is associated with improvements in executive functions, including inhibitory control, working memory, and cognitive flexibility (Moreno et al., 2011).
- Social-Emotional Development: Group musical activities can foster social skills, empathy, and emotional regulation in children (Kirschner & Tomasello, 2010).
- Brain Plasticity: Early musical training can lead to structural and functional brain changes, including enhanced connectivity between auditory and motor regions (Hyde et al., 2009).

10.2. Teens and Adults

For teenagers and adults, music continues to play a significant role in emotional regulation, social bonding, and cognitive enhancement:

- Emotional Regulation: Adolescents often use music to regulate mood and cope with stress. Music listening activates emotion regulation networks in the brain, including the prefrontal cortex and limbic system (McFerran & Saarikallio, 2014).
- Identity Formation: Music plays a crucial role in identity development during adolescence, engaging neural networks in self-referential processing and social cognition. Self-referential processing refers to the brain's ability to relate information to oneself, which is crucial for forming a coherent sense of identity. This phenomenon involves areas such as the medial prefrontal cortex (mPFC) and the posterior cingulate cortex (PCC), which are part of the default mode network (DMN). These regions are activated when individuals reflect on their thoughts, feelings, and experiences, and music can enhance this self-reflection by providing a medium through which adolescents explore and express their identities (North & Hargreaves, 1999).
- Cognitive Enhancement: Musical training in adulthood can improve cognitive functions, including attention, processing speed, and executive functions (Hanna-Pladdy & MacKay, 2011).
- Social Bonding: Shared musical experiences, such as attending concerts or playing in ensembles, can enhance social cohesion by activating neural networks involved in empathy and synchrony (Tarr et al., 2014).

- Stress Reduction: Music listening and making can reduce stress and anxiety in adults by modulating activity in the hypothalamic-pituitary-adrenal (HPA) axis and autonomic nervous system (Thoma et al., 2013).

These findings underscore the lifelong impact of music on brain function and plasticity, highlighting its potential for enhancing cognitive, emotional, and social well-being across the lifespan.

11. Music's Influence on Neuroplastic Mechanisms

The relationship between music and neuroplasticity is a fascinating area of neuroscience research. It demonstrates how musical experiences shape the brain across various domains. This relationship encompasses microstructural, macrostructural, functional, and physiological changes, illustrating the brain's remarkable adaptability to musical stimuli and training.

11.1. Microstructural Changes

Microstructural changes refer to alterations at the cellular and synaptic levels. Studies have shown that musical training can increase dendritic branching and synaptogenesis, enhancing neural connectivity. For instance:

- Synaptic Density: Musicians often exhibit greater synaptic density in areas related to auditory and motor processing (Elbert et al., 1995). This increased density allows for more efficient information processing and integration.
- Neurotrophic Factors: The experience-dependent plasticity facilitated by music can lead to changes in the expression of neurotrophic factors such as brain-derived neurotrophic factor (BDNF), which supports the growth and differentiation of new neurons and synapses (Zatorre et al., 2007). A study by Angelucci et al. (2007) found that music exposure increased BDNF levels in the hippocampus of rats, suggesting a potential mechanism for music-induced cognitive enhancement.
- Myelination: Musical training has been associated with increased myelination of white matter tracts, particularly those connecting auditory and motor regions. This enhanced myelination can lead to faster and more efficient neural transmission (Bengtsson et al., 2005).

11.2. Macrostructural Changes

At the macrostructural level, musical training and active engagement with music can lead to increased gray matter volume in various brain regions:

- Auditory Cortex: Musicians often have larger Heschl's gyri, the site of the primary auditory cortex, which may contribute to enhanced pitch perception (Schneider et al., 2002).
- Motor Areas: Increased gray matter density is observed in motor regions, reflecting the fine motor control required for playing instruments (Amunts et al., 1997).
- Corpus Callosum: The anterior portion of the corpus callosum, which connects the two hemispheres, is more significant in musicians, facilitating interhemispheric communication (Schlaug et al., 1995).
- Cerebellum: Musicians show increased cerebellar volume, particularly in areas involved in fine motor control and timing (Hutchinson et al., 2003).

Longitudinal studies have demonstrated that even short-term musical training can increase gray matter density in young children (Hyde et al., 2009). A study by Habibi et al. (2018) found that children who received music training for two years showed accelerated cortical development in auditory and motor areas compared to controls.

11.3. Functional Changes

Functional changes refer to alterations in brain activity patterns in response to musical stimuli:

- Network Activation: fMRI studies have shown that listening to music activates extensive neural networks, including areas involved in emotion, memory, and motor control (Koelsch, 2014).
- Enhanced Connectivity: Musicians often exhibit enhanced connectivity between auditory and motor regions, reflecting their ability to integrate sensory and motor information more efficiently (Bangert & Altenmüller, 2003).
- Predictive Processing: Recent research suggests that musical training enhances predictive processing in the brain, allowing for more efficient anticipation and processing of auditory stimuli (Koelsch et al., 2019).

- Cross-Modal Integration: Musical training has enhanced multisensory integration, improving the brain's ability to combine information from different sensory modalities (Paraskevopoulos et al., 2012).

11.4. Physiological Changes

Physiological changes in response to music include alterations in neurotransmitter levels and brainwave patterns:

- Dopamine Release: Listening to music we enjoy can increase the release of dopamine in the nucleus accumbens, part of the brain's reward system (Salimpoor et al., 2011). This release is associated with the pleasure and motivation derived from musical experiences.
- Cortisol Reduction: Music listening has been shown to reduce cortisol levels, indicating its potential for stress reduction (Thoma et al., 2013).
- Brainwave Synchronization: EEG studies have shown that music can enhance brainwave synchronization, particularly in the alpha and theta frequency bands, which are associated with relaxation and focused attention (Bhattacharya & Petsche, 2005).
- Autonomic Nervous System: Music can modulate activity in the autonomic nervous system, influencing heart rate, blood pressure, and respiratory rate (Koelsch & Jäncke, 2015).

12. Conclusion

The interplay between music and neuroplasticity demonstrates how musical experiences shape the brain at multiple levels, from microstructural to macrostructural, functional, and physiological changes. Music drives neuroplastic adaptations that enhance brain function and adaptability through listening, playing instruments, or singing. These changes underscore music's profound impact on overall brain health, highlighting its potential as a tool for cognitive and emotional resilience throughout the lifespan.

The multifaceted nature of music-induced neuroplasticity suggests that music could be a powerful intervention for various neurological and psychiatric conditions and a means of promoting healthy cognitive aging. Future research in this field may lead to innovative music-based therapies and educational strategies that leverage the brain's remarkable capacity for change in response to musical experiences.

13. Takeaways

1. Music engages multiple brain regions simultaneously, including the auditory, prefrontal, motor, and limbic systems, demonstrating its powerful influence on neural networks.
2. Listening to enjoyable music activates the brain's reward system, releasing dopamine and contributing to pleasurable experiences and emotional regulation.
3. Musical training induces significant structural and functional changes in the brain, including increased gray matter volume in auditory, motor, and multimodal integration areas.
4. The brain's processing of musical elements (rhythm, timbre, pitch, and lyrics) involves complex neural mechanisms that showcase the brain's remarkable plasticity.
5. Genetic factors and environmental experiences influence individual differences in musical preferences and abilities, highlighting the interplay between nature and nurture in music perception.
6. Active engagement with music, such as playing instruments or singing, leads to more pronounced neuroplastic changes than passive listening, enhancing cognitive functions and motor skills.
7. Music therapy has shown promising results in treating various neurological disorders and mental health conditions and managing pain, leveraging the brain's plasticity for therapeutic benefits.
8. Early exposure to music in childhood can enhance language development, cognitive skills, and social-emotional growth, underscoring the importance of music education.
9. Music plays a crucial role in emotional regulation, stress reduction, and cognitive enhancement throughout adolescence and adulthood.
10. The relationship between music and neuroplasticity extends from microstructural changes at the cellular level to macrostructural alterations in brain regions, demonstrating music's profound impact on brain health and adaptability across the lifespan.

Abbreviations

DMN: Default mode network
EEG: Electroencephalography
fMRI: Functional magnetic resonance imaging
HPA: Hypothalamic-Pituitary-Adrenal
mPFC: Medial prefrontal cortex
PCC: Posterior cingulate cortex
SMA: Supplementary motor area

Bibliography

1. Alluri, V., Toiviainen, P., Burunat, I., Kliuchko, M., Vuust, P., & Brattico, E. (2017). Connectivity patterns during music listening: Evidence for action-based processing in musicians. Human Brain Mapping, 38(6), 2955–2970. https://doi.org/10.1002/hbm.23565
2. Alluri, V., Toiviainen, P., Jääskeläinen, I. P., Glerean, E., Sams, M., & Brattico, E. (2012). Large-scale brain networks emerge from dynamic musical timbre, key, and rhythm processing. NeuroImage, 59(4), 3677–3689. https://doi.org/10.1016/j.neuroimage.2011.11.019
3. Altenmüller, E., Schürmann, K., Lim, V. K., & Parlitz, D. (2002). Hits to the left, flops to the right: different emotions during listening to music are reflected in cortical lateralization patterns. Neuropsychologia, 40(13), 2242–2256. https://doi.org/10.1016/S0028-3932(02)00107-0
4. Amunts, K., Schlaug, G., Jäncke, L., Steinmetz, H., Schleicher, A., Dabringhaus, A., & Zilles, K. (1997). Motor cortex and hand motor skills: structural compliance in the human brain. Human Brain Mapping, 5(3), 206–215. https://doi.org/10.1002/(SICI)1097-0193(1997)5:3<206::AID-HBM5>3.0.CO;2-7
5. Angelucci, F., Fiore, M., Ricci, E., Padua, L., Sabino, A., & Tonali, P. A. (2007). Investigating the neurobiology of music: brain-derived neurotrophic factor modulation in the hippocampus of young adult mice. Behavioural Pharmacology, 18(5–6), 491–496. https://doi.org 10.1097/FBP.0b013e3282d28f50

6. Bangert, M., & Altenmüller, E. O. (2003). Mapping perception to action in piano practice: a longitudinal DC-EEG study. BMC Neuroscience, 4(1), 26. https://doi.org/10.1186/1471-2202-4-26
7. Bendor, D., & Wang, X. (2005). The neuronal representation of pitch in primate auditory cortex. Nature, 436(7054), 1161–1165. https://doi.org/10.1038/nature03867
8. Bengtsson, S. L., Nagy, Z., Skare, S., Forsman, L., Forssberg, H., & Ullén, F. (2005). Extensive piano practice has regionally specific effects on white matter development. Nature Neuroscience, 8(9), 1148–1150. https://doi.org/10.1038/nn1516
9. Bhattacharya, J., & Petsche, H. (2005). Phase synchrony analysis of EEG during music perception reveals changes in functional connectivity due to musical expertise. Signal Processing, 85(11), 2161–2177. https://doi.org/10.1016/j.sigpro.2005.07.007
10. Blood, A. J., & Zatorre, R. J. (2001). Intensely pleasurable responses to music correlate with activity in brain regions implicated in reward and emotion. Proceedings of the National Academy of Sciences, 98(20), 11818–11823. https://doi.org/10.1073/pnas.191355898
11. Blood, A. J., Zatorre, R. J., Bermudez, P., & Evans, A. C. (1999). Emotional responses to pleasant and unpleasant music correlate with activity in paralimbic brain regions. Nature Neuroscience, 2(4), 382–387. https://doi.org/10.1038/7299
12. Bradt, J., Magee, W. L., Dileo, C., Wheeler, B. L., & McGilloway, E. (2010). Music therapy for acquired brain injury. Cochrane Database of Systematic Reviews, (7). https://doi.org/10.1002/14651858.CD006787.pub2
13. Caclin, A., McAdams, S., Smith, B. K., & Winsberg, S. (2006). Acoustic correlates of timbre space dimensions: A confirmatory study using synthetic tones. The Journal of the Acoustical Society of America, 118(1), 471–482. https://doi.org/10.1121/1.1929229
14. Chen, J. L., Penhune, V. B., & Zatorre, R. J. (2008). Moving on time: Rhythm complexity and musical training modulate the brain network for auditory-motor synchronization. Journal of Cognitive Neuroscience, 20(2), 226–239. https://doi.org/10.1162/jocn.2008.20018
15. Cohrdes, C., Wrzus, C., Frisch, S., & Riediger, M. (2017). Tune yourself in Valence and arousal preferences in music-listening choices

from adolescence to old age. Developmental Psychology, 53(9), 1777–1794. https://doi.org/10.1037/dev0000362

16. Eerola, T., & Vuoskoski, J. K. (2013). A review of music and emotion studies: Approaches, emotion models, and stimuli. Music Perception: An Interdisciplinary Journal, 30(3), 307–340. https://doi.org/10.1525/mp.2009.27.2.89
17. Elbert, T., Pantev, C., Wienbruch, C., Rockstroh, B., & Taub, E. (1995). Increased cortical representation of the fingers of the left hand in string players. Science, 270(5234), 305–307. https://doi.org/ 10.1126/science.270.5234.30
18. Foster, N. E. V., & Zatorre, R. J. (2010). Cortical structure predicts success in performing musical transformation judgments. NeuroImage, 53(1), 26–36. https://doi.org/10.1016/j.neuroimage.2010.06.042
19. Frühholz, S., Trost, W., & Kotz, S. A. (2016). The sound of emotions—Towards a unifying neural network perspective of affective sound processing. Neuroscience & Biobehavioral Reviews, 68, 96–110. https://doi.org/10.1016/j.neubiorev.2016.05.002
20. Gaser, C., & Schlaug, G. (2003). Brain structures differ between musicians and non-musicians. Journal of Neuroscience, 23(27), 9240–9245. https://doi.org/10.1523/JNEUROSCI.23-27-09240.2003
21. Gerry, D., Unrau, A., & Trainor, L. J. (2012). Active music classes in infancy enhance musical, communicative, and social development. Developmental Science, 15(3), 398–407. https://doi.org/10.1111/j.1467-7687.2012.01142.x
22. Giordano, B. L., McAdams, S., Zatorre, R. J., Kriegeskorte, N., & Belin, P. (2014). Abstract encoding of auditory objects in cortical activity patterns. Cerebral Cortex, 24(7), 2025–2037. https://doi.org/10.1093/cercor/bhs162
23. Grahn, J. A., & Brett, M. (2007). Rhythm and beat perception in motor areas of the brain. Journal of Cognitive Neuroscience, 19(5), 893–906. https://doi.org/10.1162/jocn.2007.19.5.893
24. Habib, M., & Besson, M. (2009). What do music training and musical experience teach us about brain plasticity? Music Perception, 26(3), 279–285. https://doi.org/10.1525/mp.2009.26.3.279

25. Habibi, A., Damasio, A., Ilari, B., Veiga, R., Joshi, A. A., Leahy, R. M., … & Damasio, H. (2018). Childhood music training induces a change in micro and macroscopic brain structure: results from a longitudinal study. Cerebral Cortex, 28(12), 4336–4347. https://doi.org/10.1093/cercor/bhx286

26. Herholz, S. C., & Zatorre, R. J. (2012). Musical training as a framework for brain plasticity: behavior, function, and structure. Neuron, 76(3), 486–502. https://doi.org/ 10.1016/j.neuron.2012.10.011

27. Huron, D. B. (2006). Sweet anticipation: Music and the psychology of expectation. MIT Press.

28. Hutchinson, S., Lee, L. H. L., Gaab, N., & Schlaug, G. (2003). The cerebellar volume of musicians. Cerebral Cortex, 13(9), 943–949. https://doi.org/10.1093/cercor/13.9.943

29. Hyde, K. L., Lerch, J., Norton, A., Forgeard, M., Winner, E., Evans, A. C., & Schlaug, G. (2009). Musical training shapes structural brain development. Journal of Neuroscience, 29(10), 3019–3025. https://doi.org/10.1523/JNEUROSCI.5118-08.2009

30. Janata, P. (2009). The neural architecture of music-evoked autobiographical memories. Cerebral Cortex, 19(11), 2579–2594. https://doi.org/10.1093/cercor/bhp008

31. Janata, P., Tomic, S. T., & Rakowski, S. K. (2007). Characterization of music-evoked autobiographical memories. Memory, 15(8), 845–860. https://doi.org/10.1080/09658210701734593

32. Jäusovec, N., & Habe, K. (2003). The "Mozart effect": an electroencephalographic analysis employing induced event-related desynchronization/synchronization and event-related coherence. Brain Topography, 16(2), 73–84. https://doi.org/10.1023/B:BRAT.0000006331.10425.4b

33. Kleber, B., Veit, R., Birbaumer, N., Gruzelier, J., & Lotze, M. (2010). The brain of opera singers: experience-dependent changes in functional activation. Cerebral Cortex, 20(5), 1144–1152. https://doi.org/10.1093/cercor/bhp177

34. Koelsch, S. (2010). Towards a neural basis of music perception – A review and updated model. Frontiers in Psychology, pp. 1, 25. https://doi.org/10.3389/fpsyg.2011.00110

35. Koelsch, S. (2014). Brain correlates of music-evoked emotions. Nature Reviews Neuroscience, 15(3), 170–180. https://doi.org/10.1038/nrn3666

36. Koelsch, S., & Jäncke, L. (2015). Music and the heart. European Heart Journal, 36(44), 3043–3049. https://doi.org/10.1093/eurheartj/ehv430

37. Koelsch, S., Fritz, T., Müller, K., & Friederici, A. D. (2006). Investigating emotion with music: An fMRI study. Human Brain Mapping, 27(3), 239–250. https://doi.org/10.1002/hbm.20180

38. Koelsch, S., Vuust, P., & Friston, K. (2019). Predictive processes and the peculiar case of music. Trends in Cognitive Sciences, 23(1), 63–77. https://doi.org/10.1016/j.tics.2018.10.006

39. Kung, S. J., Chen, J. L., Zatorre, R. J., & Penhune, V. B. (2013). Interacting cortical and basal ganglia networks underlying finding and tapping to the musical beat. Journal of Cognitive Neuroscience, 25(3), 401–420. https://doi.org/10.1162/jocn_a_00325

40. Ludke, K. M., Ferreira, F., & Overy, K. (2014). Singing can facilitate foreign language learning. Memory & Cognition, 42(1), 41–52. https://doi.org/10.3758/s13421-013-0342-5

41. Mas-Herrero, E., Dagher, A., & Zatorre, R. J. (2018). Modulating musical reward sensitivity up and down with transcranial magnetic stimulation. Nature Human Behaviour, 2(1), 27–32. https://doi.org/10.1038/s41562-017-0241-z

42. McLachlan, N., & Wilson, S. J. (2017). The central role of recognition in auditory perception: a neurobiological model. Psychological Review, 124(1), 72–103. https://doi.org/10.1037/a0018063

43. Merchant, H., Grahn, J., Trainor, L., Rohrmeier, M., & Fitch, W. T. (2015). Finding the beat: A neural perspective across humans and non-human primates. Philosophical Transactions of the Royal Society B: Biological Sciences, 370(1664), 20140093. https://doi.org/10.1098/rstb.2014.0093

44. Moerel, M., De Martino, F., & Formisano, E. (2014). An anatomical and functional topography of human auditory cortical areas. Frontiers in Neuroscience, 8, 225. https://doi.org/10.3389/fnins.2014.00225

45. Moreno, S., Bialystok, E., Barac, R., Schellenberg, E. G., Cepeda, N. J., & Chau, T. (2011). Short-term music training enhances verbal

intelligence and executive function. Psychological Science, 22(11), 1425–1433. https://doi.org/10.1177/0956797611416999

46. Morrison, S. J., & Demorest, S. M. (2009). Cultural constraints on music perception and cognition. Progress in Brain Research, 178, 67–77. https://doi.org/10.1016/S0079-6123(09)17805-6
47. Özdemir, E., Norton, A., & Schlaug, G. (2006). Shared and distinct neural correlates of singing and speaking. Neuroimage, 33(2), 628–635. https://doi.org/10.1016/j.neuroimage.2006.07.013
48. Pantev, C., Engelien, A., Candia, V., & Elbert, T. (2001). Representational cortex in musicians. Annals of the New York Academy of Sciences, 930, 300–314. https://doi.org/10.1111/j.1749-6632.2001.tb05740.x
49. Pantev, C., Ross, B., Fujioka, T., Trainor, L. J., Schulte, M., & Schulz, M. (2003). Music and learning-induced cortical plasticity. Annals of the New York Academy of Sciences, 999(1), 438–450. https://doi.org/10.1196/annals.1284.054
50. Paraskevopoulos, E., Kuchenbuch, A., Herholz, S. C., & Pantev, C. (2012). Musical expertise induces audiovisual integration of abstract congruency rules. Journal of Neuroscience, 32(50), 18196–18203. https://doi.org/10.1523/JNEUROSCI.1947-12.2012
51. Patel, A. D. (2010). Music, language, and the brain. Oxford University Press.
52. Patel, A. D., & Iversen, J. R. (2014). The evolutionary neuroscience of musical beat perception: the Action Simulation for Auditory Prediction (ASAP) hypothesis. Frontiers in Systems Neuroscience, 8, 57. https://doi.org/10.3389/fnsys.2014.00057
53. Penhune, V. B., Zattore, R. J., & Evans, A. C. (1998). Cerebellar contributions to motor timing: a PET study of auditory and visual rhythm reproduction. Journal of Cognitive Neuroscience, 10(6), 752–765. https://doi.org/10.1162/089892998563149
54. Peretz, I., Gaudreau, D., & Bonnel, A. M. (1998). Exposure effects on music preference and recognition. Memory & Cognition, 26(5), 884–902. https://doi.org/10.3758/BF03201171
55. Rentfrow, P. J., & Gosling, S. D. (2003). The do re mi's of everyday life: The structure and personality correlates of music preferences. Journal of Personality and Social Psychology, 84(6), 1236–1256.

56. Rouijel, I., Gür-özmen, S., & Karahan-Özcan, R. (2023). Music therapy in acute stroke. Department of Neuroscience, Bahçeşehir University Faculty of Health Science, İstanbul, Türkiye; Department of Neurology, Gebze Fatih State Hospital, İstanbul, Türkiye. . https://doi.org/10.14744/bauh.2023.65375

57. Sachs, M. E., Ellis, R. J., Schlaug, G., & Loui, P. (2016). Brain connectivity reflects human aesthetic responses to music. Social Cognitive and Affective Neuroscience, 11(6), 884–891. https://doi.org/10.1093/scan/nsw009

58. Salimpoor, V. N., Benovoy, M., Larcher, K., Dagher, A., & Zatorre, R. J. (2011). Anatomically distinct dopamine is released during anticipation and the experience of peak emotion in music. Nature Neuroscience, 14(2), 257–262.

59. Salimpoor, V. N., van den Bosch, I., Kovacevic, N., McIntosh, A. R., Dagher, A., & Zatorre, R. J. (2013). Interactions between the nucleus accumbens and auditory cortices predict music reward value. Science, 340(6129), 216–219. https://doi.org/ 10.1126/science.123105

60. Schlaug, G., Jäncke, L., Huang, Y., Staiger, J. F., & Steinmetz, H. (1995). Increased corpus callosum size in musicians. Neuropsychologia, 33(8), 1047–1055. https://doi.org/10.1016/0028-3932(95)00045-5

61. Schneider, P., Scherg, M., Dosch, H. G., Specht, H. J., Gutschalk, A., & Rupp, A. (2002). The morphology of Heschl's gyrus reflects enhanced activation in the auditory cortex of musicians. Nature Neuroscience, 5(7), 688–694. https://doi.org/10.1038/nn871

62. Slater, J., Ashley, R., Tierney, A., & Kraus, N. (2018). Got rhythm? Better inhibitory control is linked with more consistent drumming and enhanced neural tracking of the musical beat in adults and nonpercussionists. Journal of Cognitive Neuroscience, 30(1), 14–24. https://doi.org/10.1162/jocn_a_01189

63. Thaut, M. H., McIntosh, G. C., & Holmberg, V. (2015). Neurobiological foundations of neurologic music therapy: rhythmic entrainment and the motor system. Frontiers in Psychology, p. 5, 1185. https://doi.org/10.3389/fpsyg.2014.01185

64. Thoma, M. V., La Marca, R., Brönnimann, R., Finkel, L., Ehlert, U., & Nater, U. M. (2013). The effect of music on the human stress response. PloS One, 8(8), e70156. https://doi.org/10.1371/journal.pone.0070156

65. Warren, J. D. (2008). How does the brain process music? Clinical Medicine, 8(1), 32–36. https://doi.org/10.7861/clinmedicine.8-1-32

66. Wilkins, R. W., Hodges, D. A., Laurienti, P. J., Steen, M., & Burdette, J. H. (2014). Network science and the effects of music preference on functional brain connectivity: from Beethoven to Eminem. Scientific Reports, 4(1), 1–6. https://doi.org/10.1038/srep06130

67. Zatorre, R. J., Chen, J. L., & Penhune, V. B. (2007). When the brain plays music: auditory-motor interactions in music perception and production. Nature Reviews Neuroscience, 8(7), 547–558 https://doi.org/10.1038/nrn2152

Artistic Engagement and Its Effects on Brain Plasticity

Hacer Doğan

1. Introduction

Art is an essential tool for improving brain health and functionality in individuals. Recent research has shown that engaging in art increases brain plasticity, defined as brain cells' ability to change their shape and function under various conditions (Zaidel, 2010). The neuroplasticity created by art refers to the brain's ability to reshape itself and express the effects of creative processes and art activities on brain structures. Our experiences and environmental interactions are the most critical factors shaping this process (Bolwerk et al., 2014). People use art to express their experiences and the world around them. Art forms encompass different expressions of human creativity and imagination, with primary forms including visual, auditory, performance, applied, literary, and cinematic. Each of these has its distinct mode of expression. Over time, art forms can expand and intersect with one another, leading to the emergence of new forms of expression. Visual arts, for instance, encompass various disciplines such as painting, sculpture, ceramics, design, crafts, photography, film, and architecture. Visual artists develop their artistic skills and talents through education in various aesthetic elements like construction, composition, and abstraction to bring their visions and conceptual ideas to life (Tavin, 2020).

It has also been found that when art is used for advertising, it can positively influence consumers' perceptions of a product or brand, helping them make quicker and more favorable decisions. A study by Kim and colleagues concluded that consumers exposed to visual art pieces tend to have more profound emotional reactions and are more inclined to make positive decisions (Kim, 2020). Art is essential for society because it allows people to express, share, and understand their thoughts, emotions, and experiences. It helps preserve and spread cultural values, reflecting the identity and history. Art also strengthens empathy among people and offers an opportunity to understand different perspectives. In society, it promotes aesthetic appreciation and critical thinking, contributing to the development of individuals' creativity. The purpose of this section is to deeply explore how art is perceived in the brain and the effects of this perception process on neuroplasticity. From a scientific perspective, it aims to explain how art contributes to the brain undergoing structural and functional changes beyond merely being an aesthetic experience.

2. Art and Creativity

Creativity arises from the collaboration of different regions of the brain and the activation of their complex neural connections. Different fields like music, painting, and literature emerge with unique brain activation patterns (Abraham, 2018). In their study, Chen and colleagues explored the common brain regions active in all artistic creative processes. They found that regions like the prefrontal and posterior parietal cortex play significant roles in all creative processes. These regions are also associated with cognitive functions such as problem-solving, planning, and abstract thinking (Chen et al., 2020).

1. Musical Creativity: Regions such as the temporal lobes and motor cortex play essential roles in musical creativity. These areas are related to auditory processing, rhythm perception, and motor skills (Limb & Braun, 2008). (see chapter on music and neuroplasticity)
2. Drawing Creativity: Drawing creativity, or more precisely, visual artistic creativity, is the process of producing new and visually appealing visual forms, like paintings, drawings, and graphic designs. Drawing creativity is the term used to describe the mental and neurological processes that support the ability to create visually appealing, unique, and novel representations through drawing (Zeki, 2001). According to science, drawing creativity involves several interrelated brain processes, such as:
   i. Cognitive Processes: Divergent thinking is the capacity to come up with several answers to an issue, which is essential to the creative process. Divergent thinking helps artists in develop original concepts or methods when drawing (Runco, 2014).
   ii. Problem-Solving: Creative drawing frequently requires solving aesthetic and conceptual problems (such as depicting a notion fresh).

Planning and decision-making are high-level executive functions involved in drawing problem-solving (Ward et al., 1997).

iii. Mental Imagery: Artists use much mental imagery when depicting situations, objects, or abstract ideas. The ability to alter and modify mental images is crucial to creative drawing (Kosslyn & Thompson, 2003).

2.1. The Neuroscience of Drawing Creativity

Drawing activates the visual and motor cortex. It involves processing visual information, hand-eye coordination, and fine motor skills (Abraham, 2018). Key brain regions in creative drawing include:

- Prefrontal Cortex: This area handles self-monitoring, planning, and complex decisions. It is crucial for guiding drawing, evaluating it, and making changes (Abraham, 2018).
- Parietal Lobe: It aids in creating accurate shapes and understanding space (Aziz-Zadeh et al., 2013).
- Default Mode Network (DMN): Active during rest, sparks spontaneous creativity, and helps generate drawing ideas (Beaty et al., 2016).
- Motor Cortex: It manages the fine motor skills needed for drawing (Dietrich, 2004a).
- Reward System: This system, linked to dopamine, drives the pleasure and motivation in drawing (Schott, 2014).

2.2. Creative Drawing Stages

The artist prepares by gathering information, ideas, and sources of inspiration. Subconscious processing occurs during "incubation," when the artist may pause and ideas start to "incubate" independently without conscious effort. A flash of an insight or innovative breakthrough is called illumination.

- Execution: The artist uses technical skills to turn the idea in their head into an actual drawing (Dietrich, 2004b).
- The Function of Feelings: Emotional states also impact creative drawing. While negative emotions can occasionally result in more concentrated, specific labor, positive emotions can sometimes boost creative productivity by expanding thought-action repertoires. Drawing as a means of expressing

emotions can be therapeutic as they activate both brain hemispheres (Solso, 2001).

2.3. Research and Studies

fMRI studies show that creative tasks, such as painting, engage more brain regions. The default mode network (DMN) and executive control network interact crucially. This balance is critical for focused goals and spontaneous ideas. The right lateral prefrontal cortex (DLPFC) is vital in creative drawing. It manages internal focus and visualization, blocking distractions. Interestingly, amateur artists rely more on the left frontal brain during creativity. The dual model of creativity suggests that early drawing taps into the default network for idea generation. Later, it connects with the control network to evaluate these ideas (Arden et al., 2010).

The medial temporal lobe (MTL), including the hippocampus, parahippocampus, and fusiform gyrus, plays a more significant role in creative drawing than the prefrontal cortex, especially in visual art. The fusiform gyrus activates when we create mental images without outside help. Research shows that MTL activity during creative thought helps link and recall memories, aiding in new idea formation. A method called episodic specificity induction boosts activity in the left anterior hippocampus, enhancing memory and fostering creative thinking, linking MTL areas to creativity (Rominger et al., 2020).

Neuropsychology highlights the right hemisphere's role in visual creativity, especially in spatial and visual tasks, as shown in studies of brain-damaged individuals (Dietrich, 2004b). Solso (2001) and others (Aziz-Zadeh et al., 2013) examined how the brain works in art creation. It found that when drawing faces, professional artists show more activity in the right-middle frontal area but less in the right posterior parietal cortex, which processes facial information. This suggests that artists focus more on the meaning of faces than their physical traits (Rominger et al., 2020).

2.3.1. Cognitive Science

Research in cognitive science has examined the link between the brain's ability to produce creative artwork and factors such as experience, training, and drawing skills (Zeki, 2001).

2.3.2. Therapeutic Effects of Drawing

Drawing aids emotional expression and solves psychological issues in art therapy. It activates brain areas linked to emotions and fosters neuroplasticity, helping people cope with stress and trauma (Gantt & Tinnin, 2009).

Literary Creativity: It activates brain regions involved in language and symbolic thinking, mainly the left temporal and prefrontal cortex. These areas are crucial for language, storytelling, and writing. Literary creativity blends cognitive, emotional, and social factors. It is studied in neuroscience, psychology, and cognitive science.

2.3.3. Neuroscience of Literary Creativity

- Brain Areas: Studies using fMRI and EEG show that literary creativity activates several brain regions. The dorsolateral prefrontal cortex (DLPFC) is vital for planning and abstract thinking. Furthermore, the default mode network (DMN), active during daydreaming, supports creative thinking (Arden et al., 2010).
- Hemispheric Role: It is a myth that creativity is solely in the right hemisphere. This hemisphere aids in metaphorical thinking and visualization. Meanwhile, the left hemisphere, crucial for language and analytical thinking, is equally important. Effective communication between both hemispheres fuels literary creativity (Beaty et al., 2015; Dietrich, 2004b).
- Thinking Types: Literary creativity blends divergent and convergent thinking. Divergent thinking generates many ideas. In contrast, convergent thinking refines and selects the best ones. Associative networks drive divergent thinking, while executive functions aid editing and revising (Jung & Haier, 2007).

2.3.4. Cognitive Psychology of Literary Creativity

2.3.4.1. Stages of Creativity: Literary creativity follows a set of stages, often conceptualized as:

- Preparation: The writer absorbs knowledge, ideas, and experiences.
- Incubation: The writer's mind unconsciously processes this information, often during periods of rest or mind-wandering.
- Illumination: The "Aha!" moment when the creative idea emerges.
- Verification: The writer evaluates, refines, and revises the work.

- Flow and Creativity: Creative writers frequently enter a flow state, marked by intense concentration and total immersion in their work. Flow is linked to intrinsic motivation and improved performance. Research indicates that during flow, activity in the prefrontal cortex diminishes, which reduces self-criticism and boosts spontaneity (Csikszentmihalyi, 1997).

2.3.4.2. *The Role of Emotion in Creative Writing*

- Emotions and Creativity: Emotions are crucial to creativity. Positive feelings spark more ideas. Meanwhile, negative feelings lead to deeper narratives, enhancing creativity.
- Mood and Writing: A writer's mood affects their work. It shapes the tone and themes. Writers in certain moods focus on specific themes, matching their emotional state (Kaufmann, 2003).

2.3.4.3. *Creative Writing as Problem-Solving*

- Writing and Problem-Solving: Writing creatively is like solving a problem. Writers face limits, such as genre rules and personal goals. They must find innovative solutions. This aligns with the Geneplore model of creativity (Finke et al., 1992), which involves generating and exploring ideas.
- Generative Phase: The writer produces novel ideas, characters, and scenarios.
- Exploratory Phase: The writer refines these ideas, testing their coherence and fit within the narrative structure.

2.3.5. Creativity and Language

- Linguistic Flexibility: Creativity in writing hinges on using language inventively. Writers employ metaphors, analogies, and complex stories to forge new meanings. Research shows that creative writers excel in these techniques. They can bend or break traditional writing rules (Sawyer, 2012).
- Embodied Cognition: This theory posits that language and creativity stem from sensory and motor experiences. Writers often draw on sensory memories or imagine physical actions to create vivid stories (Gibbs, 2006).

2.3.6. Conclusion

Scientific studies reveal that literary creativity is complex. It involves the brain networks, cognitive steps, emotional control, problem-solving, and linguistic creativity. This understanding comes from merging insights across fields, highlighting the interplay of neurological, psychological, and social factors in creative writing.

3. Effects of Art on Neuroplasticity

3.1. Emotional and Cognitive Functions

Art increases emotional and cognitive functions, creative thinking, and problem-solving skills. Creating a piece of art or observing art pieces triggers emotional responses in the limbic system. These experiences activate neurons and strengthen connections, thereby increasing neuroplasticity. For example, being in a creative process or emotionally interacting with an artwork can activate specific neuronal groups in the brain, increasing the formation of new neural pathways. Especially while working on an artistic project, the formation of new neural pathways in the brain can be triggered by working on manual skills, visual perception, and emotional responses (Bolwerk et al., 2014).

3.2. Visual Memory

Visual arts develop the ability to remember visible details and visual information. Artists pay more attention to visual details, which can lead to stronger visual memories (Tyler & Likova, 2012).

3.3. Emotional Memory

Art-related activities can trigger emotional memories and experiences. We generally remember the information we care about emotionally better, and the emotions felt while creating art can be more lasting in memory (Mauri et al., 2020).

3.4. Focus and Concentration

Engaging in art improves attention and concentration. Increased focus can make encoding and recalling information easier (Demarin et al., 2016).

3.5. Motor Memory

Engaging in arts such as painting and sculpting enhances hand-eye coordination and fine motor skills. These motor activities can also improve memory-related brain regions (Fisher et al., 2014).

3.6. Aging Process

As we age, the ability for neuroplasticity can decrease, but creative activities like art can slow down or even reverse this process. Engaging in art helps maintain and improve cognitive flexibility and mental functions in older adults (Fioranelli et al., 2023).

3.7. Hormonal and Chemical Factors

Engaging in art also affects hormone levels and the release of chemicals in the brain. For example, it can trigger the release of happiness hormones such as endorphins and dopamine due to its stress-reducing effects. These hormones enhance neuroplasticity by strengthening the connections between nerve cells (Pilozzi et al., 2020).

3.8. Conclusion

The neuroplastic effects of art on the brain improve mental flexibility and overall brain health. Therefore, integrating artistic activities into daily life supports neuroplasticity and aesthetic and emotional health.

More research in neuroscience and psychology is needed to better understand art's complex effects on the brain.

4. Brain Activation and Art Perception

No single brain area is solely for art perception. Various regions and networks work together to achieve this. Art involves visual, auditory, emotional, and cognitive elements (Chatterjee & Vartanian, 2014).

Different brain areas are activated by different art forms. Visual arts, like painting and sculpture, engage the visual cortex. Meanwhile, music activates auditory regions. This cross-talk boosts neuroplasticity (Shah et al., 2013).

When viewing art, the brain's visual cortex is the first to activate. This area processes visual information, perceiving colors and patterns. Emotional and cognitive responses occur in the limbic system and prefrontal cortex. Art perception is complex, needing the cooperation of a vast network. Individual and cultural differences in responses show that art perception is not limited to one brain region. It involves multiple areas and connections (Miller & Limb, 2013; Chatterjee, 2011).

4.1. Frontal Lobe

The frontal lobe is critical for problem-solving, planning, decision-making, and creative thinking. It also helps develop ideas for art. This lobe shapes our art experiences through decision-making, emotional control, and abstract thinking. It helps us judge and interpret visual cues and is involved in evaluating art (Dietrich, 2004a).

Key frontal lobe areas linked to aesthetic impressions include:

- *Orbitofrontal Cortex:* This area processes pleasure and rewards. It helps us judge the emotional and enjoyable aspects of art. Our responses to art determine our enjoyment. The brain's reward system boosts this enjoyment (Vartanian & Skov, 2014).
- *Prefrontal Cortex:* This part is for higher thinking. It helps us understand art's meaning, symbolism, and beauty. We judge art by linking it to our experiences, knowledge, values, and memories (Cattaneo et al., 2014).
- *Inferior Frontal Gyrus:* This area is involved in reasoning and language and is activated when we explain our emotional responses to art. It connects feelings from the amygdala with our interpretations (Cattaneo et al., 2014).

4.2. Parietal Lobe

The parietal lobe is crucial to visual arts such as painting and sculpture. It aids in spatial processing, attention, and integrating senses. This lobe helps us navigate and understand artworks' structures (Grefkes & Fink, 2005). It also helps us perceive the relationships and orientations of objects in art.

- Spatial Awareness and Visual Perception: The parietal lobe processes spatial details in art, especially its back part. It helps us grasp depth, perspective, and distance. This awareness is critical to understanding an artwork's composition and emotional impact (Cattaneo et al., 2014).
- Integration of Sensory Information: The parietal lobe combines inputs from the visual cortex and others. It creates a complete view of the artwork, including motion and orientation (Binkofski & Buxbaum, 2013).
- Attention and Focus: This lobe directs our focus to specific art features, like a painting's focal point. It encourages exploring different parts, deepening our engagement and appreciation (Binkofski & Buxbaum, 2013).

4.3. Occipital Lobe

The occipital lobe is vital for visual processing, especially in art activities like painting (Zeki, 1999).

- Processing by the Primary Visual Cortex (V1): V1, in the occipital lobe, is the first stop for visual data. It breaks down what we see into shape, color, contrast, and orientation. These elements are crucial for recognizing patterns, colors, and edges in art (Vartanian & Skov, 2014).
- Higher-Order Visual Processing: The occipital lobe goes beyond essential visual input. It collaborates with other brain parts to understand complex visual art. For instance, it aids in recognizing objects and scenes by relaying information to the temporal and parietal lobes. This process helps us grasp what we see, whether a clear subject or an abstract piece (Vartanian & Skov, 2014).
- Perception of Depth, Movement, and Patterns: The occipital lobe also helps us see depth, movement, and patterns in art. Artists use shading or perspective to create depth, and the occipital lobe interprets these cues. Furthermore, the middle temporal area (MT or V5), part of the occipital lobe, aids in detecting motion in art (Vartanian & Skov, 2014).
- Visual Aesthetic Judgments: The occipital lobe is crucial in forming opinions about visual art. It first analyzes visual structure before the amygdala or frontal lobe considers emotional or cognitive aspects. This analysis shapes our immediate reactions to symmetry, balance, and patterns, which are critical factors in finding art pleasing (Orban, 2008).

4.4. Amygdala

The amygdala is key to how we feel about art. It processes emotions and reacts to art, making us feel joy, awe, sadness, or discomfort, depending on the artwork. When we experience art, the emotional content activates the amygdala. This, in turn, shapes our understanding and connection to the piece. For instance, viewing emotionally charged art makes the amygdala more active. This boosts the emotional link between the viewer and the artwork. Hence, our emotional processing and valuation are vital.

The widespread effects of art on the brain are highly beneficial for cognitive and emotional health (Bolwerk et al., 2014).

5. Art Experiences in the Brain

Recent studies show beautiful artworks activate the brain's reward systems and boosts dopamine level. Personal preference and experience shape aesthetic perception, which is subjective and varies between individuals (Magsamen, 2023).

Researchers have used fMRI scans to study how we perceive and experience art. The V1 cortex, crucial for aesthetic perception, is activated first when viewing art. This area responds to visual awareness, attention, and organization—different parts of the visual cortex process aspects like color and shape. The midbrain and orbitofrontal cortex (OFC) show high activity during the perception of beauty.

In a study by Kawabata and Zeki, the OFC was most active when participants found a painting beautiful. Its activity dropped for ugly paintings and was moderate for neutral ones. This shows how personal feelings influence the experience of beauty. The insula was highly active when people judged art, linking it to emotional control and expression. This suggests that empathy and personal involvement are vital in understanding art (Kawabata & Zeki, 2004).

6. Reward Neural Systems

The brain's reward system boosts art, creativity, and curiosity. Understanding these processes help us understand the brain mechanisms behind human motives and behaviors (Kenett et al, 2023).

Creativity allows people to invent new ideas. During creative tasks, the prefrontal cortex is very active. This area helps with problem-solving, flexible thinking, and seeking new experiences. The creative process depends on the brain's reward systems and memory networks working together (Dietrich, 2004b). The dopamine system and reward centers are key players, like the ventral tegmental area and nucleus accumbens. These are crucial for curiosity. When new, exciting information is found, dopamine is released. This makes curiosity feel rewarding (Jung & Vartaanian, 2018).

Aesthetic experiences trigger emotional and cognitive responses to art and beauty. Key brain areas, like the orbitofrontal cortex and insula, evaluate these experiences and manage emotions (Chatterjee, 2014; Ishizu & Zeki, 2011).

Memory and reward systems interact during these experiences, influencing responses. This interaction, involving the hippocampus and reward networks, fuels our desire for knowledge, creativity, and appreciation of beauty (Vessel et al., 2012). It boosts learning, creativity, and enjoyment of aesthetics.

Additional brain areas, including the prefrontal cortex, anterior insula, and striatum, also activate during these experiences. They are crucial for processing rewards, evaluating emotions, making decisions, and valuing experiences.

7. Neuroaesthetics

Neuroaesthetics studies how art affects the brain to understand aesthetic perception. It aims to uncover brain mechanisms behind fundamental human experiences (Zeki, 2001). Additionally, it explores the links between morality, law, and brain function. It also delves into the brain bases of creativity, achievement, and religious belief.

Recent studies have highlighted the impact of the visual arts on perception and thinking. These arts use color, shape, perspective, and composition, which are vital to human perception (Solso, 1994). Art can trigger strong emotions and activate the brain's reward system (Chatterjee & Vartanian, 2014).

Both conscious and unconscious processes shape how we perceive beauty. Understanding these processes is vital. Semir Zeki, a professor at The University College London coined "neuroaesthetics." He sees artists as informal neuroscientists, exploring the brain's capabilities through their work (Zeki, 2001). Zeki proposed two laws: consistency and abstraction. These laws explain how artists tap into the visual brain's mechanisms (Ishızu & Zeki, 2011).

Consistency involves the brain seeking to understand the constant traits of objects. It filters out changing properties. For example, artists exclude details that are not seen as constant when painting. This creates a static image, distinct from a moving view. The brain engages with the concept and emotions of the artwork rather than analyzing it (Zeki, 1999). Artists need only showcase an object's constant traits to convey its essence. Observed objects exist only in the brain and are linked to prior beliefs. Recognizing faces from different angles illustrates this. The brain focuses on key features, like shape and placement, to identify a face, even without other details (Tsao & Livingstone, 2008). The visual brain prioritizes fixed traits, filtering out dynamic ones. This also applies to artists. They exclude inconsistent details and work with fixed information (Gregory, 1997).

Abstraction is creating categories by spotting patterns in materials. It helps us analyze art by applying general rules to specific cases. Essential elements like color and shape lead to more profound concepts tied to our senses. Artists can play with these elements to evoke specific emotions. However, the brain's response to abstract versus representational art remains unclear.

Certain brain areas become more active in beautiful landscapes, including regions linked to rewards and emotions. Zhao and the team found that moving elements in art engage more brain areas than still ones. They noted that individual preferences also affect brain activity. More active responses were seen in those who preferred moving landscapes. This suggests that judging art involves emotional and motor processes, not just visual ones (Zhao et al., 2020).

Aesthetic perception works alongside cognitive processes like emotions, attention, and decision-making, making it a key topic in cognitive science. Wassiliwizky and Menninghaus explored the link between aesthetics and cognitive processes. They argued that understanding aesthetic experiences is vital for uncovering emotional and motivational mechanisms (Wassiliwizky & Menninghaus, 2021).

Aesthetic experiences also boost creativity. Cognitive science can illuminate this by studying creative thinking and its relationship with aesthetic judgment. Understanding the brain's response to aesthetic experiences is crucial for neuroscience (Wassiliwizky & Menninghaus, 2021).

8. Effects of Art at the Cellular Level

Synaptogenesis: Creating or viewing art strengthens synapses. This boosts memory and learning, leading to complex synaptic networks. Art activities increase neurotransmitters like dopamine and serotonin, enhancing feelings of reward and happiness (Malenka & Bear, 2004).

1. Neuronal Reorganization: Art engagement or observation reorganizes brain connections (Zatorre et al., 2012). Visual arts and music foster new connections across brain regions, aiding in learning and skill acquisition (Schlaug et al., 2005).

2. Long-Term Potentiation (LTP): Artistic activities boost LTP (Lynch, 2004). They strengthen synapses by enhancing neuron signals (Toga & Thompson, 2005).
3. Neurogenesis: Art stimulates new neuron growth, especially in the hippocampus, aiding learning and memory (Kempermann & Gage, 1999).
4. Dendritic Branches and Spines: Creative activities expand dendritic branches and spines (Bosch & Hayashi, 2007).
5. Neurotrophic Factors: Art increases neurotrophic factors like BDNF, promoting cell growth and neuroplasticity (Rios, 2013).
6. Glial Cells: Art enhances glial cells, aiding neuroplasticity. It boosts astrocytes and myelin production, speeding up nerve transmission (Allen & Barres, 2009).

9. Differences Between the Right and Left Brain Hemispheres in Art Processing

The right and left hemispheres of the brain are activated in different ways during the processing of art.

9.1. Right Brain

- Creativity and Imagination: The right brain is associated with creative and imaginative processes (Torrance, 1979).
- Spatial Awareness: It handles spatial awareness and understanding of visual and spatial relationships (Corballis, 2014).
- Emotional Processing: The right brain is involved in the emotional aspects of art (Gainotti, 2019).

9.2. Left Brain

Analytical and Logical Thinking: The left brain is linked to analytical and logical thinking (Goldberg, 2001).

Language Skills: It is also involved in language skills and the verbal explanation of art (Price, 2012).

During the creation of art, the left brain becomes more active when analyzing technical details and structures. It is also engaged during linguistic tasks such as explaining or writing about art. For instance, the left brain is activated when describing a piece of art or writing an essay about it (Corballis, 2014).

It is important to note that art is a complex activity involving both brain hemispheres. For example, a painter might use the right brain to imagine the overall composition and colors while focusing on details and technical skills with the left brain. Although the right brain often plays a more significant role in artistic activities, the left brain contributes significantly. The coordinated work of both hemispheres nurtures and develops both art's creative and analytical aspects.

10. The Effects of Hemiplegia on Art Perception

- Visual-Spatial Perception: Damage to the right hemisphere affects the ability to visually and spatially perceive art. For instance, the perception of perspective, depth, and composition may change (Zaidel, 2013).
- Emotional Response: Damage to the right hemisphere can alter emotional processing capabilities, thereby changing emotional responses to art (Gainotti, 2015).
- Verbal Expression: Damage to the left hemisphere affects the ability to express and analyze art verbally. Individuals may have difficulty describing or discussing the art they see or create (Zeki, 1999).
- Detail and Analysis: Damage to the left hemisphere can impact the ability to focus on artistic details. This may make it difficult for a person to perceive art's fine details and technical aspects (Benton & Tranel, 2000).
- Creative Process: The coordinated work of both hemispheres of the brain is crucial for the creative process. Hemiplegia disrupts this coordination, potentially affecting artistic creativity and productivity.

10.1. Case Studies About Artistic Transformation after Hemiplegia

Jon Sarkin, an artist from Massachusetts, underwent a major change after getting right hemiplegia. This condition shifted his career and revealed his artistic talent. Post-stroke, Sarkin suddenly became passionate about art. Before, he was a businessman. Now, he creates complex, colorful works on canvas. His art features intricate patterns, vibrant colors, and abstract figures. Some have even been showcased in galleries (Rhodes, 2023).

Similarly, Dr. Franco Magnani, an Italian engineer and watchmaker, also turned to art after hemiplegia. Following a brain infection in the 1960s, his interest in art sparked. Despite memory issues from brain damage, he began drawing vivid pictures of his childhood Italian village. His works blend memory and imagination, showcasing the brain's adaptability. This shows that, after injury, inactive brain areas can awaken, leading to new skills and interests. Often, this involves right-brain areas linked to creativity and visual memory (Treib, 2013).

10.2. Art Perception After Dementia

Dementia leads to impairments in memory, thinking, language, and judgment. An artist's work can be affected by these cognitive changes, yet the creative process can also lead to unique approaches.

Willem de Kooning, one of the prominent abstract expressionist painters of the twentieth century, was diagnosed with Alzheimer's disease in the late 1980s. As his disease progressed, his work style noticeably changed. De Kooning's early works were rich in complex and intense compositions. He used color and form in a highly dynamic and intense manner. Before the onset of dementia, de Kooning's works still maintained a distinct structure and composition. His complex and intense figurative abstraction was evident in his paintings.

When the first signs of the disease appeared, simplicity and minimalism began to be observed in his works. Lines and forms became less complex. As the disease progressed, de Kooning's works became simpler, more abstract, and more minimal. The color palette narrowed, and the forms became softer and more fluid.

During this period, although the artist's works were less detailed, they still carried aesthetic and emotional intensity. These changes can be seen not as the result of the artist's conscious choices, but rather as a reflection of changes in his cognitive capacity (Marcus et al., 2009).

11. Differences in the Brains of People Interested in Art and Those Not Interested in Art

People interested in art show different brain activation and connections than those who aren't (Leder et al, 2004).

- Brain Activation: Art lovers have more activity in the visual cortex during art-related activities, like painting or viewing art (Vartanian & Skov, 2014).
- Emotional Responses: They also have stronger emotional reactions to art, linked to heightened activity in the limbic system (Ishizu & Zeki, 2011).
- Cognitive Functions: Art enthusiasts differ in interpreting, analyzing, and expressing art. These tasks involve the prefrontal cortex (Chatterjee, 2014).
- Brain Plasticity: Engaging with art boosts brain plasticity. It strengthens and forms new neural connections (Draganski et al, 2004; Vartanian & Skov, 2014).
- Motivation and Focus: People interested in art have been observed to show more motivation and focus on art activities. This can be related to the activation of dopaminergic and other motivational systems in specific brain regions (Wise, 2004).

People not interested in art might show less brain activity or differences. However, interest and experience in art vary widely due to personal and cultural factors. So, it's vital to conduct more studies across different groups to better understand art's impact on the brain.

12. Art Therapy

Art therapy uses creativity to reduce stress and provide emotional relief. It promotes healthy brain changes in stressed individuals. This therapy improves emotional expression, stress management, communication, and psychological health (Stuckey & Nobel, 2010). Here's how it helps with various conditions:

- Stress and Mental Health: Art reduces stress and boosts mental health. It lowers cortisol, aiding brain flexibility. This improves memory, emotional control, and mental health (Bolwerk et al, 2014).
- Psychological Disorders: Art therapy aids in treating disorders like depression, anxiety, PTSD, and eating disorders. It helps express emotions and resolve conflicts (Carr et al, 2012).
- Post-Trauma Situations: For trauma survivors, art therapy aids in expressing and processing experiences. It facilitates emotional healing (Gantt & Tinnin, 2009).

- Chronic Illnesses and Pain: Art therapy reduces stress and improves life quality in chronic illness patients. It also helps manage pain by diverting attention (Baker et al., 2018).
- Neurological Diseases and Dementia: Art therapy aids in managing Alzheimer's and similar diseases. It stimulates memory, strengthens connections, and enhances quality of life (Chandra & Liddle, 2014).
- Children and Adolescents: Art therapy helps with emotional regulation, behavior issues, and boosting confidence (Rubin, 2016).
- Chronic and Cancer Illnesses: Art therapy alleviates pain and emotional burdens in chronic illnesses (Slayton et al., 2010).
- Schizophrenia and Psychoses: Art therapy helps manage symptoms in schizophrenia and other psychotic disorders (Ferszt & Mularski, 2012).

The benefits of art therapy stem from creative engagement and brain plasticity (Stuckey & Nobel, 2010). It significantly boosts well-being and quality of life when combined with other treatments.

Only certified art therapists should conduct art therapy. They help individuals tap into creativity and support emotional healing. Each person's needs are unique, so a tailored approach is vital.

13. Impact of Art Education on Performance in Other Subjects

- Improves Academic Performance: Art education boosts students' grades, especially in math and language arts.
- Enhances Motivation: Art classes spark interest in school, boosting motivation and success.
- Develops Memory and Concentration: Art activities sharpen memory and focus, aiding learning in other subjects.
- Builds Discipline and Patience: Art projects teach focus, patience, and perseverance, improving overall academic performance.

Art education is vital. It enhances creativity, analysis, emotional, and social skills. These improvements carry over to other subjects and life. Its benefits make it essential in education.

14. Stendhal Syndrome

Stendhal Syndrome causes strong emotional and physical reactions to art. It's often seen in art galleries, museums, or historic sites. Symptoms include rapid heartbeat, dizziness, fainting, nausea, hallucinations, confusion, and extreme emotions like crying or joy. Named after the French writer Stendhal, the syndrome was highlighted by his experience in 1817 after visiting Florence's Basilica of Santa Croce. He noted, "When I left Santa Croce, my heart was beating fast, life seemed to be drained from me, and I feared I might fall." While not officially recognized in today's psychiatry, some psychologists and psychiatrists acknowledge it (Palacios-Sánchez et al, 2018).

For art lovers, art is more than beauty; it's a deep emotional and psychological journey. Stendhal Syndrome exemplifies this intense reaction to art. It helps us understand art's influence on the mind and the depth of aesthetic experiences.

15. Göbeklitepe and Art

Göbeklitepe, in Turkey's Şanlıurfa province, is the oldest known human settlement. Its Neolithic structures feature T-shaped pillars. These pillars display engravings of humans, animals, hands, arms, and abstract symbols. Experts believe these dense engravings form a cohesive message (UNESCO World Heritage List, The Archaeological Site of Göbeklitepe). Early art, like this, sparked advances in abstract thought, creativity, and symbolic thinking. It engaged brain regions such as the prefrontal cortex and limbic system, promoting neuroplasticity (Fuster, 2008; Güner, 2019).

The rituals and art at Göbeklitepe fostered symbolic thinking, social bonds, and collective awareness. Engaging with art involved higher cognitive functions such as abstraction and creativity. This likely formed new neural connections, enhancing neuroplasticity (Kolb & Whishaw, 2015). Creating art at Göbeklitepe likely activated the prefrontal cortex for design and execution. Planning, interpreting symbols, and engraving them on stone would have restructured neural circuits in this region. Moreover, rituals and social interactions may have fostered hippocampal plasticity, strengthening social memory pathways. Creating and viewing art likely activated mirror neurons, enhancing shared understanding and social cognition. Strong emotional responses in the limbic system, triggered by art, further supported neuroplasticity through increased dopamine (Ramachandran, 2011).

In conclusion, Göbeklitepe's art sheds light on the evolution of cognitive processes linked to symbolic thinking and abstract art. This art likely influenced neuroplasticity, aiding cognitive, motor, and emotional growth. It played a crucial role in the brain's development, enhancing cognitive capacity (Saxe & Kanwisher, 2003).

15.1. Aesthetic Perception at Göbeklitepe

The art and aesthetic perception at Göbeklitepe indicate that early humans had advanced cognitive processes. These included symbolic thinking, aesthetic judgment, and collective consciousness. This insight is vital for understanding the role of art in the development of human cognition, culture, and society (Ayaz et al., 2022).

15.1.1. Symbolic Thinking and Abstraction

Göbeklitepe's art shows its creators had advanced symbolic thinking and abstraction. They used symbols to understand the world and think abstractly. This laid the groundwork for modern skills in language, mathematics, philosophy, and science. Early art allowed people to share complex emotions and ideas (Ramachandran & Blakeslee, 1998; Schirmer & Adolphs, 2017).

15.1.2. The Evolution of Aesthetic Perception

Göbeklitepe's artworks served both practical and aesthetic purposes. This shows early humans could appreciate beauty, order, and form. A better sense of beauty influenced cultural values and responses. Today, this perception is key in art, architecture, design, and fashion. It shapes self-expression and cultural identity. Göbeklitepe's aesthetic likely influenced its time's culture, suggesting our current sense of beauty evolved from it (Kolb and Gibb, 2017).

15.1.3. The Social and Ritual Role of Art

At Göbeklitepe, art was part of rituals. It united people, built identity, and fostered belonging. This strengthened cultural ties. Today, art still connects people. Cinema, music, and painting play key roles (Renfrew and Morley, 2009).

Neuroplasticity and Cognitive Development Through Art

Art at Göbeklitepe boosted intelligence and emotional understanding. It helped people adapt and collaborate. This shows art's vital role in brain growth. Modern art therapy supports this. Art fosters emotional control, creativity, and flexibility. It aids brain adaptability (Tattersall, 2014).

15.1.4. Architecture and Collective Building Consciousness

Göbeklitepe's structures and art show early societies' collaboration. This spirit continues in today's architecture and projects. They reflect our ability to unite and share visions (Vygotsky, 1978).

16. Nobel Laureates with Artistic Talents

- Richard Feynman (1918–1988): Feynman won the Nobel Prize in Physics in 1965. Besides science, he loved painting and playing the bongo drums. He signed his art as "Ofey." (Feynman, 1985).
- Santiago Ramón y Cajal (1852–1934): Cajal earned the Nobel Prize in Medicine in 1906. He founded modern neuroanatomy and created beautiful drawings of nerve cells. These works are both scientific and artistic (Ramón y Cajal, 1999).
- Niels Bohr (1885–1962): Bohr received the Nobel Prize in Physics in 1922. He had a passion for poetry. He used poetic language to explain complex ideas, believing that art and literature enhance scientific understanding (Pais, 1991).
- Alexis Carrel (1873–1944): Carrel was awarded the Nobel Prize in Medicine in 1912. He was a notable scientist and a prolific writer. His works, both scientific and philosophical, were rich and deep. He explored the impact of science on life, often in a poetic and reflective style (Carrel, 1935).

17. Exercises to Boost Artistic Talent and Right-Brain Function

- Drawing Skills
  - Free Drawing: Regular practice boosts creative thinking and spatial awareness.
  - Color Exercises: Experimenting with colors enhances perception and memory.

- o Nature Drawing: Drawing from nature sharpens observation and attention to detail.

- · Mental Visualization Exercises
  - o Meditation and Visualization: Imagining scenes improves memory and creativity.
  - o Mind Maps: Creating helps organize thoughts and make connections.

- · Problem-Solving and Creative Thinking
  - o Brainstorming: Group sessions enhance creativity and problem-solving.
  - o Alternative Uses Tests: Finding new uses for objects sparks creativity.

- · Physical Exercises and Dance
  - o Dance: Exploring various dance styles boosts body awareness and skills.
  - o Yoga and Tai Chi: These practices strengthen the mind-body link and boost creativity.

- · Suggestions for Right-Brain Development
  - o Artistic Activities: Regular art engagement stimulates the right brain.
  - o Nature Interaction: Time in nature enhances sensory and emotional processing.
  - o New Experiences: Hobbies, travel, and cultural interactions improve the right-brain's flexibility.

18. Takeaways

1. The neuroplastic effects of art on the brain improve mental flexibility and overall brain health. Therefore, integrating artistic activities into daily life is important for supporting neuroplasticity and for aesthetic and emotional health.
2. Scientific approaches to literary creativity reveal it to be a multifaceted process involving brain networks, cognitive stages, emotional regulation, problem-solving skills, and linguistic creativity. The integration of different disciplines provides a more comprehensive understanding

of how creative writing emerges from a blend of neurological, psychological, and social factors.
3. Art increases neurotransmitter release in neurons, supports the formation of new connections in the brain, and promotes neuroplasticity, which strengthens learning and memory processes.
4. Aesthetic experiences foster creativity. Cognitive science can expand knowledge in this field by examining the processes of creative thinking and the impact of aesthetic judgment on creativity. Understanding the effects of aesthetic experiences on the brain greatly contribute to neuroscience research.
5. The art utilized by early humans played a pivotal role in shaping the world we know today.
6. The art employed at Göbeklitepe contributed to brain development and evolution by supporting neuroplasticity. The artifacts at Göbeklitepe are intricate and aesthetic, conveying a narrative.
7. Art education can enhance cognitive abilities, creativity, and problem-solving skills, thereby improving students' academic performance and motivation.
8. The artifacts at Göbeklitepe are intricate and aesthetic, conveying a narrative.
9. The ritualistic use of art fostered societal cohesion and collaboration.

Abbreviations

DMN: Default mode network
MTL: Medial temporal lobe
DLPFC: The dorsolateral prefrontal cortex
MT: Middle temporal area
OFC: Orbitofrontal cortex
LTP: Long-term potentiation
BDNF: Brain-derived neurotrophic factor

Bibliography

1. Abraham, A. (2018). The neuroscience of creativity. Cambridge, United Kingdom: Cambridge University Press.

2. Allen, N. J., & Barres, B. A. (2009). "Glia - more than just brain glue." Nature, 457(7230), 675–677. https://doi.org/10.1038/457675a

3. Arden, R., Chavez, R. S., Grazioplene, R., & Jung, R. E. (2010). "Neuroimaging creativity: A psychometric view." Behavioural Brain Research, 214(2). https://doi.org/10.1016/j.bbr.2010.05.015

4. Ayaz, O., Çelik, B., & Çakmak, F. (2022). "Status Society": Sociological Thinking Of Göbekli Tepe And Karahan Tepe In The Context Of Social Stratification. Karadeniz Uluslararası Bilimsel Dergi, 1(56), 122–142. https://doi.org/10.17498/kdeniz.1186376

5. Aziz-Zadeh, L., Liew, S. L., & Dandekar, F. (2013). I am exploring the neural correlates of visual creativity. Social cognitive and affective neuroscience, 8(4), 475–480. https://doi.org/10.1093/scan/nss021

6. Baker, F. A., Metcalf, O., Varker, T., & O'Donnell, M. (2018). A Systematic Review of the Efficacy of Creative Arts Therapies in the Treatment of Adults with PTSD. Psychological Trauma: Theory, Research, Practice, and Policy, 10(6), 643–651. https://doi.org/10.1037/tra0000353

7. Beaty, R. E., Benedek, M., Silvia, P. J., & Schacter, D. L. (2016). "Creative cognition and brain network dynamics." Trends in Cognitive Sciences, 20(2). https://doi.org/ 10.1016/j.tics.2015.10.004

8. Benton, A. L., & Tranel, D. (2000). "Visuoperceptual, visuospatial, and visuoconstructive disorders." The Clinical Neuropsychologist, 14(4).

9. Binkofski, F., & Buxbaum, L. J. (2013). Two action systems in the human brain. Brain and Language, 127(2), 222–229. https://doi.org/10.1016/j.bandl.2012.07.007

10. Bolwerk, A., Mack-Andrick, J., Lang, F. R., Dörfler, A., & Maihöfner, C. (2014). How Art Changes Your Brain: Differential Effects of Visual Art Production and Cognitive Art Evaluation on Functional Brain Connectivity. PLOS ONE, 9(7) https://doi.org/10.1371/journalpone.0101035

11. Bosch, M., & Hayashi, Y. (2012). Structural plasticity of dendritic spines. Current opinion in neurobiology, 22(3), 383–388. https://doi.org/10.1016/j.conb.2011.09.002

12. Carrel, A. (1935). Man, the Unknown. Harper & Brothers.
13. Cattaneo, Z., Lega, C., Gardelli, C., Merabet, L. B., Cela-Conde, C. J., & Nadal, M. (2014). The role of prefrontal and parietal cortices in aesthetic appreciation of representational and abstract art: a TMS study. Neuroimage, 99, 443–450. https://doi.org/10.1016/j.neuroimage.2014.05.037
14. Csikszentmihalyi, M. (1997). Flow and the psychology of discovery and invention. HarperPerennial, New York, 39, 1–16.
15. Çelik, B. (2019). Göbeklitepe: World's first temple and its symbolic significance. Archaeological Heritage Journal, 45(2), 112–134. http://www.gobeklitepeturkey.org/pdf/G%C3%B6beklit<wbr></wbr>epeKitap.pdf
16. Chandra, M. S., & Liddle, P. F. (2014). Art Therapy in Dementia Care: Harnessing the Creativity of Patients with Alzheimer's Disease. Aging & Mental Health, 18(5), 601–611.
17. Chatterjee, A. (2011). Art and the brain: effects of art on the viewer's brain. Progress in Brain Research, pp. 204, 177–194.
18. Chatterjee, A. (2014). "The aesthetic brain: How we evolved to desire beauty and enjoy art." Oxford University Press. ISBN: 978-0199811809.
19. Chatterjee, A., & Vartanian, O. (2014). Neuroaesthetics. Trends in Cognitive Sciences, 18(7), 370–375. https://doi.org/ 10.1016/j.tics.2014.03.003
20. Chen, Q., Beaty, R. E., & Qiu, J. (2020). Mapping the artistic brain: Common and distinct neural activations associated with musical, drawing, and literary creativity. Human Brain Mapping, 41(12), 3403–3419. https://doi.org/10.1002/hbm.25025
21. Corballis, M. C. (2014). "Left brain, right brain: facts and fantasies." PloS biology, 12(1), e1001767. https://doi.org/10.1371/journal.pbio.1001767
22. Csikszentmihalyi, M. (1996). Creativity: Flow and the Psychology of Discovery and Invention. HarperCollins.
23. Demarin, V., Roje Bedeković, M., Bosnar Puretić, M., & Bošnjak Pašić, M. (2016). Arts, brain and cognition. Psychiatria Danubina, 28(4), 343–348

24. Dietrich, A. (2004a). "Neurocognitive mechanisms underlying the experience of flow." Consciousness and Cognition, 13(4). https://doi.org/10.1016/j.concog.2004.07.002
25. Dietrich, A. (2004b). "The cognitive neuroscience of creativity." Psychonomic Bulletin & Review, 11(6), 1011–1026. https://doi.org/10.3758/BF03196731
26. Draganski, B., Gaser, C., Busch, V., Schuierer, G., Bogdahn, U., & May, A. (2004). "Neuroplasticity: Changes in grey matter induced by training." Nature, 427(6972). https://doi.org/10.1038/427311a
27. Ferszt, G. G., & Mularski, K. (2012). Art Therapy with Patients Who Have Schizophrenia and Other Psychoses. Art Therapy: Journal of the American Art Therapy Association, 29(1), 38–42. https://doi.org/10.1080/07421656.1993.10759011
28. Feynman, R. P. (1985). Surely You're Joking, Mr. Feynman! Adventures of a Curious Character. W. W. Norton & Company.
29. Finke, R. A., Ward, T. B., & Smith, S. M. (1992). Creative Cognition: Theory, Research, and Applications. MIT Press.
30. Fioranelli, M., Roccia, M. G., & Garo, M. L. (2023). The role of arts engagement in reducing cognitive decline and improving quality of life in healthy older people: a systematic review. Frontiers in Psychology, 14, 1232357. https://doi.org/10.3389/fpsyg.2023.1232357
31. Fisher, B. E., Morton, S. M., & Lang, C. E. (2014). From motor learning to physical therapy and back again: the state of the art and science of motor learning rehabilitation research. Journal of Neurologic Physical Therapy, 38(3), 149–150. https://doi.org/10.1097/NPT.0000000000000043
32. Fuster, J. M. (2008). The prefrontal cortex (4th ed.). Academic Press.
33. Gainotti, G. (2015). "Emotional behavior and hemispheric side of the lesion." Cortex, 72. https://doi.org/10.1016/S0010-9452(72)80026-1
34. Gainotti, G. (2019). "The role of the right hemisphere in emotional and behavioral disorders of patients with unilateral brain damage." Advances in Clinical Neuroscience & Rehabilitation, 19(1), 4–7. https://doi.org/10.3389/fnagi.2019.00055
35. Gantt, L., & Tinnin, L. W. (2009). Support for a Neurobiological View of Trauma with Implications for Art Therapy. The Arts in Psychotherapy, 36(3), 148–153. https://doi.org/10.1016/j.aip.2008.12.005

36. Gibbs, R. W. (2006). Embodiment and Cognitive Science. Cambridge University Press.
37. Goldberg, E. (2001). "The Executive Brain: Frontal Lobes and the Civilized Mind." Oxford University Press.
38. Grefkes, C., & Fink, G. R. (2005). "The functional organization of the inferior parietal lobule in humans." Neuroscientist, 11(5), 432–439. https://doi.org/10.1177/1073858405276326
39. Gregory, R. L. (1997). "Knowledge in perception and illusion." Philosophical Transactions of the Royal Society of London. Series B: Biological Sciences, 352(1358), 1121–1127. https://doi.org/10.1098/rstb.1997.0095
40. Güner, A. (2019). Göbeklitepe: The world's first temple. In Ancient History uncovered (pp. 24–37). History Press.
41. Ishizu, T., & Zeki, S. (2011). "Toward A Brain-Based Theory of Beauty." PloS One, 6(7), e21852. https://doi.org/10.1371/journal.pone.0021852
42. Jung, R. E., & Haier, R. J. (2007). "The Parieto-Frontal Integration Theory (P-FIT) of intelligence: Converging neuroimaging evidence." https://doi.org/10.1017/S0140525X07001185
43. Jung, R. E., & Vartanian, O. (2018). "The Cambridge Handbook of the Neuroscience of Creativity." Cambridge University Press. https://doi.org/10.1017/9781316556238
44. Kaufmann, G. (2003). The Effect of Mood on Creativity in the Innovative Process. Creativity Research Journal, 15(2–3), 131–135.
45. Kawabata, H., & Zeki, S. (2004). Neural correlates of beauty. Journal of Neurophysiology, 91(4), 1699–1705. https://doi.org/10.1152/jn.00696.2003
46. Kempermann, G., & Gage, F. H. (1999). "Experience-dependent regulation of adult hippocampal neurogenesis: effects of long-term stimulation and stimulus withdrawal." Hippocampus, 9(3), 321–332. https://doi.org/10.1002/(SICI)1098-1063(1999)9:3<321::AID-HIPO11>3.0.CO;2-C
47. Kenett, Y. N., Humphries, S., & Chatterjee, A. (2023). A thirst for knowledge: Grounding curiosity, creativity, and aesthetics in memory and reward neural systems. Creativity Research Journal, 35(3), 412–426. https://doi.org/10.1080/10400419.2023.2165748

48. Kim, Y., Park, K., Kim, Y., Yang, W., Han, D., & Kim, W. S. (2020). The impact of visual art and high affective arousal on heuristic decision-making in consumers. Frontiers in Psychology, 11, 565829. https://doi.org/10.3389/fpsyg.2020.565829
49. Kolb, B., & Gibb, R. (2017). Brain plasticity and behaviour in the developing brain. Journal of Clinical and Experimental Neuropsychology, 39(7), 667–676. https://doi.org/10.1080/13803395.2016.1253665
50. Kolb, B., & Whishaw, I. Q. (2015). An introduction to brain and behavior (5th ed.). Worth Publishers.
51. Kosslyn, S. M., & Thompson, W. L. (2003). "The Case for Mental Imagery." Oxford University Press.
52. Limb, C. J., & Braun, A. R. (2008). Neural substrates of spontaneous musical performance: An fMRI study of jazz improvisation. PLoS one, 3(2), e1679. https://doi.org/10.1371/journal.pone.0001679
53. Leder, H., Belke, B., Oeberst, A., & Augustin, D. (2004). A model of aesthetic appreciation and aesthetic judgments. British journal of psychology, 95(4), 489–508. https://doi.org/10.1348/0007126042369811
54. Lynch, G. (2004). "Long-term potentiation and memory." Physiological Reviews, 84(1), 87–136. https://doi.org/10.1152/physrev.00014.2003.
55. Magsamen, S., & Ross, I. (2023). Your brain on art: How the arts transform us. Random House.
56. Malenka, R. C., & Bear, M. F. (2004). "LTP and LTD: an embarrassment of riches." Neuron, 44(1), 5–21. https://doi.org/1 10.1016/j.neuron.2004.09.012
57. Marcus, E. L., Kaufman, Y., & Cohen-Shalev, A. (2009). Creative work of painters with Alzheimer's disease. Harefuah, 148(8), 548–53. PMID: 19899261.
58. Mauri, M., Crippa, A., Bacchetta, A., Grazioli, S., Rosi, E., Gazzola, E., … & Nobile, M. (2020). The utility of NIRS technology for exploring emotional processing in children. Journal of Affective Disorders, 274, 819–824. https://doi.org/10.1016/j.jad.2020.06.004
59. Miller, B. L., & Limb, C. J. (2013). Art and the brain: How imagery alters neural networks. Cerebrum, 2013, 10.

60. Orban, G. A. (2008). Higher order visual processing in macaque extrastriate cortex. Physiological Reviews, 88(1), 59–89. https://doi.org/10.1152/physrev.00008.2007

61. Pais, A. (1991). Niels Bohr's Times: In Physics, Philosophy, and Polity. Clarendon Press.

62. Palacios-Sánchez, L., Botero-Meneses, J. S., Pachón, R. P., Hernández, L. B. P., Triana-Melo, J. D. P., & Ramírez-Rodríguez, S. (2018). Stendhal syndrome: a clinical and historical overview. Arquivos de neuro-psiquiatria, 76(2), 120–123. https://doi.org/10.1590/0004-282X20170189

63. Pilozzi, A., Carro, C., & Huang, X. (2020). Roles of β-endorphin in stress, behavior, neuroinflammation, and brain energy metabolism. International journal of molecular sciences, 22(1), 338. https://doi.org/10.3390/ijms22010338

64. Price, C. J. (2012). "A review and synthesis of the first 20 years of PET and fMRI studies of heard speech, spoken language and reading." Neuroimage, 62(2), 816–847. https://doi.org/10.1016/j.neuroimage.2012.04.062

65. Ramachandran, V. S. (2011). The tell-tale brain: A neuroscientist's quest for what makes us human. W. W. Norton & Company. . https://doi.org/110.1172/JCI57214

66. Ramachandran, V. S., & Blakeslee, S. (1998). Phantoms in the brain: Probing the mysteries of the human mind. William Morrow.

67. Ramón y Cajal, S. (1999). Advice for a Young Investigator. MIT Press. (Originally published in Spanish in 1897 as Reglas y consejos sobre investigación científica).

68. Renfrew, C., & Morley, I. (2009). Becoming human: Innovation in prehistoric material and spiritual culture. Cambridge University Press.

69. Rhodes, C. (2023). Jon Sarkin: the eternal now. Epidemiology and Psychiatric Sciences, 32, e28. https://doi.org/10.1017/S2045796023000227

70. Rios, M. (2013). BDNF and the central control of feeding: accidental bystander or essential player? Trends in neurosciences, 36(2), 83–90. https://doi.org/ 10.1016/j.tins.2012.12.00

71. Rominger, C., Papousek, I., Perchtold, C. M., Benedek, M., Weiss, E. M., Weber, B., … & Fink, A. (2020). Functional coupling of brain networks during creative idea generation and elaboration in the figural domain. NeuroImage, 207, 116395. https://doi.org/10.1016/j.neuroimage.2019.116395

72. Rubin, J. A. (2016). Child Art Therapy (25th Anniversary Edition). Routledge.

73. Runco, M. A. (2014). "Creativity: Theories and Themes: Research, Development, and Practice." Academic Press. http://hdl.handle.net/20.500.12010/37618

74. Sawyer, R. K. (2012). Explaining Creativity: The Science of Human Innovation. Oxford University Press.

75. Saxe, R., & Kanwisher, N. (2003). People thinking about thinking people: The role of the temporo-parietal junction in "theory of mind". NeuroImage, 19(4), 1835–1842. https://doi.org/10.1016/S1053-8119(03)00230-1

76. Schirmer, A., & Adolphs, R. (2017). Emotion perception from face, voice, and touch: Comparisons and convergence. Trends in Cognitive Sciences, 21(3), 216–228. https://doi.org/10.1016/j.tics.2017.01.001

77. Schlaug, G., Norton, A., Overy, K., & Winner, E. (2005). "Effects of music training on the child's brain and cognitive development." Annals of the New York Academy of Sciences, 1060(1), 219–230. https://doi.org/10.1196/annals.1360.015

78. Schott, B. H. et al. (2014). "Mesolimbic functional magnetic resonance imaging activations during reward anticipation correlate with reward-related ventral striatal dopamine release." https://doi.org/10.1523/JNEUROSCI.2058-08.2008

79. Shah, C., Erhard, K., Ortheil, H. J., Kaza, E., Kessler, C., & Lotze, M. (2013). Neural correlates of creative writing: an fMRI study. Human brain mapping, 34(5), 1088–1101. https://doi.org/10.1002/hbm.21493

80. Slayton, S. C., D'Archer, J., & Kaplan, F. (2010). Outcome Studies on the Efficacy of Art Therapy: A Review of Findings. Art Therapy: Journal of the American Art Therapy Association, 27(3), 108–118. https://doi.org/10.1080/07421656.2010.10129660

81. Solso, R. L. (1994). "Cognition and the Visual Arts." MIT Press. ISBN: 978-0262691867.

82. Solso, R. L. (2001). Brain activities in a skilled versus a novice artist: An fMRI study. Leonardo, 34(1), 31–34. https://doi.org/10.1162/002409401300052479

83. Stuckey, H. L., & Nobel, J. (2010). The connection between art, healing, and public health: A review of current literature. American journal of public health, 100(2), 254–263. https://doi.org/10.2105/AJPH.2008.156497

84. Tattersall, I. (2014). Masters of the planet: The search for our human origins. St. Martin's Griffin.

85. Tavin, K. (2020). Art education in the age of the experience economy: Why art education has never been about art. Studies in Art Education, 61(4), 285–296.

86. Toga, A. W., & Thompson, P. M. (2005). "Temporal dynamics of brain anatomy." Annual Review of Biomedical Engineering, 7, 77–107. https://doi.org/10.1146/annurev.bioeng.5.040202.121611

87. Torrance, E. P. (1979). "The Search for Satori and Creativity." Creative Education Foundation.

88. Treib, M. (2013). Spatial recall: Memory in architecture and landscape. Routledge.

89. Tsao, D. Y., & Livingstone, M. S. (2008). "Mechanisms of face perception." Annual Review of Neuroscience, 31, 411–437. https://doi.org/10.1146/annurev.neuro.30.051606.094238

90. Tyler, C. W., & Likova, L. T. (2012). The role of the visual arts in enhancing the learning process. Frontiers in human neuroscience, 6, 8. https://doi.org/10.3389/fnhum.2012.00008

91. UNESCO. (2018). Göbeklitepe Arkeolojik Alanı. https://whc.unesco.org/tr/liste//1578/

92. Vartanian, O., & Skov, M. (2014). "Neural correlates of viewing paintings: Evidence from a quantitative meta-analysis of functional magnetic resonance imaging studies." Brain and Cognition, 87. https://doi.org/10.1016/j.bandc.2014.03.004

93. Vessel, E. A., Starr, G. G., & Rubin, N. (2012). "The brain on art: intense aesthetic experience activates the default mode network." Frontiers in Human Neuroscience, 6, 66. https://doi.org/10.3389/fnhum.2012.00066

94. Vygotsky, L. S. (1978). Mind in society: The development of higher psychological processes. Harvard University Press.

95. Ward, T. B., Smith, S. M., & Vaid, J. (1997). "Creative thought: An investigation of conceptual structures and processes." American Psychologist, 52(4).

96. Wassiliwizky, E., & Menninghaus, W. (2021). Why and how should cognitive science care about aesthetics?. Trends in Cognitive Sciences, 25(6), 437–449. https://doi.org/10.1016/j.tics.2021.03.008

97. Wise, R. A. (2004). "Dopamine, learning and motivation." Nature Reviews Neuroscience, 5(6). https://doi.org/ 10.1038/nrn1406

98. Zaidel, D. W. (2010). "The Arts and the Brain: What Does Your Brain See in a Work of Art?." Journal of American Medical Association (JAMA), J Anat. 2010 Feb; 216(2): 177–183. https://doi.org/ 10.1111/j.1469-7580.2009.01099.x

99. Zaidel, D. W. (2013). "Neuropsychology of Art: Neurological, Cognitive, and Evolutionary Perspectives." Psychology Press.

100. Zatorre, R. J., Fields, R. D., & Johansen-Berg, H. (2012). "Plasticity in gray and white: neuroimaging changes in brain structure during learning." Nature Neuroscience, 15(4), 528–536. https://doi. org/10.1038/nn.3045

101. Zeki, S. (1999). "Art and the brain." Journal of Consciousness Studies, 6(6–7), 76–96.

102. Zeki, S. (2001). "Artistic creativity and the brain." Science, 293(5527), 51–52.

103. Zhao, G., Lu, Y., Huang, X., & Lee, T. M. C. (2020). The neural mechanism of aesthetic judgments of dynamic landscapes. Scientific Reports, 10, 19700. . https://doi.org/ 10.1038/s41598-020-77658-y

Mindfulness Practices and Neural Plasticity: Yoga and Meditation

Merve Bilgiç Koç

1. Introduction

Yoga and meditation have become increasingly popular worldwide and are considered a complementary medical practice. As a result, interest in these practices in the literature is growing daily. Although meditation practices are distinct from yoga, they are often incorporated into a yoga session, and many practitioners engage in meditation as part of their practice.

Recent studies have reported that they are effective in treating patients with conditions like depression, stress, musculoskeletal system diseases, and cardiovascular diseases. This section of the book will explain how yoga and meditation affect neuroplasticity.

2. Yoga

Yoga, a practice that dates back thousands of years, is essential in Indian culture and health system. Although its meaning varies in different sources, it has definitions such as "to unite," "to control," and "to yoke."

The practice aims to unite the body, mind, and soul trio and establish a harmony among them. This purpose is achieved through an eightfold practice path described in classical yoga philosophy. These stage include ethical restraints (Yama), personal observances (Niyama), Postures (Asana), breath Control (Pranayama), withdrawal of the senses (Pratyahara), concentration (Dharana), meditation (Dhyana) and absorption (Samadhi). The three popular types today are Bhakti (yoga of devotion), Jnana (yoga of knowledge), and Karma yoga (yoga of action) (Hawley, 2011; Easwaran, 2011).

Patanjali's teachings outline an 8-step path (Birch & Singleton, 2019):

1. Yama: Rules are against harm, lying, stealing, unfaithfulness, and greed.
2. Niyama: This encourages purity, contentment, perseverance, self-study, and devotion.

3. Asana: These are postures.
4. Pranayama: This is breath awareness.
5. Pratyahara: This means abstraction.
6. Dharana: This is concentration.
7. Dhyana: This is meditation.
8. Samadhi: This is liberation.

2.1. Therapeutic Form

Yoga has been practiced in Eastern societies for thousands of years and has recently become popular in Western societies (Ivtzan & Jegatheeswaran, 2015). A study conducted in the United States in 2018 showed that 14.3% of the population practiced yoga in the previous 12 month (Clarke et al., 2018).

Individuals aim to achieve healthy living, gain fitness and flexibility, and manage stress. This intervention is, therefore, essential for the aging population (Barnes et al., 2008).

It is also a form of physical activity. A study in the USA in 2016 examined physical energy expenditure during Hatha yoga. According to this study, some poses were quite demanding (>3METs). The American College of Sports Medicine has classified yoga as a form of "light intensity physical activity" (2-2.9 METs) (Larson-Meyer, 2016; Swain et al., 2014).

To achieve comprehensive therapeutic outcomes, studies should also focus on self-compassion, which involves adopting a lifestyle that includes breathing techniques, therapeutic/physical postures, meditation techniques, self-compassion skills, a healthy and nutritious diet, avoiding psychoactive substances such as caffeine, drugs, and alcohol, healthy sleep hygiene, and paying attention to family and environmental relations (Stephens, 2017).

An example application of the therapeutic yoga postures is presented (Figure 1).

2.2. Yoga as a Stress Control Method

Today, many individuals live in a state of intense "sympathetic overdrive," and their bodies feel the stress they experience throughout the day. The constant stress they feel activates the hypothalamic-pituitary-adrenal (HPA) axis, triggering various processes, such as suppressing the immune system and affecting energy metabolism.

The released cortisol disrupts homeostasis by affecting blood glucose levels and water and sodium retention in the kidneys. It converts protein and

Figure 1: Therapeutic Yoga Poses

carbohydrates into glucose via glycogenesis, which increases blood volume and pressure.

Because the human body has not evolved to be physiologically resistant to chronic internal stress, stress plays a significant role in the development of many diseases. According to a study by the Centers for Disease Control and Prevention, 75% of patients who consult a physician with stess-related conditions (Stephens, 2017).

When the sympathetic system is activated, the body triggers a fight-flight response. These changes include an increased heart rate, vasoconstriction in the veins, emptying of liver glycogen stores, expansion of the bronchioles, redistribution of blood circulation, slowing down the digestive system, and urinary retention.

In contrast, when the parasympathetic system is stimulated, these changes are reversed, and the body activates the "rest and digest" mechanism, reducing physiological arousal and regulating perceived physical stress (Sarla, 2020; Salleh, 2008).

These practice, balance the autonomic nervous system by stimulating the parasympathetic system.

As studies on the benefits of yoga have increased, discussions have begun regarding the integration of yoga and meditation into the healthcare system.

The therapeutic form of yoga, which is adapted for individuals who are not suitable for standart practice, provides valuable information about its benefits across different populations (Larson-Meyer, 2016).

The literature reports a wide range of physical, mental and physiological benefits. Reported physical benefits include improvements in the musculoskeletal system, delayed onset of muscle pain, enhanced flexibility and balance, reductions in chronic pain and muscle atrophy, decreased body weight and increased bone mineral density. Mental and psychologicel benefits include enhanced self-regulation, self efficacy, attentional control, and reductions in anxiety, depression, mood disturbances and perceived stress. (Chong et al., 2011).

Studies in the literature have also found positive effects on diabetes, cardiovascular problems, and musculoskeletal disorders (Aljasir et al., 2010; Chu et al., 2016). Other studies have examined its effects on depression, stress, anxiety, and general discomfort (Kirkwood et al., 2005). Lin et al. reported positive effects on stress, anxiety and depression levels in their study on individuals with cancer (Lin et al., 2011).

3. Meditation

Meditation training helps regulate emotions and attention. Often, it shifts attention from external stimuli to internal experiences (Lutz, 2008b).

The Dalai Lama sparked interest in meditation with his 2015 book. He said, "The mind can be transformed through meditation." Meditation is gaining popularity in the West despite its Eastern roots, attracting scientific research (Ricard et al., 2014; Millière et al., 2018).

Research on meditation is still new. Yet, it has benefits, such as better emotional control, attention, and flexibility (Brown & Ryan, 2003).

Meditation involves cultivating awareness of internal experiences with an attitude of opennes and nonjudgement (Kabatt, 2003). For seasoned meditators, it is a physical act. Recent studies suggest that top-down and bottom-up processes influence thinking (Wilson, 2002; Barsalou, 2010; Winkielman et al., 2015).

Much of the existing research on meditation focused on applied outcomes rather than theoretical frameworks. Nevertheless, it can be limited to four types of meditation. these can be broadly categorized into four types: 1) focused attention, 2) open monitoring, 3) loving-kindness meditation, and 4) mantra meditation.

There are many types of meditation practice, and in this book section, we will include the more common types in the literature. Our sources (Ospina et al., 2007; Sedlmeier et al., 2012) provide detailed information.

3.1. Focused Attention

This meditation is a type of attention training in which individuals develop the ability to direct and sustain their attention to the "present moment."

In this meditation, individuals focus on a target object, such as breathing, which faciliates the ability to refocus attention when distractions arise. Samatha (breath-focused) and vipassana meditation are sub-types of focused attention practice.

It is a meditation in which the practitioner focuses on the nostrils in a comfortable position. In this practice, deep breathing is done, thoughts that arise are acknowledged without engagement, and staying in the breath is done by concentrating on the path of the breath, its depth, and its duration (Guidotti et al., 2023).

3.2. Loving-Kindness Meditation

It is a Buddhist meditation that aims to cultivate unconditional kindness toward oneself and others. The individual silently repeats phrases such as "I wish you happiness" and "I wish you relief from suffering."

Targets progess from less emotionally challenging to more complex, typically starting with oneself and then moving on to loved ones, neutral people, those who are difficult to forgive, and finally, all beings.

As a result, four core attitudes develop within the practitioner:

1. Loving-kindness with friendship, devotion,
2. Compassion, which is the desire to end the suffering of those who are suffering,
3. Appreciative joy, expressing happiness for the success of others,
4. Equanimity, which expresses calmness toward the fate of others based on wisdom (Sujiva, 2007).

3.3. Open Monitoring

In this practice, individuals observe their thoughts and bodily sensations. They aim to remain non-reactive and non-judgmental throughout the experience. This form of meditation allows emotions and thoughts to surface and fade away (Guidotti et al., 2023).

3.4. Mantra Meditation

Mantra meditation involves repeating a word, phrase, or sound silently or aloud. This practice helps override mental chatter and redirect negative thoughts, alleviating psychological distress. Sub-types include Transcendental Meditation, OM Chanting, and Benson Relaxation Therapy (Lynch et al., 2018).

4. Effects of Yoga and Meditation on Neuroplasticity

Studies on neuroplasticity have often involved small sample sizes and heterogeneous methodologies. However, they consistently showed similar findings regarding brain structure and function (Field, 2011; Vancampfort et al., 2012; Villemure et al., 2015).

One of the earliest studies, conducted in 1991, used positron emission tomography (PET) to investigate brain activity.

The study measured glucose metabolism in eight meditation practitioners during rest and yoga-based relaxation. No significant differences were observed between resting and relaxation conditions. (Herzog et al., 1990).

Significant differences were found in the volumes of the left hippocampus and parahippocampal regions according to experience status. No volumetric changes observed in non-practitioners

Yoga practitioners had larger volumes in several brain areas. These included the temporal, frontal, occipital, and limbic systems. Compared with yoga practitioners, non practitioners reported a higher frequency of cognitive failures (Froeliger et al., 2012).

In another study looking at the effect of practice time on changes in brain volume; significant changes were found in the left middle insula and orbital frontal cortex. Meanwhile, these changes were also observed in other sections (Villemure et al., 2015).

A further investigation of 11 experienced Kundalini meditators found decreased Cerebral Blood Flow (CBF) in several brain areas during meditation. However, there was increased CBF in the posterior cingulate and left superior temporal gyrus (Khalsa et al., 2009).

Another study on focused Kundalini meditation showed increased CBF in the medial frontal gyrus and caudate nucleus before and after meditation. Baseline measurements also revealed decreased CBF in the inferior superior occipital gyrus and inferior parietal lobe.

Breath-based meditation increased CBF in the insula, amygdala, hippocampus, parahippocampus, and superior temporal gyrus. The CBF patterns differed between the two types of meditation. A signifficant correlation was observed between meditation depth and increased CBF in the insula, inferior frontal cortex, and temporal lobe (Wang et al., 2011).

Other studies noted decreases in amygdala density and volume (Hölzel et al., 2010). These studies also demonstrated that lower perceived stress leads to changes in the volume and shape of the hippocampus and amygdala (Roozendaal et al., 2009; Gotink et al., 2018; Mcewen et al., 2016; Mcewen, 2017).

Practitioners have more Gray Matter (GM) than controls. Meditation practice is also boosts connectivity in the brain, increasing GM density (Avery et al., 2014).

Repeated mental and behavioral activities trigger changes in brain matter through various processes (Dayan & Cohen, 2011; Zatorre et al., 2012; Lövdén et al., 2013).

Greater practice duration is associated with increased left hippocampal volume. Neural activity has been observed even during yogic sleep (Lou et al., 1999). This activity is heightened in Kundalini meditation (Lazar et al., 2005).

An 8-week mindfulness intervention was associated with increased left hippocampal GM (Holzel et al., 2011). In Vippasana meditators, GM density in the medial prefrontal cortex (MPFC) relates to meditation hours (Lazar et al., 2000).

Experienced meditators have a larger hippocampus than controls (Holzel et al., 2008; Luders et al., 2009). Hippocampal volume correlates with experience (Luders et al., 2013).

Yoga practice has been shown to reduce perceived stress levels (Goosens, 2011). Lower stress increases the hippocampal volume (Pruessner et al., 2005). In older adults, larger hippocampal volume has been associated with higher fitness levels (Pereira et al., 2007; Erickson et al., 2009).

A 2015 study showed that the human brain can adapt quickly. GM changes appear to be practice-dependent and may partially reverse following cessation of training. For instance, GM volume increased in participants who trained in visual-motor skills but returned to normal after training ended (Draganski et al., 2004).

This implies that ongoing practice is vital to maintaining GM volume. Increasing weekly practice helps maintain these changes by strengthening certain areas of the brain: the right somatosensory cortex/superior parietal lobe (S1/SPL), the left hippocampus, the midline precuneus/posterior cingulate cortex (PCC), and the right primary visual cortex (V1) (Villemure et al., 2015).

Individual focuses on body alignment, emotional state, and and breathing patterns. More practice hours increase sensory input, affecting the S1 cortex and Precuneus/PCC. These areas are crucial for integrating stimuli in emotional and personal contexts. Increased awareness explains the changes in these regions (Corbetta & Shulman, 2002; Northoff & Bermpohl, 2004; Holzel et al., 2011).

Experience that involving visualization or breath control can activate V1 (Villemure et al., 2015). Neuroimaging shows V1 active during visual mental imagery (Kosslyn & Thompson, 2003). V1 activation has also been associated with breath control practices.

The insula processes internal signals like heart rate and pain. The insula is also engaged during breathing regulation, meditation and yoga practices (Gu et al., 2012; Critchley et al., 2015). High awareness of internal signals helps manage stress, while low awareness is linked to depression (Avery et al., 2014).

A yoga group showed better pain tolerance than the controls. This result is linked to the insula's role in homeostasis and reduced pain expectation due to increased volume (Villemure et al., 2014).

After 12 weeks of practice, functional connectivity increased in specific brain areas. This study used Kritan Kriya meditation, which may boost memory by activating specific brain networks. It also showed increases in Default Mode Network (DMN) areas (Eyre et al., 2016).

These results support earlier findings on aerobic exercise. They show moderate activity activates the Anterior Cingulate Cortex (ACC) (Hillman, 2008).

Cognitive tasks like processing stimuli and updating memory involve the frontal lobe, ACC, inferior temporal lobe, and parietal cortex.

The ACC, part of the limbic system, is linked to areas that handle sensory, motor, and cognitive tasks (Carter et al., 1999).

Several of these regions overlap with core components of the DMN. After yoga, better connectivity leads to improved verbal memory recall (Buckner et al., 2008; Eyre et al., 2016).

In a Functional Magnetic Resonance (fMRI) study, researchers compared 8-year practitioners with a matched control group. The practitioners had better connectivity between the MPFC and the right angular gyrus. This was true at rest.

While aging reduces connectivity, yoga practice may partially counteract age-related declines in functional connectivity (Santaella et al., 2019).

Neural activity in the DMN is linked to mind wandering, which can lower happiness. Meditation reduces mind wandering and boosts connectivity between the DMN and cognitive control areas (Brewer et al., 2011).

A review found the frontal region most active during meditation, regardless of type (Froeliger et al., 2012; Hernández et al., 2015; Mishra et al., 2017).

Practitioners deactivate the amygdala and activate the prefrontal cortex during emotional tasks, which shows that they manage emotions using cognitive control (Gothe et al., 2018).

In a study with four mildly hypertensive patients, 12 weeks of Iyengar yoga and meditation reduced blood flow in specific brain areas. The intervention was associated with increased activation in the right prefrontal cortex and other areas. After the intervention, activity in the left ACC, dorsomedial frontal cortex, and superior temporal lobe were higher than in the right hemisphere (Cohen et al., 2009).

A study of Hatha yoga practitioners with similar backgrounds found more significant differences in the left hemisphere than the right (Froeliger et al., 2012).

Left-hemispheric activity has been associated with parasympathetic regulation and affective balance, whereas right-hemispheric activity is more strongly linked to arousal, stress responses, and survival-related process (Craig, 2009; Lou et al., 1999).

In a meditation study, participants showed brain activation in the hippocampus while using silent mantras. When chanting "Om," their vagal tone improved. Mantra-based meditation was associated with reduce activity in several brain regions: the amygdala, anterior cingulate gyrus, hippocampus, insula, orbitofrontal cortex, parahippocampal gyrus, and thalamus (Kalyani et al., 2011).

Another study on Sahaja meditation demonstrated increased neural activity in the medial frontal gyrus, anterior cingulate gyrus, inferior frontal gyrus, insula, superior temporal gyrus, and medial temporal gyrus compared to controls (Hernández et al., 2015).

In 2023, researchers examined brain activation patterns during focused attention meditation (FAM) and Open Monitoring Meditations (OMM).

OMM activates the Angular Gyrus and Inferior Parietal Lobule, which process sensory, auditory, and visual information. FAM engages the Frontal Eye Field, ACC, MPFC, and PCC, critical areas for attention and self-awareness.

Moreover, intensive meditation alters PCC connections, aiding the shift from mind wandering to focused attention (Marzetti et al., 2014; Zhang et al., 2021). This study, like others, highlighted left hemisphere dominance (Manna et al., 2010; Marzetti et al., 2014).

4.1. Cognitive Effects of Yoga and Meditation

A study on Hatha yoga found no change in participants' episodic memory (Baklouti et al., 2022). These findings contrast with earlier reports suggesting cognitive benefits of yoga practice.

However, Kundalini yoga improved verbal memory after 12 and 24 weeks (Eyre et al., 2016).

Another study using the Hopkins Verbal Learning Test found improved episodic memory (Cinalli et al., 2008).

A 6-month intervention was also associated with improvements in verbal episodic memory (Hariprasad et al., 2013).

These improvements may be partly explained by increased hippocampal volume, which is vital for memory. The DMN also aids memory processes (Kramer &Erickson, 2007).

Comprehensive meditation training enhances DMN connectivity, pottentialy contributing to the observed cognitive effects (Taylor et al., 2013).

Brain Wave Vibration (BWV) meditators show less cortical thickness in self-referential areas (Kang et al., 2013). In contrast, increased cortical thickness has been observed in the lateral prefrontal cortex, MPFC, and temporal regions, which are linked to sensory processing and emotional integration.

In BWV practitioners, DMN connectivity in the prefrontal cortex is significantly higher (Jang et al., 2011).

A 2018 BWV study noted mood improvements after practice and found increased connectivity between the prefrontal cortex, thalamus, and anterior insula (Jang et al., 2018; Manuello et al., 2016; Tang et al., 2015).

The thalamus helps create emotional awareness by relaying signals from the insula to the cortex. The anterior insula integrates sensory signals with predictions (Gu et al., 2013). Higher connectivity between these regions is linked to better emotional attention (Saggar et al., 2015).

The MFPC helps manage emotions by coordinating responses to social needs. Studies show that meditation improves emotional control (Lutz, 2008a; Jung et al., 2010).

The PFC connects with the insula (Craig, 2009). Meditation has been associated with increased activity in attentional control regions, including the dorsolateral PFC. It also enhances connectivity between the dorsolateral PFC and the insula.

This increased PFC activity strengthens links between the control and salience networks (Mooneyham et al., 2016).

The DMN has two parts: the anterior (aDMN) and posterior (pDMN). The aDMN includes the MPFC, ACC, and precuneus, while the pDMN comprises the precuneus and inferior parietal cortex (Damoiseaux et al., 2008).

In BWV meditation, aDMN areas for emotional regulation thicken, whereas pDMN areas for self-referential processing are thin (Kang et al., 2013).

The caudate is critical in the reward system and executive functions including attention regulation and impulse control, as well as anxiety modulation (Guehl et al., 2008). It plays a key role in attentional control and response inhibition (Aron & Poldrack, 2005). In fMRI studies, Parkinson's patients showed increased GM in the caudate after meditation (Pickut et al., 2013). After 8 weeks of mindfulness meditation, blood flow increased in the caudate and insula. Notably, higher blood flow in the left caudate is linked to lower anxiety (Kirk et al, 2015).

Buddhist meditation practitioners show heightened activity in the somatosensory and superior temporal cortices. This outcome helps them manage negative responses to unfairness (Kirk et al, 2011).

Meditation makes people aware of negative emotions and thoughts. Accepting these without reaction aids in managing anxiety and making decisions (Jung et al., 2016).

4.2. Brain Neurotransmitter Studies

A study investigating endogenous striatal dopamine release during yogic sleep (nidra) meditation using PET observed an 8% decrease in binding potential in the ventral striatum.

It was associated with the workers' experience of being in the moment and less ready for action (Kjaeret al., 2002). In another study using only asanas, yoga practice was associated with increas Gamma-aminobutyric acid (GABA) levels (Streeter et al, 2007). It has been reported that the increased GABA level may be due to increased PFC activation during and after yoga.

This situation is thought to increase glutamate and result in increased GABA release via the reticular nucleus of the thalamus (Guglietti et al., 2013).

5. Conclusion

Many studies have consistently demonstrated that regular yoga and meditation practices induce significant anatomical changes in the brain. However, it is essential to note that the extent of these changes is directly proportional to the duration and consistency of one's yoga practice. The more frequent and prolonged the practice, the more pronounced the alterations in brain structure.

Notably, yoga and meditation have been found to enhance executive functions, including attention, verbal memory, and visual memory, by strengthening connections between specific brain regions. These findings hold considerable promise for the prevention and delayed progresion of age related neurodegeneratvie disorders, such as Alzheimer's and dementia. By incorporating yoga into one's lifestyle, individuals can delay or even reverse the onset of these degenerative conditions, leading to a healthier and more agile brain well into old age.

6. Takeaways

1. Yoga practice brings about anatomical changes in the frontal cortex, hippocampus, anterior cingulate cortex, and insula.
2. Increased connectivity in the default mode network leads to improved verbal memory performance.
3. Studies on yoga, meditation and neuroplasticity have used varied interventions and small sample sizes.
4. Anatomical changes are more pronounced in the left than in the right brain.
5. Different types of meditation have distinct effects on the brain, resulting in unique psychological benefits.
6. Changes in gray matter volume vary depending on the duration of weekly yoga practice and are reversed when practice stops.
7. The frontal region shows the most neural activity during meditation.
8. Yoga activates the prefrontal cortex and deactivates the amygdala.
9. Experienced meditators show a more significant increase in hippocampal gray matter volume than beginners.
10. A decrease in the volume and density of the amygdala has been reported in yoga and meditation practitioners.

Abbreviations

ACC: Anterior cingulate cortex
BWV: Brain wave vibration
CBW: Cerebral blood flow
DMN: Default mode network
FAM: Focused attention meditation
fMRI: Functional magnetic resonance
GABA: Gamma-aminobutyric acid
GM: Gray matter
HPA: Hypothalamus-Pituitary-Adrenal axis
MPFC: Medial prefrontal cortex
OMM: Open monitoring meditation
PET: Positron emission tomography
PCC: Posterior cingulate cortex
S1: Somatosensory cortex
SPL: Superior parietal lobe
V1: Primary visual cortex

Bibliography

1. Aljasir, B., Bryson, M., & Al-Shehri, B. (2010). Yoga practice for the management of type II diabetes mellitus in adults: a systematic review. Evidence-Based Complementary and Alternative Medicine, 7(4), 399–408. https://doi.org/10.1093/ecam/nen027
2. Aron, A. R.,& Poldrack, R. A. (2005). The cognitive neuroscience of response inhibition: relevance for genetic research in attention-deficit/hyperactivity disorder. Biological psychiatry, 57(11), 1285–1292. https://doi.org/10.1016/j.biopsych.2004.10.026
3. Avery, J. A., Drevets, W. C., Moseman, S. E., Bodurka, J., Barcalow, J. C., & Simmons, W. K. (2014). Major depressive disorder is associated with abnormal interoceptive activity and functional connectivity in the insula. Biological psychiatry, 76(3), 258–266. https://doi.org/10.1016/j.biopsych.2013.11.027
4. Baklouti, S., Aloui, A., Baklouti, H., Souissi, N., & Jarraya, M. (2022). Effects of Hatha yoga on cognitive functions in the elderly: a cross-sectional study. Libyan Journal of Medicine, 17(1). eISSN: 1819-6357

5. Barnes, P. M., Bloom, B., & Nahin, R. L. (2008). Complementary and alternative medicine use among adults and children: United States, 2007. https://stacks.cdc.gov/view/cdc/5266/cdc_5266_DS1.pdf
6. Barsalou, L. W. (2010). Grounded cognition: Past, present, and future. Topics in cognitive science, 2(4), 716–724. https://doi.org/10.1111/j.1756-8765.2010.01115.x
7. Birch, J.,& Singleton, M. (2019). The yoga of the Haṭhābhyāsapaddhati: Haṭhayoga on the cusp of modernity. Journal of Yoga Studies, 2, 3–70. https://journalofyogastudies.org/index.php/JoYS/article/view/2019.v2
8. Brewer, J. A., Worhunsky, P. D., Gray, J. R., Tang, Y. Y., Weber, J., & Kober, H. (2011). Meditation experience is associated with differences in default mode network activity and connectivity. Proceedings of the National Academy of Sciences, 108(50), 20254–20259. https://doi.org/10.1073/pnas.1112029108
9. Brown, K. W.,& Ryan, R. M. (2003). The benefits of being present: mindfulness and its role in psychological wellbeing. Journal of personality and social psychology, 84(4), 822.
10. Buckner, R. L., Andrews-Hanna, J. R., & Schacter, D. L. (2008). The brain's default network: anatomy, function, and relevance to disease. Annals of the new York Academy of Sciences, 1124(1), 1–38. https://doi.org/10.1196/annals.1440.011
11. Carter, C. S., Botvinick, M. M., & Cohen, J. D. (1999). The contribution of the anterior cingulate cortex to executive processes in cognition. *Reviews in the Neurosciences, 10*(1), 49–58. https://doi.org/10.1515/REVNEURO.1999.10.1.49
12. Ch, H. (2008). Be smart, exercise your heart: Exercise affects the brain and cognition. *Nat Rev Neurosci*, *9*, 58–65. https://doi.org/10.1038/nrn2298
13. *Chong, C. S., Tsunaka, M., & Chan, E. P. (2011). Effects of yoga on stress management in healthy adults: a systematic review. Alternative therapies in health and medicine, 17(1), 32.*
14. Chu, P., Gotink, R. A., Yeh, G. Y., Goldie, S. J., & Hunink, M. M. (2016). The effectiveness of yoga in modifying risk factors for cardiovascular disease and metabolic syndrome: A systematic review and meta-analysis of randomized controlled trials. *European journal of preventive cardiology*, *23*(3), 291–307. https://doi.org/10.1177/2047487314562741

15. Cinalli, F. Z., Serrao, V. T., Campanholo, K. R., Jungerman, F. S., Lúcia, M. C. S. D., Scaff, M., & Miotto, E. C. (2008). Influence of yoga practice on cognitive functioning in normal aging. *Psicologia Hospitalar, 6*(2), 88–99. ISSN 2175-3547

16. Clarke, T. C., Barnes, P. M., Black, L. I., Stussman, B. J., & Nahin, R. L. (2018). *Use of yoga, meditation, and chiropractors among US adults aged 18 and over* (pp. 1–8). Washington, DC, USA: US Department of Health and Human Services, Centers for Disease Control and Prevention, National Center for Health Statistics.

17. Cohen, D. L., Wintering, N., Tolles, V., Townsend, R. R., Farrar, J. T., Galantino, M. L., & Newberg, A. B. (2009). Cerebral blood flow effects of yoga training: preliminary evaluation of 4 cases. *The Journal of Alternative and Complementary Medicine, 15*(1), 9–14. https://doi.org/10.1089/acm.2008.0008

18. Corbetta, M., & Shulman, G. L. (2002). Control of goal-directed and stimulus-driven attention in the brain. *Nature Reviews Neuroscience, 3*(3), 201–215. https://doi.org/10.1038/nrn755

19. Craig, A. D. (2009). How do you feel—now? The anterior insula and human awareness. *Nature Reviews Neuroscience, 10*(1), 59–70. https://doi.org/10.1038/nrn2555

20. Critchley, H. D., Nicotra, A., Chiesa, P. A., Nagai, Y., Gray, M. A., Minati, L., & Bernardi, L. (2015). Slow breathing and hypoxic challenge: cardiorespiratory consequences and their central neural substrates. PloS one, 10(5), e0127082. https://doi.org/10.1371/journal.pone.0127082

21. Damoiseaux, J. S., Beckmann, C. F., Arigita, E. S., Barkhof, F., Scheltens, P., Stam, C. J., … & Rombouts, S. A. R. B. (2008). It reduced resting-state brain activity in the "default network" in normal aging. *Cerebral cortex, 18*(8), 1856–1864.

22. Dayan, E., & Cohen, L. G. (2011). Neuroplasticity subserving motor skill learning. *Neuron, 72*(3), 443–454. https://doi.org/ 10.1016/j.neuron.2011.10.008

23. Draganski B., Gaser C., Busch V., Schuierer G., Bogdahn U., May A. (2004). Neuroplasticity: changes in grey matter induced by training. Nature 427 311–312 10.1038/427311a https://doi.org/10.1038/427311a

24. Easwaran, E. (2011). *The essence of the Bhagavad Gita is a contemporary guide to yoga, meditation, and Indian philosophy* (Vol. 2). Nilgiri Press.

25. Erickson, K. I., Prakash, R. S., Voss, M. W., Chaddock, L., Hu, L., Morris, K. S., … & Kramer, A. F. (2009). Aerobic fitness is associated with hippocampal volume in elderly humans. *Hippocampus*, *19*(10), 1030–1039. https://doi.org/10.1002/hipo.20547

26. Eyre, H. A., Acevedo, B., Yang, H., Siddarth, P., Van Dyk, K., Ercoli, L., … & Lavretsky, H. (2016). Changes in neural connectivity and memory following a yoga intervention for older adults: a pilot study. *Journal of Alzheimer's disease*, *52*(2), 673–684. https://doi.org/10.3233/JAD-1506

27. Field, T. (2011). Yoga clinical research review. *Complementary therapies in clinical practice*, *17*(1), 1–8. https://doi.org/10.1016/j.ctcp.2010.09.007

28. Froeliger, B., Garland, E. L., & McClernon, F. J. (2012). Yoga Meditation Practitioners Exhibit Greater Gray Matter Volume, and Fewer Reported Cognitive Failures: Results of a Preliminary Voxel-Based Morphometric Analysis: *Evidence-Based Complementary and Alternative Medicine*, *2012*(1), 821307. https://doi.org/10.1155/2012/821307

29. Goosens, K. A. (2011). Hippocampal regulation of aversive memories. *Current opinion in neurobiology*, *21*(3), 460–466. https://doi.org/10.1016/j.conb.2011.04.003

30. Gothe, N. P., Hayes, J. M., Temali, C., & Damoiseaux, J. S. (2018). Differences in brain structure and function among yoga practitioners and controls. *Frontiers in integrative neuroscience*, *12*, 26 https://doi.org/10.3389/fnint.2018.00026

31. Gotink, R. A., Vernooij, M. W., Ikram, M. A., Niessen, W. J., Krestin, G. P., Hofman, A., … & Hunink, M. M. (2018). Meditation and yoga practice are associated with smaller right amygdala volume: the Rotterdam study. *Brain imaging and behavior*, *12*, 1631–1639 *https://doi.org/10.1007/s11682-018-9826-z*

32. Gu, X., Gao, Z., Wang, X., Liu, X., Knight, R. T., Hof, P. R., & Fan, J. (2012). Anterior insular cortex is necessary for empathetic pain perception. *Brain*, *135*(9), 2726–2735. https://doi.org/10.1093/brain/aws199

33. Gu, X., Hof, P. R., Friston, K. J., & Fan, J. (2013). Anterior insular cortex and emotional awareness. *Journal of Comparative Neurology*, *521*(15), 3371–3388. https://doi.org/10.1002/cne.23368

34. Guehl, D., Benazzouz, A., Aouizerate, B., Cuny, E., Rotgé, J. Y., Rougier, A., … & Burbaud, P. (2008). Neuronal correlates of

obsessions in the caudate nucleus. *Biological psychiatry*, *63*(6), 557–562. https://doi.org/10.1016/j.biopsych.2007.06.023

35. Guglietti, C. L., Daskalakis, Z. J., Radhu, N., Fitzgerald, P. B., & Ritvo, P. (2013). Meditation-related increases in GABAB modulated cortical inhibition. *Brain stimulation*, *6*(3), 397–402 https://doi.org/10.1016/j.brs.2012.08.005

36. Guidotti, R., D'Andrea, A., Basti, A., Raffone, A., Pizzella, V., & Marzetti, L. (2023). Long-term and meditation-specific modulations of brain connectivity revealed through multivariate pattern analysis. *Brain Topography*, *36*(3), 409–418. https://doi.org/10.1007/s10548-023-00950-3

37. Hariprasad, V. R., Koparde, V., Sivakumar, P. T., Varambally, S., Thirthalli, J., Varghese, M., … & Gangadhar, B. N. (2013). Randomized clinical trial of yoga-based intervention in residents from elderly homes: Effects on cognitive function. *Indian journal of psychiatry*, *55*(Suppl 3), S357-S363 https://doi.org/1 10.4103/0019-5545.116308

38. Hernández, S. E., Suero, J., Rubia, K., & Gonzalez-Mora, J. L. (2015). Monitoring the neural activity of the state of mental silence while practicing Sahaja yoga meditation. *The Journal of Alternative and Complementary Medicine*, *21*(3), 175–179. https://doi.org/10.1089/acm.2013.0450

39. Herzog, H., Lele, V. R., Kuwert, T., Langen, K. J., Kops, E. R., & Feinendegen, L. E. (1990). Changed pattern of regional glucose metabolism during yoga meditative relaxation. *Neuropsychobiology*, *23*(4), 182–187. https://doi.org/10.1159/000119450

40. Holzel B. K., Ott U., Gard T., Hempel H., Weygandt M., Morgen K., et al. (2008). Investigation of mindfulness meditation practitioners with voxel-based morphometry. Soc. Cogn. Affect. Neurosci. 3 55–61 10.1093/scan/nsm038 https://doi.org/10.1093/scan/nsm038

41. Hölzel, B. K., Carmody, J., Evans, K. C., Hoge, E. A., Dusek, J. A., Morgan, L., … & Lazar, S. W. (2010). Stress reduction correlates with structural changes in the amygdala. *Social cognitive and affective neuroscience*, *5*(1), 11–17. https://doi.org/10.1093/scan/nsp034

42. Hölzel, B. K., Carmody, J., Vangel, M., Congleton, C., Yerramsetti, S. M., Gard, T., & Lazar, S. W. (2011). Mindfulness practice leads

to increases in regional brain gray matter density. *Psychiatry research: neuroimaging*, *191*(1), 36–43. https://doi.org/10.1016/j.pscychresns.2010.08.006

43. Ivtzan, I.,& Jegatheeswaran, S. (2015). The yoga boom in Western society: Practitioners' spiritual vs. physical intentions and their impact on psychological wellbeing. *Journal of Yoga & Physical Therapy*, *5*(03). http://dx.doi.org/10.4172/2157-7595.1000204
44. Jack Hawley (2011). The Bhagavad Gita, ISBN 978-1-60868-014-6, pages 50, 130
45. Jang, J. H., Jung, W. H., Kang, D. H., Byun, M. S., Kwon, S. J., Choi, C. H., & Kwon, J. S. (2011). Increased default mode network connectivity associated with meditation. *Neuroscience letters*, *487*(3), 358–362. https://doi.org/10.1016/j.neulet.2010.10.056
46. Jang, J. H., Kim, J. H., Yun, J. Y., Choi, S. H., An, S. C., & Kang, D. H. (2018). Differences in functional connectivity of the insula between brain wave vibration in meditators and non-meditators. *Mindfulness*, *9*, 1857–1866. https://doi.org/10.1007/s12671-018-0928-x
47. Jung, Y. H., Ha, T. M., Oh, C. Y., Lee, U. S., Jang, J. H., Kim, J., … & Kang, D. H. (2016). The effects of an online mind-body training program on stress, coping strategies, emotional intelligence, resilience, and psychological state. *PloS one*, *11*(8), e0159841.. https://doi.org/10.1371/journal.pone.0159841
48. Jung, Y. H., Kang, D. H., Jang, J. H., Park, H. Y., Byun, M. S., Kwon, S. J., … & Kwon, J. S. (2010). The effects of mind-body training on stress reduction, positive affect, and plasma catecholamines. *Neuroscience Letters*, *479*(2), 138–142. https://doi.org/10.1016/j.neulet.2010.05.048
49. Kabat-Zinn, J. (2003). Mindfulness-based interventions in context: past, present, and future. https://doi.org/10.1093/clipsy.bpg016
50. Kalyani, B. G., Venkatasubramanian, G., Arasappa, R., Rao, N. P., Kalmady, S. V., Behere, R. V., … & Gangadhar, B. N. (2011). Neurohemodynamic correlates of 'OM' chanting: A pilot functional magnetic resonance imaging study. *International journal of yoga*, *4*(1), 3–6. https://doi.org/10.4103/0973-6131.78171
51. Kang, D. H., Jo, H. J., Jung, W. H., Kim, S. H., Jung, Y. H., Choi, C. H., … & Kwon, J. S. (2013). The effect of meditation on

brain structure: cortical thickness mapping and diffusion tensor imaging. *Social cognitive and affective neuroscience*, *8*(1), 27–33. https://doi.org/10.1093/scan/nss056

52. Khalsa, D. S., Amen, D., Hanks, C., Money, N., & Newberg, A. (2009). Cerebral blood flow changes during chanting meditation. *Nuclear medicine communications*, *30*(12), 956–961. https://doi.org/10.1097/MNM.0b013e32832fa26c
53. Kirk, U., Brown, K. W., & Downar, J. (2015). Adaptive neural reward processing during anticipation and receipt of monetary rewards in mindfulness meditators. *Social cognitive and affective* neuroscience, 10(5), 752–759. https://doi.org/10.1093/scan/nsu112
54. Kirk, U., Downar, J., & Montague, P. R. (2011). Interoception drives increased rational decision-making in meditators playing the ultimatum game—*frontiers in neuroscience*, *5*, 49. https://doi.org/10.3389/fnins.2011.00049
55. Kirkwood, G., Rampes, H., Tuffrey, V., Richardson, J., & Pilkington, K. (2005). Yoga for anxiety: a systematic review of the research evidence. *British journal of sports medicine*, *39*(12), 884–891. https://doi.org/10.1136/bjsm.2005.018069
56. Kjaer, T. W., Bertelsen, C., Piccini, P., Brooks, D., Alving, J., & Lou, H. C. (2002). Increased dopamine tone during meditation-induced change of consciousness. *Cognitive Brain Research*, *13*(2), 255–259. https://doi.org/10.1016/S0926-6410(01)00106-9
57. Kosslyn, S. M., & Thompson, W. L. (2003). When is the early visual cortex activated during visual mental imagery? *Psychological bulletin*, *129*(5), 723.
58. Kramer, A. F.,& Erickson, K. I. (2007). Capitalizing on cortical plasticity: influence of physical activity on cognition and brain function. *Trends in cognitive sciences*, *11*(8), 342–348. https://doi.org/10.1016/j.tics.2007.06.009
59. Larson-Meyer, D. E. (2016). A systematic review of the energy cost and metabolic intensity of yoga. *Medicine and science in sports and exercise*, *48*(8), 1558–1569.
60. Lazar, S. W., Bush, G., Gollub, R. L., Fricchione, G. L., Khalsa, G., & Benson, H. (2000). Functional brain mapping of the relaxation response and meditation. Neuroreport, 11(7), 1581–1585.

61. Lazar, S. W., Kerr, C. E., Wasserman, R. H., Gray, J. R., Greve, D. N., Treadway, M. T., ... & Fischl, B. (2005). Meditation experience is associated with increased cortical thickness. Neuroreport, 16(17), 1893–1897 https://doi.org/10.1097/01.wnr.0000186598.66243.19

62. Lin, K. Y., Hu, Y. T., Chang, K. J., Lin, H. F., & Tsauo, J. Y. (2011). Effects of yoga on psychological health, quality of life, and physical health of patients with cancer: a meta-analysis. *Evidence-Based Complementary and Alternative Medicine*, *2011*(1), 659876. https://doi.org/10.1155/2011/659876

63. Lou, H. C., Kjaer, T. W., Friberg, L., Wildschiodtz, G., Holm, S., & Nowak, M. (1999). A 15O-H2O PET study of meditation and the resting state of ordinary consciousness. *Human brain mapping*, 7(2), 98–105. https://doi.org/10.1002/(SICI)1097-0193(1999)7:2<98::AID-HBM3>3.0.CO;2-M

64. Lövdén, M., Wenger, E., Mårtensson, J., Lindenberger, U., & Bäckman, L. (2013). Structural brain plasticity in adult learning and development. *Neuroscience & Biobehavioral Reviews*, *37*(9), 2296–2310. https://doi.org/10.1016/j.neubiorev.2013.02.014

65. Luders, E., Kurth, F., Toga, A. W., Narr, K. L., & Gaser, C. (2013). Meditation effects within the hippocampal complex revealed by voxel-based morphometry and cytoarchitectonic probabilistic mapping. *Frontiers in Psychology*, *4*, 398. https://doi.org/10.3389/fpsyg.2013.00398

66. Luders, E., Toga, A. W., Lepore, N., & Gaser, C. (2009). The underlying anatomical correlates of long-term meditation: larger hippocampal and frontal volumes of gray matter. *Neuroimage*, *45*(3), 672–678. https://doi.org/10.1016/j.neuroimage.2008.12.061

67. Lutz, A., Brefczynski-Lewis, J., Johnstone, T., & Davidson, R. J. (2008a). Regulation of the neural circuitry of emotion by compassion meditation: effects of meditative expertise. PloS one, 3(3), e1897. https://doi.org/10.1371/journal.pone.0001897

68. Lutz, A., Slagter, H. A., Dunne, J. D., & Davidson, R. J. (2008b). Attention regulation and monitoring in meditation. *Trends in cognitive sciences*, *12*(4), 163–169. https://doi.org/10.1016/j.tics.2008.01.005

69. Lynch, J., Prihodova, L., Dunne, P. J., Carroll, A., Walsh, C., McMahon, G., & White, B. (2018). Mantra meditation for mental health in the general population: A systematic review. *European*

Journal of Integrative Medicine, pp. 23, 101–108.-108 https://doi.org/10.1016/j.eujim.2018.09.010

70. Manna, A., Raffone, A., Perrucci, M. G., Nardo, D., Ferretti, A., Tartaro, A., ... & Romani, G. L. (2010). Neural correlates of focused attention and cognitive monitoring in meditation. *Brain Research Bulletin*, *82*(1–2), 46–56. https://doi.org/10.1016/j.brainresbull.2010.03.001
71. Manuello, J., Vercelli, U., Nani, A., Costa, T., & Cauda, F. (2016). Mindfulness meditation and consciousness: An integrative neuroscientific perspective. *Consciousness and Cognition*, *40*, 67–78. https://doi.org/10.1016/j.concog.2015.12.005
72. Marzetti, L., Di Lanzo, C., Zappasodi, F., Chella, F., Raffone, A., & Pizzella, V. (2014). Magnetoencephalographic alpha band connectivity reveals differential default mode network interactions during focused attention and open monitoring meditation—*frontiers in human neuroscience*, *8*, 832. https://doi.org/10.3389/fnhum.2014.00832
73. McEwen, B. S. (2017). Neurobiological and systemic effects of chronic stress. *Chronic stress*, *1*, 2470547017692328 https://doi.org/10.1177/247054701769232
74. McEwen, B. S., Nasca, C., & Gray, J. D. (2016). Stress effects on neuronal structure: hippocampus, amygdala, and prefrontal cortex. *Neuropsychopharmacology*, *41*(1), 3–23. https://doi.org/10.1038/npp.2015.171
75. Millière, R., Carhart-Harris, R. L., Roseman, L., Trautwein, F. M., & Berkovich-Ohana, A. (2018). Psychedelics, meditation, and self-consciousness. *Frontiers in Psychology*, *9*, 1475. https://doi.org/10.3389/fpsyg.2018.01475
76. Mishra, S. K., Khosa, S., Singh, S., Moheb, N., & Trikamji, B. (2017). Changes in functional magnetic resonance imaging with Yogic meditation: a pilot study. *AYU (An International Quarterly Journal of Research in Ayurveda)*, *38*(3–4), 108–112 https://doi.org/10.4103/ayu.AYU_34_17
77. Mooneyham, B. W., Mrazek, M. D., Mrazek, A. J., & Schooler, J. W. (2016). Signal or noise: brain network interactions underlying the experience and training of mindfulness. *Annals of the New York Academy of Sciences*, *1369*(1), 240–256 https://doi.org/10.1111/nyas.13044

78. Northoff, G., & Bermpohl, F. (2004). Cortical midline structures and the self. *Trends in cognitive sciences*, *8*(3), 102–107. https://doi.org/10.1016/j.tics.2004.01.004

79. Ospina, M. B., Bond, K., Karkhaneh, M., Tjosvold, L., Vandermeer, B., Liang, Y., … & Klassen, T. P. (2007). Meditation practices for health: state of the research. *Evidence report/technology assessment*, (155), 1–263.

80. Pereira, A. C., Huddleston, D. E., Brickman, A. M., Sosunov, A. A., Hen, R., McKhann, G. M., … & Small, S. A. (2007). An in vivo correlate of exercise-induced neurogenesis in the adult dentate gyrus. *Proceedings of the National Academy of Sciences*, *104*(13), 5638–5643. https://doi.org/10.1073/pnas.0611721104

81. Pickut, B. A., Van Hecke, W., Kerckhofs, E., Mariën, P., Vanneste, S., Cras, P., & Parizel, P. M. (2013). Mindfulness-based intervention in Parkinson's disease leads to structural brain changes on MRI: a randomized controlled longitudinal trial. *Clinical neurology and neurosurgery*, *115*(12), 2419–2425. https://doi.org/10.1016/j.clineuro.2013.10.002

82. Pruessner, J. C., Baldwin, M. W., Dedovic, K., Renwick, R., Mahani, N. K., Lord, C., … & Lupien, S. (2005). Self-esteem, locus of control, hippocampal volume, and cortisol regulation in young and old adulthood. *Neuroimage*, *28*(4), 815–826. https://doi.org/10.1016/j.neuroimage.2005.06.014

83. Ricard, M., Lutz, A., & Davidson, R. J. (2014). The mind of the meditator. *Scientific American*, *311*(5), 38–45. https://www.jstor.org/stable/26041821

84. Roozendaal, B., McEwen, B. S., & Chattarji, S. (2009). Stress, memory, and the amygdala. *Nature Reviews Neuroscience*, *10*(6), 423–433. https://doi.org/10.1038/nrn2651

85. Saggar, M., Zanesco, A. P., King, B. G., Bridwell, D. A., MacLean, K. A., Aichele, S. R., … & Miikkulainen, R. (2015). Mean-field thalamocortical modeling of longitudinal EEG acquired during intensive meditation training. *NeuroImage*, *114*, 88–104 https://doi.org/10.1016/j.neuroimage.2015.03.073

86. Salleh, M. R. (2008). Life events, stress, and illness. *The Malaysian Journal of Medical Sciences: MJMS*, *15*(4), 9 PMCID: PMC3341916 PMID: 22589633

87. Santaella, D. F., Balardin, J. B., Afonso, R. F., Giorjiani, G. M., Sato, J. R., Lacerda, S. S., … & Kozasa, E. H. (2019). More excellent anteroposterior default mode network functional connectivity in long-term elderly yoga practitioners. *Frontiers in Aging Neuroscien*ce, 11, 158 https://doi.org/10.3389/fnagi.2019.00158

88. Sarla, G. S. (2020). The Therapeutic Effects of Yoga on Health and Related Diseases: An Analytical Review. *Journal of Scientific Research in Medical and Biological Sciences*, *1*(2), 109–115. https://doi.org/10.47631/jsrmbs.v1i2.133

89. Sedlmeier, P., Eberth, J., Schwarz, M., Zimmermann, D., Haarig, F., Jaeger, S., & Kunze, S. (2012). The psychological effects of meditation: a meta-analysis. *Psychological bulletin*, *138*(6), 1139. https://doi.org/10.1037/a0028168

90. Stephens, I. (2017). Medical yoga therapy. *Children*, *4*(2), 12. *https://doi.org/10.3390/children4020012*

91. Streeter, C. C., Jensen, J. E., Perlmutter, R. M., Cabral, H. J., Tian, H., Terhune, D. B., … & Renshaw, P. F. (2007). Yoga Asana sessions increase brain GABA levels: a pilot study. *The journal of alternative and complementary medicine*, *13*(4), 419–426. https://doi.org/10.1089/acm.2007.6338

92. Sujiva, V. (2007). Loving-kindness meditation. *Buddha Dharma Education Association Inc.*

93. Swain, D. P., Brawner, C. A., & American College of Sports Medicine. (2014). *ACSM's resource manual for guidelines for exercise testing and prescription*. Wolters Kluwer Health/Lippincott Williams & Wilkins.

94. Tang, Y. Y., Hölzel, B. K., & Posner, M. I. (2015). The neuroscience of mindfulness meditation. *Nature Reviews Neuroscience*, *16*(4), 213–225. https://doi.org/10.1038/nrn3916

95. Taylor, V. A., Daneault, V., Grant, J., Scavone, G., Breton, E., Roffe-Vidal, S., … & Beauregard, M. (2013). Impact of meditation training on the default mode network during a restful state. *Social cognitive and affective neuroscience*, *8*(1), 4–14. https://doi.org/10.1093/scan/nsr087

96. Vancampfort, D., Vansteelandt, K., Scheewe, T., Probst, M., Knapen, J., De Herdt, A., & De Hert, M. (2012). Yoga in schizophrenia: a systematic review of randomized controlled

trials. *Acta Psychiatrica Scandinavica*, *126*(1), 12–20. https://doi.org/10.1111/j.1600-0447.2012.01865.x

97. Villemure, C., Čeko, M., Cotton, V. A., & Bushnell, M. C. (2014). The insular cortex mediates increased pain tolerance in yoga practitioners. *Cerebral cortex*, *24*(10), 2732–2740. https://doi.org/10.1093/cercor/bht124
98. Villemure, C., Čeko, M., Cotton, V. A., & Bushnell, M. C. (2015). Neuroprotective effects of yoga practice: age-, experience-, and frequency-dependent plasticity. *Frontiers in human neuroscience*, pp. 9, 281. https://doi.org/10.3389/fnhum.2015.00281
99. Wang, D. J., Rao, H., Korczykowski, M., Wintering, N., Pluta, J., Khalsa, D. S., & Newberg, A. B. (2011). Cerebral blood flow changes are associated with different meditation practices and perceived depth of meditation. *Psychiatry Research: Neuroimaging*, *191*(1), 60–67. https://doi.org/10.1016/j.pscychresns.2010.09.011
100. Wilson, M. (2002). Six views of embodied cognition. *Psychonomic bulletin & review, pp. 9*, 625–636. https://doi.org/10.3758/BF03196322
101. Winkielman, P., Niedenthal, P., Wielgosz, J., Eelen, J., & Kavanagh, L. C. (2015). Embodiment of cognition and emotion. https://doi.org/10.1037/14341-004
102. Zatorre, R. J., Fields, R. D., & Johansen-Berg, H. (2012). Plasticity in gray and white: Neuroimaging changes in brain structure during learning. *Nature Neuroscience*, *15*(4), 528–536. https://doi.org/10.1038/nn.3045
103. Zhang, Z., Luh, W. M., Duan, W., Zhou, G. D., Weinschenk, G., Anderson, A. K., & Dai, W. (2021). Longitudinal effects of meditation on brain resting-state functional connectivity. *Scientific reports*, *11*(1), 11361. https://doi.org/10.1038/nn.3045

Psychedelic Substances and Their Impact on Neuroplasticity

Selin Sert Yurdakul

1. Introduction to Neuroplasticity and Psychedelic Substances

Psychedelics are hallucinogenic substances that alter perceptions of the mind, senses, and environment. The term "psychedelic" was coined by the psychiatrist Humphry Osmond in 1956. Incorrespondence with the renowned author Aldous Huxley comes from "psyche" ("mind") and "delos" ("manifest"). The word means "mind-manifesting" (Lukasiewicz et al., 2021). Describes substances that expand consciousness. Induced changes from drugs are temporarily reversible. They include acute perceptual shifts, cognitive distortions, irrational thoughts, ego dissolution, and psychosis. A key theory on psychedelics' long-term effects labels them "psychoplastogens." These substances quickly boost neuroplasticity and cause lasting changes (Calder & Hasler, 2023).

2. The Historical Context of Psychedelic Substance Use

Psychedelic substances have a long history, starting with ancient Mesoamerican cultures. For instance, North American tribes used peyote for visions. Meanwhile, the Aztecs valued psilocybin mushrooms. European colonizers, however, suppressed these practices, leading to bans and persecution. Yet, in the nineteenth century, Western ethnologists studying these cultures found sacred objects with extraordinary properties. This discovery reignited interest in these substances and their effects on the mind and body.

Hallucinogens, particularly those derived from plants, have a long history of use in religious and spiritual practices. However, the first synthetic hallucinogen, lysergic acid diethylamide (LSD), was synthesized in 1938 by Swiss chemist Albert Hofmann while working for the pharmaceutical company Sandoz. In 1943, Hofmann accidentally came into contact with LSD, experiencing vivid visual hallucinations described as "an uninterrupted stream of fantastic pictures" with intense, kaleidoscopic colors. In 1947, Sandoz marketed LSD under the trade name "Delysid," promoting it for use in psychotherapy and research into the nature of psychoses (Lee & Shlain, 1985).

In 1960, psychologist Timothy Leary launched the Harvard Psilocybin Project to explore the potential of psilocybin as an adjunct in psychotherapy. Along with Richard Alpert, they experimented with LSD. This led to their firing from Harvard in 1963. After Sandoz's LSD patents expired in 1963, illegal LSD production surged. This raised alarms for European and U.S. governments about unsupervised use. The U.S. then banned psychedelic research with the 1970 Controlled Substances Act and declared a "War on Drugs." This ended research into psychedelics' therapeutic potential. Recently, interest in these substances has surged again. Organizations like Multidisciplinary Association for Psychedelic Studies (MAPS), the Heffter Research Institute, and the Beckley Foundation are leading this revival. They work with regulatory bodies to ensure proper protocols. Major universities, including Johns Hopkins and Imperial College London, have now set up dedicated centers to study psychedelics' effects on the mind, brain, and psychiatric disorders.

Despite the U.S. Drug Enforcement Administration (DEA) still classifying LSD, psilocybin, ayahuasca, and MDMA as Schedule I drugs, a global trend toward decriminalizing and regulating the use of psychedelics for psychotherapy is emerging. Countries like Australia have legalized MDMA and psilocybin for medical use. Canada offers limited access through its Special Access Program for medical use. In the U.S., Oregon and Colorado have decriminalized certain psychedelics. Several cities have eased enforcement on psychedelic offenses. Meanwhile, Jamaica, Brazil, and the British Virgin Islands have decriminalized psychedelics for spiritual or therapeutic use. These shifts show a growing recognition of the therapeutic potential of psychedelics, despite regulatory challenges. They bring hope for the future of mental health treatments and highlight the need for ongoing research and advocacy (UC Berkeley, 2024).

3. Different Types of Psychedelic Substances and Their Action in Molecular Level

Psychedelic substances can be broadly categorized into three main groups based on their pharmacological mechanisms of action: classical psychedelics, dissociatives, and empathogens.

3.1. Classical Psychedelics

Psychedelic substances that act as agonists on serotonin receptors, often referred to as "classical psychedelics." They've been used in spiritual, therapeutic and ceremonial settings for centuries. These substances, include tryptamines and indole alkaloids, such as psilocybin, LSD, and DMT. They bind serotonin receptors, especially 5-HT2A. Their action enhances neuroplasticity by promoting dendritogenesis and synaptogenesis. This occurs mainly in the prefrontal cortex and hippocampus. Scientists find their potential to enhance neuroplasticity is a fascinating area of research that continues to grow. (Calder & Hasler, 2022).

Classical psychedelics act primarily through agonistic interactions with the serotonin 5-HT2A receptors. These are G-protein-coupled receptors in key brain areas including the prefrontal cortex, hippocampus, and thalamus. The activation leads to modulation of neuronal activity by disrupting normal serotonin signaling pathways and trigger alterations in the axonal activity. This affects the electrical patterns in thalamocortical circuits altering perception and consciousness (Swanson, 2018).

Recent studies explored the impact of psychedelics on neuroplasticity, which leads to long- term structural and functional changes in the brain. They activate signal pathways that encourage dendritic growth and synaptogenesis and other forms of neural adaptation. This might help break reinforced patterns of thinking and behaving, explained by the "entropic brain" hypothesis and the "Relaxed Beliefs Under Psychedelics" (REBUS) model (Carhart-Harris & Friston, 2019).

Psilocybin closely resembles the structure of serotonin and quickly metabolized into psilocin, a 4-hydroxyl derivative of serotonin which activates 5-HT2A receptors (de Vos et al., 2021). Similarly, LSD and DMT interact with these serotonergic pathways, suggesting shared mechanisms of action (Cruzat et al., 2022).

When psychedelics bind to 5-HT2A receptors, they activate several intracellular signaling pathways that lead to their psychoactive and therapeutic effects.

3.1.1. Activation of G-Protein Pathways

Upon binding to 5-HT2ARs, psychedelics preferentially activate the Gq protein pathway, leading to the stimulation of phospholipase C (PLC). This enzyme catalyzes the breakdown of phosphatidylinositol 4,5-bisphosphate

(PIP2) into inositol triphosphate (IP3) and diacylglycerol (DAG). IP3 then mobilizes calcium release from intracellular stores, while DAG activates protein kinase C (PKC). This cascade is crucial for the regulation of synaptic plasticity and neurogenesis, which are central to the potential therapeutic effects of psychedelics in treating psychiatric disorders (Banks et al., 2021).

3.1.2. Biased Agonism and Pathway Selectivity

Psychedelics display biased agonism at the 5-HT2AR, selectively activating specific signaling pathways while avoiding others. For example, psychedelics may favor G-protein signaling over β-arrestin pathways, resulting in different downstream effects that may reduce adverse effects and enhance therapeutic benefits. This bias in receptor signaling contributes to the unique psychoactive properties of psychedelics, such as alterations in perception, mood, and cognition, while minimizing side effects associated with non-selective receptor activation (Wacker et al., 2017).

3.1.3. Interplay Between Serotonin and Glutamate Systems

Psychedelics also modulate glutamate receptor signaling, mainly through the interaction between 5-HT2ARs and metabotropic glutamate receptors (mGluR2). This interaction involves serotonergic and glutamatergic signaling. It's crucial for neural plasticity and may explain the quick antidepressant effects of psychedelics. Research shows that 5-HT2ARs can form complexes with mGluR2. This changes G-protein coupling and signaling, affecting mood and cognition (Moreno et al., 2016).

Novel synthetic compounds, such as 4-phosphoryl-N, N-dimethyltryptamine, and 5-methoxy-N, N-dimethyltryptamine, have also shown similar serotonergic actions, indicating that they may share common pharmacological properties with classical psychedelics. More research is needed to compare the pharmacodynamics and therapeutic potential of these substances. Psychedelics are quickly absorbed, either orally, nasally, or by inhalation. Psilocybin peaks in the brain within 90 minutes, while LSD takes a few hours. Despite their strong effects, classical psychedelics are generally considered low-risk and unlikely to cause lethal outcomes (Banks et al., 2021).

3.1.4. Induction of Neuroplasticity

Psychedelics boost neuroplasticity by raising brain-derived neurotrophic factor (BDNF) levels. They also increase glutamate, the brain's main excitatory neurotransmitter. This action activates AMPA and NMDA receptors, enhancing synaptic plasticity and promoting long-term potentiation (LTP) (Holze et al., 2021). Studies show that treating mice with 5-HT2A receptor agonists raises BDNF mRNA and protein in the cortex. In rats, LSD also boosts BDNF, but only in the frontal cortex (Inserra et al.2022).

The study by Ly et al. (2018) investigates psychedelics like LSD and DMT. It finds these substances significantly promote structural and functional neural plasticity. This occurs through activating serotonin 5-HT2A receptors. This activation triggers pathways involving BDNF and mTOR, essential for synaptic growth and neuron survival. The researchers used rodent neurons in their experiments. They observed psychedelics increasing the complexity of dendrites. Specifically, there was a rise in the number and length of dendritic branches. Additionally, the density of dendritic spines, crucial for neuron communication, also increased. These findings suggest psychedelics could treat mood disorders by enhancing neural resilience and plasticity. Notably, the effects were blocked by ketanserin, a 5-HT2A receptor antagonist. This highlights the importance of this receptor in the observed changes (Ly et al., 2018).

3.2. Dissociative Psychedelics

Dissociative psychedelics such as ketamine, phencyclidine (PCP), and dextromethorphan are known for inducing a trance-like state where users feel detached from their bodies and surroundings. These substances primarily act as antagonists at the N-methyl-D-aspartate (NMDA) receptors, which play a crucial role in synaptic plasticity and cognitive function. Dissociatives block NMDA receptors, disrupting the glutamate system crucial for normal brain function. This leads to hallucinations and altered perceptions. They also induce dissociation, making people feel detached from their bodies and surroundings, and changing their sense of time and space. Ketamine, initially developed as an anesthetic, has shown quick antidepressant effects. Now, it's used in clinics for treatment-resistant depression (Marguilho, 2023).

Ketamine is the only legal psychedelic used in clinics, mainly for its quick antidepressant effects. It works by blocking NMDA receptors, a type of glutamate receptor in the brain. This action blocks the excitatory neurotransmitter glutamate, lowering neuronal excitability and altering synaptic plasticity (Zanos et al., 2018).

Ketamine blocks NMDA receptors at the PCP site. This prevents calcium from entering cells, crucial for synaptic transmission and plasticity. The drug lowers overall excitatory brain activity and alters neural circuits, especially in mood-regulating areas like the prefrontal cortex and hippocampus. Additionally, ketamine interacts with muscarinic, opioid, and adrenergic receptors. It also blocks hyperpolarization-activated cyclic nucleotide-gated (HCN1) channels and the transporters for serotonin and norepinephrine. Moreover, it stops the reuptake of serotonin and dopamine and affects AMPA, kainate, and GABA-A receptors. This accounts for its varied effects on the central nervous system (Marguilho, 2023).

Beyond its effects on NMDA receptors, ketamine promotes rapid neuroplastic changes. It enhances the release of brain-derived neurotrophic factor (BDNF). It activates the mammalian target of the rapamycin (mTOR) pathway, leading to increased synaptogenesis, dendritic spine density, and overall synaptic strength. These changes are believed to underlie ketamine's fast-acting antidepressant effects, often seen within hours of administration. This is much faster than traditional antidepressants, which usually take weeks to show results (Zanos et al., 2018). This neuroplasticity is essential for restoring normal neuronal function and connectivity in mood-related brain regions (Deka et al., 2021).

Research indicates that ketamine's antidepressant effects are not solely due to the parent compound but also involve its metabolites, particularly hydroxynorketamine (HNK). HNK does not significantly bind to NMDA receptors but contributes to the antidepressant effects through mechanisms involving AMPA receptor activation and increased synaptic signaling. This finding suggests that both ketamine and its metabolites may work synergistically to exert rapid and sustained antidepressant effects (Zanos et al., 2018).

Ketamine's unique action and ability to induce synaptic plasticity make it effective for major depressive disorder, particularly when other treatments fail. Its legal status enables further research to uncover its full benefits and deepen our understanding of pharmacology. (Drozdz et al., 2022).

Phencyclidine (PCP) was first made as an anesthetic. It blocks NMDA receptors, similar to ketamine. However, its use is rare due to its unpredictable and potentially dangerous psychological effects. Dextromethorphan, found in cough syrups, shows dissociative and hallucinogenic effects at high doses. It works by blocking NMDA receptors and activating sigma receptors. This drug, while showing some protective benefits in neurodegenerative diseases, mainly serves to suppress coughs. Both PCP and dextromethorphan have less therapeutic value than ketamine. They are not approved for use as psychedelic therapies (Olson, 2022).

3.3. Empathogen Psychedelics

Empathogens are psychoactive substances that include MDMA, MDEA, and MMDA. They produce both empathogenic and entactogenic effects increasing empathy, emotional closeness, and sociability (Hysek et al., 2014). MDMA, or "Ecstasy," is the most researched. It's popular in social settings like raves. Yet, its potential therapeutic effects are also being studied, especially for PTSD. MDMA enhances interpersonal communication and emotional insights, making it promising for therapy (Lukasiewicz et al., 2021).

MDMA mainly works by releasing serotonin, dopamine, and norepinephrine. This action makes it a stimulant and boosts mood. It increases serotonin by reversing the serotonin transporter (SERT). This causes a serotonin overflow in the synaptic cleft and prolonged receptor activation. This release fosters emotional empathy, trust, and better social interactions. Additionally, MDMA activates dopamine D2 receptors in the nucleus accumbens, a region linked to reward and social bonding. This further enhances its prosocial and mood-enhancing effects (Rein et al., 2024).

MDMA promotes social bonding and empathy by increasing oxytocin levels. Studies show that MDMA raises plasma oxytocin, enhancing prosocial feelings of closeness and empathy toward others. However, MDMA's impact on emotional empathy also involves serotonin 2A and vasopressin 1A receptors, not just oxytocin (Kuypers et al., 2014).

MDMA has been studied for its therapeutic effects in various settings. Research shows it can reduce social anxiety, increase emotional openness, and improve comfort with emotional disclosure. This makes MDMA a promising aid for treating PTSD, anxiety, and other psychological conditions (Mitchell et.al., 2021). Its Schedule I classification reflects concerns about its abuse and neurotoxicity, especially with high or repeated doses. These can cause serotonergic neurotoxicity, similar to amphetamines (Capela et al., 2022).

While these categories provide a helpful framework, psychedelics' realm extends beyond these boundaries, with new psychoactive substances continually emerging and expanding therapeutic possibilities. Continued research is necessary to understand fully the mechanisms through which psychedelics impact neuroplasticity and how these effects can be harnessed for therapeutic purposes (Hipólito et al., 2023).

4. Psychedelics Effect on Critical Period of Neuroplasticity

Psychedelics create a regulated state of plasticity, often driven by oxytocin. This process involves the nucleus accumbens metaplasticity, or the ability to modify plasticity thresholds, induced by psychedelics. Psychedelics prime the brain for synaptic changes rather than driving hyperplasticity (excessive synaptic growth). This fine-tuned form of plasticity allows neurons to adapt their synaptic strength more flexibly in response to their environment. This mechanism explains the long-term benefits of psychedelics in therapy.

Research on mice showed that substances like MDMA, psilocybin, LSD, ketamine, and ibogaine reopened the critical period for social reward learning in adults. The length of this "open state" matched the drugs' effects length in humans. Ketamine's effect lasted 48 hours. Psilocybin and MDMA kept the period open for two weeks. LSD extended it to three weeks, while ibogaine pushed it to four weeks (Neill et al., 2021).

The extracellular matrix (ECM) is vital for synaptic plasticity. It supports or limits synaptic growth. Psychedelics remodel the ECM by changing levels of certain components, like fibronectin (FN1) and matrix metalloproteinases (MMP-16). This creates a supportive environment for plasticity, affecting proteoglycans and collagen. In groups treated with psychedelics, genes linked to angiogenesis (blood vessel growth), endothelial cell development, and tissue morphogenesis were significantly regulated. This included Trpv4, Tinagl1, and Robo4. It suggests ECM modulation is a critical factor in psychedelics ability to induce long-lasting neural changes. The 5-HT2AR receptor is crucial for LSD and psilocybin. However, psychedelics like MDMA and ketamine operate through different pathways. This shows a common ability to unlock critical periods, but through distinct upstream signaling cascades (Nardau et al., 2023).

5. Psychedelics and Their Action on Functional Level in Brain Connectivity

Psychedelics have gained interest for their unique effects on consciousness and perception. Research is also looking into their potential to treat mood, anxiety, and substance use disorders. Recent studies using neuroimaging have shown how psychedelics affect brain connectivity and activity. This helps us understand their impact on consciousness and possible therapeutic benefits.

Several models explain these effects. One key model focuses on modulation of the cortico-striato-thalamo-cortical loop involving the prefrontal cortex, thalamic nuclei, ventral striatum, pallidum, sensory cortex, and posterior parietal cortex. Psychedelics activate serotonin 5-HT2A receptors in these areas, leading to changes in brain information flow. Specifically, they modulate the activity of PFC layer five pyramidal neurons that control important thalamic nuclei, such as the reticular nucleus and the mediodorsal thalamus, influencing both cortical and subcortical circuits (Dosss et al., 2022).

A meta-analysis by Castelhano et al. (2021) shows that tryptamine psychedelics mainly affect brain areas rich in serotonin receptors, especially 5-HT2A and 5-HT1A. These receptors are plentiful in the visual cortex, fusiform gyrus, cingulate cortex, and medial prefrontal cortex, vital for visual processing, emotional regulation, and social cognition. Activation of 5-HT2A receptors here leads to changes in consciousness and perception typical of psychedelics. The analysis also notes significant effects in the right amygdala, essential for emotional processing. Changes in this region may explain the altered emotional experiences during psychedelic use and highlight their potential in treating anxiety and depression (Castelhano et al., 2021).

Another model, called the "relaxed beliefs under psychedelics" model, suggests that psychedelics inhibit top-down regulation by higher-order cortical regions, like the PFC and PPC. This leads to prediction errors and rapid updates in "prior beliefs," resulting in increased brain entropy. By this way it makes it possible to have more comprehensive range of possible brain states. The default mode network (DMN), which includes the PFC, PPC, and hippocampus, is key in this model. All these areas have high levels of 5-HT2ARs (Carhart-Harris & Friston, 2019).

Research shows psychedelics consistently reduce connectivity in the DMN and simultaneously promotes global brain integration. These pattern of connectivity changes likely lead to ego dissolution and expanded states of consciousness typical of psychedelic experiences. Notably, such shifts in brain network dynamics are linked to better treatment results, especially for mood and anxiety disorders (Gattuso et al., 2023).

A third model, the "cortical-claustral-cortical" model, highlights the claustrum's role in psychedelics' effects. This brain area, rich in 5-HT2ARs, integrates sensory input and coordinates activity across different regions of the brain. It may explain the altered consciousness states linked to psychedelics. However, more research is needed. Specifically, we need to understand psychedelics' impact on 5-HT2AR neurons and their networks, especially regarding neuroplasticity over time (De Gregorio et al., 2021).

Nasretdinov et al. (2023) found that LSD and ketamine affect brain activity differently in rodents. Using local field potential (LFP) recordings, they showed ketamine boosts neuronal activity significantly across most brain regions. In contrast, LSD doesn't increase overall activity but shifts the balance between excitation and inhibition in brain networks. Both drugs reduce functional connectivity. However, they do so in different ways. Ketamine decouples neural networks globally. Meanwhile, LSD affects connectivity by targeting serotonin 2A (5-HT2A) receptors, enhancing glutamatergic transmission without raising firing rates. These findings highlight the unique neurophysiological effects of ketamine and LSD, linking them to their specific pharmacological targets. This offers insights into their potential use in treating psychiatric disorders.

Recent fMRI studies have shown how ketamine, an NMDA receptor blocker, impacts functional brain networks. After ketamine infusion, in overall brain connectivity, especially in the frontal cortex, increases. This change may be linked to altered prefrontal cortex and hippocampus connectivity. Moreover, large-scale brain networks also reorganize. They shift from less integrated and more compartmentalized information processing. This change lowers global efficiency and small-worldness, indicating reduced long-range communication. Yet, it enhances local processing. Notably, cyclic motifs in the networks significantly increase. This suggests network organization move toward localized recurrent processing over extended linear communication. These findings suggest a loss of small-world network properties and reduced functional integration with NMDA receptor dysfunction. Both human and animal studies show these connectivity changes. They imply that ketamine's antidepressant effects are tied to shifts in brain networks, moving away from overly rigid patterns to more flexible dynamic configurations (Becker et al., 2016).

A study found that ketamine negatively impacts dynamic resting-state functional connectivity. This is especially true for the left visual network and between different visual networks. These changes match ketamine's effects on visual processing, using which are linked to its role in psychosis and suggest stable modifications in visual processing networks. (Spies et al., 2019).

Another study by Sahib et al. (2020) found that repeated ketamine treatments significantly reduced activity in the inhibitory control network, which includes the prefrontal and parietal regions. The study also observed changes in the supplementary motor area (SMA) during response inhibition. These changes could predict treatment outcomes, suggesting they might serve as biomarkers for treatment response in major depression.

6. Psychedelic Substances and Cognitive Functions

Psychedelics affect cognitive functions differently based on the type of the psychedelic substance, usage context which it is administered, and duration of its use. It's also vital to distinguish effects during active use in acute phase from those that appear in long term. We'll explore these in separate sections for clarity.

6.1. Classical Psychedelics Effect on Cognition

Psychedelics might boost cognitive function by increasing neuroplasticity, especially in the prefrontal cortex (PFC). This brain area is vital for memory, attention, and emotional regulation. The growth of dendrites and synapses in PFC neurons could explain better mood and cognitive function. The PFC regulates other areas, like the amygdala, which are involved in emotions (Calder, 2023).

In animal studies, chronic LSD treatment is linked to improved learning. In humans, LSD enhanced memory retrieval and increased reward sensitivity. This might boost reinforcement learning. Psychedelics like ayahuasca and psilocybin also improve cognitive flexibility. This is linked to circuits originating in the prefrontal cortex (PFC). Regular ayahuasca users excel in tasks requiring behavioral inhibition, cognitive flexibility, working memory, and executive function (Cardoso, 2021).

Psychedelics might boost mindfulness a form of attentional regulation by affecting the prefrontal cortex (PFC) and anterior cingulate cortex (ACC). It's suggested that growth in PFC and ACC neurons leads to these cognitive improvements. Often, such changes are linked to increased neuroplasticity in the cortex, vital for enhancing brain functions (Cardoso, 2021).

Psychedelics boost learning, memory, and cognitive flexibility. They enhance executive function and mindfulness. These properties make them potentially useful for cognitive improvement and rehabilitation when combined with supportive therapy.

6.2. Dissociative Psychedelics Effect on Cognition

Recent studies have revealed nuanced insights into both the benefits and risks of ketamine. In adults with treatment-resistant depression (TRD), ketamine often improves cognitive function. However, it can temporarily impair executive functions and attention due to its anesthetic effects. These impairments are short-lived. A 2021 study showed intravenous ketamine significantly improved sustained attention, working memory, and response control. This was linked to a reduction in depressive symptoms (McIntyre et al., 2021). Another review found that low doses of ketamine enhanced working memory and visual learning in TRD patients without impairing cognitive function (Gill et al., 2020).

Ketamine offers therapeutic benefits, but also risks cognitive impairment. Its effects on cognitive functions are complex. In stressed animals, it improves memory in some cases but not in others. A study showed temporary attention and executive function impairment in patients with major depressive disorder (MDD) and posttraumatic stress disorder (PTSD) after ketamine use. However, these issues were resolved within a day. Clinicians should note this temporary impairment. It suggests patients might struggle with tasks needing focus or problem-solving right after treatment (Davis et al., 2021).

Ketamine's antidepressant effects are linked to specific neural activation patterns. This includes the dorsolateral prefrontal cortex (DLPFC) and the default mode network (DMN). A study showed that higher activation in the DLPFC and reduced DMN deactivation predicted reduced cognitive symptoms in depressed patients. This suggests the neurobiological basis for ketamine's cognitive benefits. However, while these benefits are promising, especially for depression, they need to be weighed against the risk of cognitive impairment in chronic users. (Stippl et al., 2021).

Chronic ketamine use can lead to cognitive decline. However, users showed recovery after stopping. A study found significant memory and executive function improvements after 12 weeks. This suggests that the impairments from long-term use might be reversible with sustained abstinence. (Tang et al., 2019).

6.3. Emphatogen Psychedelics Effect on Cognition

MDMA's impact on cognition is mixed. Many studies report specific impairments. For example, verbal memory impairments are consistently observed during use. It also negatively affects sustained attention and visuospatial memory. However, visual perception, planning, and semantic memory retrieval remain unaffected (Kuypers et al., 2014).

MDMA is also known for its emotional and empathogenic effects, such as increased emotional empathy and social behavior (Rein et al., 2014). However, it doesn't improve cognitive empathy or social cognitive inferences. It's linked to a reduced recognition of negative emotions, especially fear. This could hinder emotional processing in social situations. In psychedelic-assisted therapy, acute cognitive impairment might challenge interventions. But the "afterglow" period offers more therapeutic support opportunities. MDMA's memory impairments could help PTSD treatment by disrupting traumatic memory reconsolidation (Hysek et al., 2014).

MDMA is associated with positive mood and an increase in hormones like oxytocin and prolactin. However, it doesn't reliably enhance cognition. Studies caution against using it for cognitive improvement. The risks include long-term memory loss, reduced appetite, and attention deficit problems. In summary, psychedelics and MDMA have different cognitive and therapeutic effects. This makes them suitable for different therapeutic approaches.

7. Psychedelics in Mental Health Therapies

Psychedelics like psilocybin, LSD, MDMA, and ayahuasca are gaining scientific attention for treating psychiatric disorders. This interest is mainly in psychedelic-assisted psychotherapy. Their research had stalled after being banned in the 1970s. Now, studies show promise for conditions like PTSD, depression, anxiety, and substance abuse. They suggest psychedelics boost neuroplasticity, neurogenesis, and neuronal growth, and connectivity. This raises the possibility of using them for mental health treatment. Mental illness is a growing concern. Currently, doctors mainly prescribe two types of medications: selective serotonin reuptake inhibitors and atypical antipsychotics. While these can help, they often take weeks to work and don't address the root causes of mental illness. Recent research shows psychedelics, however, can have quick anti-addictive, anti-anxiety, and antidepressant effects. These effects can kick in within minutes or hours (De Gregorio et al., 2021).

Psychedelic-assisted psychotherapy is a structured process. First, a drug is given in a safe, supportive setting. This often catalyzes to emotional breakthroughs. Therapists then explore and integrate these in later sessions. MDMA and psilocybin are beneficial in enhancing the therapeutic process. They lower psychological barriers and deepen the patient-therapist connection. This makes therapy more meaningful. (Reiff et al., 2020).

Psychedelics reopen critical periods for social reward learning. This significantly boosts psychotherapeutic interventions. They enhance metaplasticity, mainly through oxytocin in the nucleus accumbens. Moreover, they reshape the extracellular matrix (ECM). Such insights clarify how psychedelics achieve lasting brain changes. This makes them effective for treating PTSD, depression, and addiction (Nardou et al., 2023).

Psychedelics tap into the brain's plasticity. They help patients to revisit and reshape deep-seated emotional and behavioral patterns. These patterns often resistant to standard treatments. This supports integrating the use of psychedelics in therapy. It also sparks the creation of new drugs. These drugs aim to mimic therapeutic effects. Importantly, they seek to avoid inducing the altered states of consciousness of psychedelic experiences. Such progress could make psychedelic therapies more widely available and applicable (Reiff et al., 2020).

MDMA shows promise in treating PTSD. It reduces the amygdala's response to fear and improves emotional processing. By affecting serotonin, dopamine, and oxytocin systems, MDMA creates emotional openness. This allows patients to face traumatic memories without intense fear or anxiety. Clinical trials confirm that MDMA therapy significantly reduces PTSD symptoms. Many participants enjoy lasting benefits after just a few sessions (Mitchell et al., 2021).

Psilocybin is being studied for its potential to ease depression and existential anxiety, especially in terminally ill patients. It works by activating serotonin receptors, enhances neural plasticity, and disrupting overactive rigid networks, such as the default mode network (DMN) linked to depression. Research shows that therapy with psilocybin can quickly and significantly reduce depression. This often leads to significant emotional and cognitive shifts. In therapy, psilocybin helps people face and change deep-seated emotions and cognitive patterns. Such experiences are often seen as "mystical" or "spiritual" and are linked to better therapeutic outcomes (Reiff et al., 2020).

Ketamine is now recognized for its rapid-acting antidepressant effects on mental health. When used in low doses, it alleviates depression, PTSD, and substance use issues, especially alongside therapy. Its quick mood improvements in tough cases show ketamine's promise in therapy. This approach, called Ketamine-Assisted Psychotherapy (KAP), leverages ketamine's ability to boost brain adaptability, aiding emotional and cognitive restructuring during therapy processes (Drozdz et al., 2022)

The drug can alleviate severe depressive symptoms and reduce suicidal thoughts. This opens a window for patients to engage in therapy without being overwhelmed by depression or trauma-related symptoms. Moreover, ketamine boosts neuroplasticity, enhancing its effects, especially with psychotherapy. This increased plasticity helps patients restructure negative thought patterns and emotional responses leading to lasting improvements in therapeutic outcomes (Davis et al., 2021).

KAP has shown promise in treating substance use disorders. It reduces cravings and promotes abstinence. Patients on high-dose ketamine with psychotherapy had better results in stopping heroin use. This was compared to those on low doses or standard treatments. Ketamine's dissociative effects help patients see their addiction from new perspectives. This fosters motivation for long-term change (Krupitsky et al., 2002).

Ketamine, combined with psychotherapy, quickly eases severe symptoms and boosts treatment success by fostering emotional connection and facilitating the therapeutic process. KAP sessions have three parts: preparation, ketamine administration, and integration. Preparation sets therapeutic intentions, while integration applies insights from ketamine experiences to real life. Studies show patients using ketamine with psychotherapy maintained symptom relief for months. Yet, how long this relief lasts depends on the condition, session frequency, and therapy type. Because of differences in studies and patients, more research is vital to create standard KAP protocols. Continued investigation is also necessary to find the optimal ketamine dose and frequency for lasting KAP benefits (Drozdz et al., 2022).

Psychedelics show promise in therapy, especially for patients unresponsive to standard treatments. They induce neuroplasticity and improve emotional processing, making them ideal for complex psychiatric health issues.

However, psychedelic-assisted therapies (PAP) face significant legal and ethical challenges for broader use. Most psychedelics, like psilocybin and MDMA, are still illegal except in trials or with special permits. Ketamine is more accessible as it's Schedule III classification. But, the slow approval process and concerns about profit-driven practices raise issues about future access and affordability. This could lead to a situation where only wealthier individuals can afford these therapies, widening the gap in healthcare.

On the ethical side, ensuring patient safety is crucial due to the intense effects of psychedelics. Patients must be fully informed about the risks and benefits. Therapists need to keep professional boundaries, especially when patients are vulnerable in altered states. All should have access to these therapies, not just those who can pay. It's also vital to respect the cultural roots of psychedelics, particularly those from Indigenous traditions, to avoid misuse. Addressing these challenges is critical to safely bringing PAP into mainstream mental health care.

8. Takeaways

1. Psychedelics, particularly classical ones like psilocybin and LSD, enhance neuroplasticity by promoting the growth of dendrites, synapses, and neurons.
2. The use of psychedelics dates back to ancient cultures and was later revived in the twentieth century for both recreational and therapeutic purposes.
3. Psychedelics act on serotonin 5-HT2A receptors, which leads to a cascade of intracellular signaling that influences synaptic plasticity and neurogenesis.
4. Psychedelics can reopen critical periods for neuroplasticity, regulated by mechanisms like oxytocin particularly for social reward learning.

5. Psychedelics alter brain-wide connectivity, particularly within the default mode network (DMN), reducing its overactivity in conditions like depression.
6. Ketamine, a dissociative psychedelic, acts on NMDA receptors and promotes neuroplasticity via brain-derived neurotrophic factor (BDNF). Its rapid antidepressant effects have made it a valuable treatment option.
7. MDMA, often used in the treatment of PTSD, enhances emotional empathy and social bonding through serotonin and oxytocin release.
8. Recent studies show that psychedelics can effectively treat psychiatric disorders such as PTSD, depression, and substance abuse. Their ability to induce neuroplasticity opens new possibilities for treating conditions resistant to conventional therapies.
9. While psychedelics show promise in therapy, they come with risks such as potential neurotoxicity (particularly with MDMA) and transient cognitive impairment, which necessitates controlled, supportive environments during their use.
10. The integration of psychedelics into mainstream mental health care is complicated by their Schedule I classification, slow regulatory approval, and the commercialization of treatment, which raises concerns about equitable access and ethical use.

Abbreviations

ACC: Anterior cingulate cortex
AMPA: α-Amino-3-Hydroxy-5-Methyl-4-Isoxazolepropionic acid
BDNF: Brain-derived neurotrophic factor
DEA: Drug enforcement administration
DLPFC: Dorsolateral prefrontal cortex
DMN: Default mode network
DMT: N-Dimethyltryptamine
ECM: The extracellular matrix
fMRI: Functional magnetic resonance imaging
KAP: Ketamine-assisted psychotherapy
LFP: Local field potential
LSD: Lysergic acid diethylamide
MAPS: Multidisciplinary Association for Psychedelic Studies

MDD: Major depressive disease
MDEA: 3,4-methylenedioxymethamphetamine
MDMA: 3,4-Methylenedioxymethamphetamine
MMDA: 5-methoxy-3,4 methylenedioxyamphetamine
mTOR: Mammalian target of the rapamycin
NMDA: N-Methyl-D-Aspartate
PAP: Psychedelic-assisted psychotherapy
PCP: Phencyclidine
PFC: Prefrontal cortex
PPC: Posterior parietal cortex
PTSD: Posttraumatic stress disorder
REBUS: Relaxed beliefs under psychedelics
TRD: Treatment-resistant depression
5-HT2A: 5-Hydroxytryptamine 2A receptor

Bibliography

1. Banks, M., Zahid, Z., Jones, N. T., Sultan, Z. W., & Wenthur, C. J. (2021). *Catalysts for change: the cellular neurobiology of psychedelics.* Molecular Biology of the Cell, 32(11), 1350–1361. https://doi.org/10.1091/mbc.E20-05-0340
2. Becker, R., Braun, U., Schwarz, A., Gass, N., Schweiger, … , & Meyer-Lindenberg, A. (2016). *Species-conserved reconfigurations of brain network topology induced by ketamine.* Translational Psychiatry, 6(4), e843. https://doi.org/10.1038/tp.2016.53
3. Calder, A. E., & Hasler, G. (2023). *Towards An Understanding Of Psychedelic-İnduced Neuroplasticity.* Springer Nature Neuropsychopharmacology 2023 48:104–112 https://doi.org/10.1038/s41386-022-01389-z
4. Capela, J. P., & Carvalho, F. D. (2022). *A review on the mitochondrial toxicity of "ecstasy" (3,4-methylenedioxymethamphetamine, MDMA).* Current Research in Toxicology, *3*, 100075. https://doi.org/10.1016/j.crtox.2022.100075
5. Cardoso, C. G., & Viana, M. de B. (2023). *Psychedelics and cognitive function: A systematic review.* Research, Society and Development, 12(14), e16121444436. https://doi.org/10.33448/rsd-v12i14.44436

6. Carhart-Harris, R., & Friston, K. J. (2019). *REBUS and the Anarchic Brain: Toward a Unified Model of the Brain Action of Psychedelics.* Pharmacological Reviews, 71(3), 316–344. https://doi.org/10.1124/pr.118.017160

7. Castelhano, J., Lima, G., Teixeira, M., Soares, C., Pais, M., & Castelo-Branco, M. (2021). *The Effects Of Tryptamine Psychedelics İn The Brain: A Meta-Analysis Of Functional And Review Of Molecular İmaging Studies.* Frontiers in Pharmacology, 12, 739053. https://doi.org/10.3389/fphar.2021.739053

8. Cruzat, J., Perl, Y., Escrichs, A., Vohryzek, J., Timmermann, C., Roseman, L., … & Kringelbach, M. (2022). *Effects Of Classic Psychedelic Drugs On Turbulent Signatures İn Brain Dynamics.* Network Neuroscience, 6(2), 178–198. https://doi.org/10.1162/netn_a_00250

9. Davis, M. T., DellaGiogia, N., Maruff, P., Pietrzak, R. H., & Esterlis, I. (2021). *Acute Cognitive Effects of Single-Dose Intravenous Ketamine in Major Depressive and Posttraumatic Stress Disorder.* Translational Psychiatry, 11, 1–16. https://doi.org/10.1038/s41398-021-01327-5

10. De Gregorio, D., Aguilar-Valles, A., Preller, K. H., Heifets, B. D., Hibicke, M., Mitchell, J., & Gobbi, G. (2021). *Hallucinogens in Mental Health: Preclinical and Clinical Studies On LSD, Psilocybin, MDMA, And Ketamine.* The Journal Of Neuroscience, *41*(5), 891–900. https://doi.org/10.1523/JNEUROSCI.1659-20.2020

11. Deka, B., Dash, B., Bharali, A., & Ahmed, A. (2021). *Ketamine: More than Just an NMDA Blocker.* IntechOpen, 17(6) https://doi.org/10.5772/intechopen.101113

12. Doss M. K, Madden M. B., Gaddis A., M. B. Nebel, Griffiths R., Mathur B. N., Barrett F.S., (2022). Models of psychedelic drug action: modulation of cortical-subcortical circuits, *Brain*, Volume 145, Issue 2, February 2022, Pages 441–456 https://doi.org/10.1093/brain/awab406

13. Drozdz, S. J., Goel, A., McGarr, M. W., Katz, J., Ritvo, P., Mattina, G. F., Bhat, V., Diep, C., & Ladha, K. S. (2022). *Ketamine Assisted Psychotherapy: A Systematic Narrative Review of the Literature.* Journal of Pain Research 2022:15 1691–1706

14. Gattuso, J. J., Perkins, D., Ruffell, S., Lawrence, A. J., Hoyer, D., Jacobson, L. H., … & Sarris, J. (2023). *Default Mode Network Modulation By Psychedelics: A Systematic Review.* International Journal

of Neuropsychopharmacology, 26(3), 155–188. https://doi.org/10.1093/ijnp/pyac074

15. Gill, H., Gill, B., Rodrigues, N. B., Lipsitz, O., Rosenblat, J., El-Halabi, S., … , & McIntyre, R. (2020). *The Effects Of Ketamine On Cognition İn Treatment-Resistant Depression: A Systematic Review.* Neuroscience & Biobehavioral Reviews, 120, 78–85. https://doi.org/10.1016/j.neubiorev.2020.11.020
16. Hipólito, I., Mago, J., Rosas, F., & Carhart-Harris, R. (2023). *Pattern Breaking: A Complex Systems Approach To Psychedelic Medicine.* Neuropsychopharmacology, 48(2), 232–245. https://doi.org/10.1093/nc/niad017
17. Holze, F., Vizeli, P., Ley, L., Müller, F., Dolder, P. C., Stocker, M., … & Liechti, M. E. (2021). *Acute Dose-Dependent Effects Of Lysergic Acid Diethylamide İn A Double-Blind Placebo-Controlled Study İn Healthy Subjects.* Neuropsychopharmacology, 46(3), 537–544. https://doi.org/10.1038/s41386-020-00883-6
18. Hysek, C., Schmid, Y., Simmler, L., Domes, G., Heinrichs, M., Eisenegger, C., Preller, K., Quednow, B., & Liechti, M. (2014). *MDMA Enhances Emotional Empathy And Prosocial Behavior.* Social Cognitive and Affective Neuroscience, 9(11), 1645–1652. https://doi.org/10.1093/scan/nst161
19. Krupitsky E, Burakov A, Romanova T, Dunaevsky I, Strassman R, Grinenko A. (2002). Ketamine psychotherapy for heroin addiction: immediate effects and two-year follow-up. J Subst Abuse Treat. 23(4):273–283. https://doi.org/10.1016/S0740-5472(02)00275-1
20. Kuypers, K., de la Torre, R., Farré, M., Yubero-Lahoz, S., Dziobek, I., van den Bos, W., & Ramaekers, J. (2014). *No Evidence that MDMA-Induced Enhancement of Emotional Empathy Is Related to Peripheral Oxytocin Levels or 5-HT1a Receptor Activation.* Plos One, 9(6). https://doi.org/10.1371/journal.pone.0100719
21. Lee, M. A., & Shlain, B. (1985). *Acid Dreams: The Complete Social History of LSD: The CIA, the Sixties, and Beyond*. Grove Press.
22. Lukasiewicz, K., Baker, J. J., Zuo, Y., & Lu, J. (2021). *Serotonergic Psychedelics in Neural Plasticity*. Frontiers in Molecular2023Neuroscience, 14, 748359.

23. Ly, C., Greb, A. C., Cameron, L. P., Wong, J. M., Barragan, E. V., Wilson, P. C., … & Olson, D. E. (2018). *Psychedelics promote structural and functional neural plasticity.* Cell Reports, 23(11), 3170–3182. https://doi.org/ 10.1016/j.celrep.2018.05.022

24. Marguilho, M., Figueiredo, I., & Castro-Rodrigues, P. (2023). *A Unified Model Of Ketamine's Dissociative And Psychedelic Properties.* Journal of Psychopharmacology 2023, Vol. 37(1) 1432. https://doi.org/10.1177/0269881122114001

25. McIntyre, R., Rosenblat, J., Rodrigues, N. B., Lipsitz, O., … , & Lee, Y. (2021). *The Effect of İntravenous Ketamine on Cognitive Functions in Adults with Treatment-Resistant Major Depressive or Bipolar Disorders: Results From the CRTCE.* Psychiatry Research, 302, 113993. https://doi.org/10.1016/j.psychres.2021.113993

26. Mitchell, J. M., Bogenschutz, M., Lilienstein, A., …. &, Van der Kolk, B., Doblin, R. (2021). *MDMA-assisted therapy for severe PTSD: a randomized, double-blind, placebo-controlled phase 3 study. Nature Medicine*, 27(6), 1025–1033. https://doi.org/10.1038/s41591-021-01336-3

27. Moreno, J. L., Miranda-Azpiazu, P., García-Bea, A., Younkin, J., Cui, M., Kozlenkov, A., González-Maeso, J. (2016). *Allosteric signaling through an mGlu2 and 5-HT2A heteromeric receptor complex and its potential contribution to schizophrenia.* Science Signaling, 9(410). https://doi.org/10.1126/scisignal.aab0467

28. Nardou, R., Sawyer, E., Song, Y. J., Wilkinson, M., … , & Dölen, G. (2023). *Psychedelics Reopen the Social Reward Learning Critical Period.* Nature, *618*(7966), 790–798. https://doi.org/10.1038/s41586-023-06204-3

29. Neill, J. C., Khundakar, A. A., Imbimbo, B. pietro, … , & Mason, N. L. (2021). *Psychedelics and Neuroplasticity: A Systematic Review Unraveling the Biological Underpinnings of Psychedelics.* Frontiers in Psychiatry 12, 724606. https://doi.org/10.3389/fpsyt.2021.724606

30. Olson, D. E. (2022). *Biochemical Mechanisms Underlying Psychedelic-Induced Neuroplasticity.* Biochemistry, 61(2), 123–137. https://doi.org/10.1021/acs.biochem.1c00812

31. Reiff, C. M., Richman, E. E., Nemeroff, C. B., Carpenter, L. L., … , & McDonald, W. M. (2020). *Psychedelics And Psychedelic-Assisted*

Psychotherapy. American Journal of Psychiatry, 177(5), 391–410. https://doi.org/10.1176/appi.ajp.2019.1901003

32. Rein, B., Raymond, K., Boustani, C., Tuy, S., Zhang, J., Laurent, R., … , & Malenka, R. C. (2024). *MDMA enhances empathy-like behaviors in mice via 5-HT release in the nucleus accumbens.* Science Advances (Vol. 10). eadl6554 https://doi.org/ 10.1126/sciadv.adl6554
33. Stippl, A., Scheidegger, M., Aust, S., Herrera, A., … , & Grimm, S. (2021). *Ketamine Specifically Reduces Cognitive Symptoms İn Depressed Patients: An İnvestigation Of Associated Neural Activation Patterns.* Journal Of Psychiatric Research, 136, 402–408. https://doi.org/10.1016/j.jpsychires.2021.02.028
34. Swanson, L. (2018). *Unifying Theories of Psychedelic Drug Effects.* Frontiers in Pharmacology, pp. 9, 172. https://doi.org/10.3389/fphar.2018.00172
35. Tang, W., Lau, C. G., Ungvari, G. S., Lin, S., & Lane, H. Y. (2019). *Recovery Of Cognitive Functioning Following Abstinence From Ketamine.* Addictive Behaviors, 99, 106081. https://doi.org/10.1016/j.addbeh.2019.106081
36. UC Berkeley Center for the Science of Psychedelics. (2024). *Psychedelic Legalization & Decriminalization Tracker.* Retrieved from https://psychedelics.berkeley.edu
37. Wacker, D., Wang, S., McCorvy, J. D., Betz, R. M., Venkatakrishnan, A. J., … , & Roth, B. L. (2017). *Crystal structure of an LSD-bound human serotonin receptor.* Cell, 168(3), 377–389. https://doi.org/10.1016/j.cell.2016.12.033
38. Zanos, P., Moaddel, R., Morris, P. J., Riggs, L. M., Highland, J. N., … , & Gould, T. D. (2018). *Ketamine and Ketamine Metabolite Pharmacology: Insights into Therapeutic Mechanisms.* Pharmacological Reviews, 70(3), 621–660. https://doi.org/10.1124/pr.117.015198

Olfactory Stimulation and Neuroplasticity: The Science of Aromatherapy

Tuğba Duygu Özmet

1. Introduction

Modern research focuses on treating brain ailments and boosting memory. Neuroplasticity plays a crucial role in combating Alzheimer's, dementia, anxiety, and depression. Various factors influence this process, including aromatherapy—a centuries-old practice.

Aromatherapy enhances and accelerates neuroplasticity. The interplay between these two fields merits attention. Their relationship offers promising avenues for brain health and cognitive function, highlighting the potential of holistic approaches in neuroscience.

2. Factors Increasing and Decreasing Neuroplasticity

Neuroplasticity fluctuates with age, a non-modifiable factor. As we grow older, our brain's ability to adapt diminishes. Young dogs learn faster than old ones, mirroring human experience. Aging alters brain composition and neurotransmitter balance, impacting neuroplasticity.

However, elderly individuals can still enhance memory and motor skills through targeted training. Childhood education may boost neuroplasticity, suggesting a lifelong potential for brain adaptation.

Lifestyle choices significantly influence neuroplasticity. Regular physical activity prevents cognitive decline and dementia. Aerobic exercise bolsters brain-derived neurotrophic factors, enlarges the hippocampus, and increases gray and white matter volume. Even brief daily cycling can improve memory and concentration.

Sleep quality profoundly affects neuroplasticity. Well-regulated sleep strengthens memory and accelerates learning. Treating sleep disorders can enhance neuroplasticity, promoting healthier aging.

Chronic stress damages the hippocampus and impairs neuronal communication, potentially leading to cell death. It causes structural deterioration in dendrites, hindering neuron multiplication and survival. High-stress levels induce brain atrophy, reducing neuroplasticity.

Mindfulness programs and aromatherapy can mitigate stress and anxiety, thereby fostering neuroplasticity. These ancient practices offer alternative methods to enhance brain adaptability.

3. Aromatherapy

Fragrant essences from plants form the core of aromatherapy, a burgeoning alternative treatment. This holistic approach harnesses concentrated oils extracted from various plant parts—blossoms, foliage, stems, fruits, and roots—to bolster health and well-being. Inhalation of these potent aromas is believed to stimulate brain activity. The oils may also seep through the skin into the bloodstream, fostering healing throughout the body. As a complementary therapy, aromatherapy has gained traction globally in recent years, offering a natural path to rejuvenate the mind, body, and spirit. These botanical extracts serve as the primary therapeutic agents in this increasingly popular healing modality (Lee et al., 2012).

Aromatherapy harnesses plants' pure essential oils, offering body and mind benefits (Brennan et al., 2022). This practice effectively alleviates anxiety, depression, and negative emotions, enhancing overall comfort. Aromatherapy regulates mood by linking olfaction to the limbic system, triggering endorphin and serotonin release (Li et al., 2022). These neurotransmitters induce calm, soothing effects, easing anxiety (Singewald et al., 2015). Lavender, rose, lemon, mint, and rosemary oils have proven effective in reducing stress, boosting memory, and improving cognition. Consequently, aromatherapy shows promise in treating Alzheimer's disease (Luan et al., 2023).

Essential oils, potent aromatic extracts from nature's bounty, form the heart of aromatherapy. These concentrated compounds, sourced from diverse plant parts, offer various applications. Diffusion and massage are primary delivery methods, unlocking the oils' potential. Beyond fragrant ambiance, aromatherapy's reach extends to clinical realms. It can transform a space with floral notes or potentially aid in battling neurodegenerative conditions like Alzheimer's by targeting problematic enzymes (Agatonovic-Kustrin et al., 2019).

Aromatherapy harnesses essential oils to boost physical and mental well-being. Black pepper, cypress, rose, valerian, ginger, bergamot, and lavender yield potent extracts. Black pepper oil enhances circulation, eases pain, and combats mental fatigue. It lifts mood, restores motivation, and relieves tension. Users can apply it topically or inhale its vapors for optimal benefits (Shutes & Galper, 2022).

Rose's healing properties span various ailments. Applied topically or inhaled, it soothes varicose veins, eases menstrual discomfort, and aids fertility. This versatile flower combats sleeplessness, calms nerves, and lifts postpartum spirits. Available as creams or oils, rose blends well with citrus scents and ylang-ylang. Its aromatic baths offer a luxurious remedy for the body and mind (Shutes & Galper, 2022).

Goat's rue calms and regulates sleep, easing digestive issues, insomnia, and stress-induced fatigue. It can be used as an inhaler, roll-on, or in baths.

Bergamot, an Italian-grown plant, yields a premium essential oil with a rich, exotic scent. Its calming properties aid anxiety, insomnia, focus, and memory. Apply in baths, inhalers, diffusers, or roll-ons. Mixed with eucalyptus, it stimulates.

Ginger's spicy aroma stimulates digestion, eases menstrual pain, and relieves muscle aches. It boosts motivation, focus, and sexual function. Apply it as an inhaler, oil, cream, or lotion. It pairs well with jasmine, rose, black pepper, and ylang-ylang.

Tangerine soothes and uplifts. It treats stress-related gut problems, premenstrual syndrome (PMS), anxiety, child hyperactivity, and mild depression.

Versatile lavender improves mood, sleep, and pain relief through GABA inhibition. It reduces inflammation, eases depression via glutamate stimulation and calcium inhibition, and exhibits antiepileptic properties (Bavarsad et al., 2023).

Rheumatology is linked to sweet-scented geranium, a member of the geraniaceae family. Its oil, rich in geraniol, citronellol, terpineol, and alcohol, boasts various health benefits. These include anti-inflammatory, analgesic, antioxidant, anti-cancer, and antimicrobial effects (Gazerani et al., 2021). Also, geranium oil boosts learning and memory in the elderly by increasing neuroplasticity (Atef et al., 2022).

3.1. History of Aromatherapy

Aromatherapy has a long history dating back to 1555 BC when Hippocrates used it to treat patients (Agnihotry et al., 2024). He discovered that burning aromatic plants had beneficial effects on health. In Greece, Theophrastus wrote about oils used in aromatherapy. The fathers of aromatherapy, the Egyptians used essential oils in massage, medicine, skincare, and mummification. They believed that God had a unique scent and gave great importance to hygiene. After learning distillation, they used essential oils in aromatherapy and sanitation. The Kyphi formula, containing 16 plants, was used for sleep and anxiety and as an antiseptic. Aromatherapy was also used in Chinese, Indian, and Roman cultures. In ancient times, plants were used for medicinal purposes, hygiene, and beauty products (Shutes & Galper, 2022; Yousefsani et al., 2021). Today, eight nurse theorists recognize aromatherapy as a method for patient care, promoting relaxation, adaptation, and holistic harmony (Farrar & Farrar, 2020; Gnatta et al., 2016).

3.2. Pathophysiological Process in Aromatherapy Application

When essential oils used in aromatherapy are inhaled, the olfactory nerve is stimulated, and the respiratory and gastrointestinal systems are stimulated. The molecules formed during aromatherapy secrete neurotransmitters. This affects blood pressure, pulse, body temperature, and pupil dilation. The process can be listed as follows,

- The olfactory nerve is stimulated.
- The stimulus affects the brain and stimulates the amygdala. This affects the emotional state.
- Inhaled aromatherapy molecules pass from the upper respiratory tract to the lower respiratory tract.
- Molecules enter the circulation from the pulmonary blood vessels and spread to tissues and organs (Farrar & Farrar, 2020).

3.3. Safety Guidelines for the Use of Aromatherapy

The use of essential oils is widespread worldwide. Aromatherapy has beneficial effects, for example, in the after-work bath. Aromatherapy can be applied by people who do not have enough information about safety precautions.

However, if used excessively or by people who have not received training, it can lead to toxic effects. This toxic effect can result in burns or even death. The precautions to be taken can be listed as follows.

- This information should include environment, diet and existing health problems of the individual.
- If an individual has a known allergy or is suspected to be sensitive, a skin patch test should be performed before application.
- If the individual has a chronic disease such as diabetes or epilepsy, precautions should be taken. For example, fennel, hyssop, and rosemary oils should not be used in patients with epilepsy, and angelica oil should not be used in patients with diabetes.
- After obtaining the information, essential oils that are appropriate for the individual should be preferred.
- High-quality oils that can be checked from other sources should be used.
- It should be kept out of sight when applied. It should be stored in a cold and dark environment. It should be protected from sunlight.
- It should be ensured that the treatment room has adequate ventilation. (National Association of Holistic Aromatherapy, 2024).
- Caution should be taken when using it on patients who are not knowledgeable about other essential oils.
- Essential oils should not be used internally.
- For long-term use, it should be used under the supervision of a health professional. Undiluted essential oils should not be used. The use of undiluted oils can cause inflammation. Therefore, it should be applied with lotion or cream.
- If female patients are suspected of being pregnant, precautions should be taken. Low doses should be used in children, the elderly, and pregnant women.
- Nurses must observe the principle of beneficence when using aromatherapy. Failure to comply with this principle may harm the individual and lead to legal consequences (Farrar & Farrar, 2020). To avoid this situation, healthcare personnel who have received advanced aromatherapy training should apply aromatherapy (Alliance of International Aromatherapists, 2024).

4. Neuroplasticity and Aromatherapy

Essential oils impact the nervous system. Mint and coffee stimulate it, as seen in brain scans. Inhaling oils like lavender, rosemary, and bergamot alters autonomic nervous system activity. This also applies to eucalyptus, rose, lemon, Meniki, Hinoki, Juniperus phoenicea gum, Copaifera officinalis resin, Aniba rosaeodora oil, Juniperus virginiana oil, grapefruit oil, and petitgrain. Massage with lavender and geranium oil lowers heart rate and blood pressure. Lavender also enhances sleep and calms. Mint and coffee are stimulating (Sattayakhom et al., 2023). It is also known that lavender oil has the effect of reducing anxiety and depression by increasing neuroplasticity (Sánchez-Vidaña et al., 2019).

Depression is the most studied clinical finding regarding the effect of essential oils. Neuroplasticity decreases with the decrease in serotonin in depression (Page et al., 2024). It is stated that the anxiolytic effect of lavender is equivalent to low-dose diazepam (Alireza et al., 2014). Lavender, chamomile, bergamot, sweet orange, anise, and geranium can reduce depressive symptoms in various groups, including the elderly, postpartum women, and patients with specific conditions. This was shown in long-term care units. In a study by Conrad and Adams (2012), they focused on postpartum anxiety in 28 women. The intervention group received aromatherapy through hand massages and inhaled rose and lavender oil. This happened for 15 minutes twice a week over four weeks. As a result, their depression and anxiety levels decreased. (Conrad & Adams, 2012). In studies conducted on women who gave birth by Cesarean section, it was found that post-cesarean depression scores decreased in both women who drank lavender tea and women who smelled it. In addition, it was found that the quality of women's sleep increased, and state anxiety decreased with the smell of lavender oil (Chen & Chen, 2015).

Lavender, bergamot, and lemon balm reduce pain and anxiety in mice, rats, neonates, premature infants, and women with painful periods. Similarly, these effects are also exhibited by extracts from Juniperus phoenicea, Copaifera officinalis, Aniba rosaeodora, virginiana oil, and citrus (Citrus spp.) (Bavarsad et al., 2023).

Another plant used in aromatherapy is patchouli. Essential oils obtained from patchouli leaves have anti-depressant, antimicrobial, and antiviral effects and effects that relax the nervous system (Farrar & Farrar, 2020).

Studies show that essential oils reduce fatigue, memory, and sleep issues, and behavioral symptoms in breast cancer patients who had breast reconstruction, children with burns, and patients over 65 with mild to moderate dementia (Pereira et al., 2022).

In the systematic analysis conducted by Terzi et al. (2020), it was stated that aromatherapy showed improvement in pain and depression symptoms in the elderly. In addition, studies found that aromatherapy was applied with acupressure and music therapy. In these studies, it was found that there was a significant decrease in the anxiety level and agitation scores of elderly individuals compared to the group that applied aromatherapy alone (Terzi et al., 2020).

When used by inhalation, essential oils affect cognitive functions and improve symptoms of Alzheimer's and dementia. Oils entering the blood reach the limbic system from the nose and stimulate the olfactory nerve. In this way, it strengthens cognitive functions and memory. In a study conducted by Lin et al. (2007), it was determined that essential oils obtained from L. angustifolia improved agitation symptoms in Alzheimer's and dementia patients (Lin et al., 2007). A study using Rosemarinus Officinalis (R. Officinalis) and vegetable plants as essential oils found that they strengthen memory in dogs and mice (Benny & Thomas, 2019).

Dried bay leaves have analgesic and neurological system-improving and calming effects (Agnihotry et al., 2024). It also treats epilepsy, Parkinson's, and neuralgia (Pilipović et al., 2023).

5. Conclusion

Individuals use aromatherapy to improve their overall well-being, which encompasses physical, mental, and social health. Research has demonstrated the positive impact of essential oils on health, including alleviating anxiety and depression, enhancing Alzheimer's treatment, and regulating sleep disorders by increasing neuroplasticity. For example, a study found that inhaling essential oils decreased anxiety scores and improved sleep quality in patients with chronic stress (Lin et al.2007).

However, the misuse of aromatherapy is a widespread concern. Many individuals use it without proper understanding, which can lead to severe side effects, even death. Essential oils can interact with medications, trigger allergic reactions, or exacerbate existing health conditions. To avoid these risks, healthcare professionals must administer aromatherapy with aromatherapy training.

It is like using a powerful tool without instructions—the results can be disastrous. Only through proper training and knowledge can healthcare professionals harness aromatherapy's therapeutic benefits, ensuring safe and effective treatment.

In summary, while aromatherapy has shown promise in promoting neuroplasticity, further research is necessary to fully understand its effects. By increasing our knowledge of aromatherapy and its relationship with neuroplasticity, we can unlock its full potential as a holistic approach to health.

6. Takeaways

1. Neuroplasticity can be defined as the self-renewal of neurons and the brain.
2. Increasing neuroplasticity prevents neurological diseases such as Alzheimer's and dementia. It has also been reported that there is an improvement in anxiety disorder, depression symptoms, as well as memory, sleep, and concentration disorders.
3. Aromatherapy is an intervention that increases neuroplasticity. It has been known and applied since human history and is used by many religious and ethnic groups. Aromatherapy can be carried out by inhalation, massage, or dermal means.
4. Trained people should apply aromatherapy. While applying, safety precautions, should be taken.
5. Applying essential oils improves many symptoms, such as pain and depression. They also have antibacterial and antiviral effects.
6. It is more effective when applied using aromatherapy, acupressure, and music therapy.

Abbreviations

PMS: Premenstrual syndrome
R. Officinalis: Rosemarinus officinalis

Bibliography

1. Agatonovic-Kustrin, S., Kustrin, E., & Morton, D. W. (2019). Essential oils and functional herbs for healthy aging. *Neural Regeneration Research*, *14*(3), 441–445. https://doi.org/10.4103/1673-5374.245467
2. Agnihotry, S., Chopra, D., Singh, J., Negi, S., Srivastav, A., Upadhyay, J., & Sharma, G. (2024). Aromatherapy Evolution and Blending Basics of Essential Oils. Aromatherapy: The Science of Essential Oils, p. 1.
3. Alireza, K., Nasrin, H. F., Mina, R., Zahra, A., Siamak, S., & Abdolrahman, S. (2014). Study the anti-anxiety effect of acute injection of Lavandula angustifolia extract in elevated plus-maze test in male rats. *International Journal of Pharmacognosy and Phytochemical Research*, *6*(4), 879–882.
4. Alliance of International Aromatherapists [AIA] (2024). *Aromatherapy safety.* https://www.alliance-aromatherapists.org/aromatherapy-safety
5. Atef, M. M., Emam, M. N., Abo El Gheit, R. E., Elbeltagi, E. M., Alshenawy, H. A., Radwan, D. A., Younis, R. L., & Abd-Ellatif, R. N. (2022). Mechanistic Insights into Ameliorating Effect of Geraniol on d-Galactose Induced Memory Impairment in Rats. *Neurochemical Research*, *47*(6), 1664–1678. https://doi.org/10.1007/s11064-022-03559-3
6. Bavarsad, N. H., Bagheri, S., Kourosh-Arami, M., & Komaki, A. (2023). Aromatherapy for the brain: Lavender's healing effect on epilepsy, depression, anxiety, migraine, and Alzheimer's disease: A review article. *Heliyon*, *9*(8), e18492. https://doi.org/10.1016/j.heliyon.2023.e18492
7. Benny, A., & Thomas, J. (2019). Essential Oils as Treatment Strategy for Alzheimer's Disease: Current and Future Perspectives. *Planta Medica*, *85*(3), 239–248. https://doi.org/10.1055/a-0758-0188
8. Brennan, S. E., McDonald, S., Murano, M., & McKenzie, J. E. (2022). Effectiveness of aromatherapy for prevention or treatment of disease, medical or preclinical conditions, and injury: protocol for a systematic review and meta-analysis. *Systematic Reviews*, *11*(1), 1–18. https://doi.org/10.1186/s13643-022-02015-1
9. Chen, S.-L., & Chen, C.-H. (2015). Effects of Lavender Tea on Fatigue, Depression, and Maternal-Infant Attachment in Sleep-Disturbed Postnatal Women. *Worldviews on Evidence-Based Nursing*, *12*(6), 370–379. https://doi.org/10.1111/wvn.12122

10. Conrad, P., & Adams, C. (2012). The effects of clinical aromatherapy for anxiety and depression in the high-risk postpartum woman - a pilot study. *Complementary Therapies in Clinical Practice*, *18*(3), 164–168. https://doi.org/10.1016/j.ctcp.2012.05.002
11. Farrar, A. J., & Farrar, F. C. (2020). Clinical Aromatherapy. *The Nursing Clinics of North America*, *55*(4), 489–504. https://doi.org/10.1016/j.cnur.2020.06.015
12. Gazerani, A., Sarchahi, Z., Hosseini, S. S., lakziyan, R., & Abavisani, M. (2021). The effect of inhalation aromatherapy of geranium on pain and physiological indices after appendectomy: A double-blind randomized clinical trial. *International Journal of Surgery* Open, 28, 44–49. https://doi.org/10.1016/j.ijso.2020.12.004
13. Gnatta, J. R., Kurebayashi, L. F. S., Turrini, R. N. T., & Da Silva, M. J. P. (2016). Aromatherapy and nursing: Historical and theoretical conception. *Revista Da Escola de Enfermagem*, *50*(1), 127–133. https://doi.org/10.1590/S0080-623420160000100017
14. Lee, M. S., Choi, J., Posadzki, P., & Ernst, E. (2012). Aromatherapy for health care: an overview of systematic reviews. *Maturitas*, *71*(3), 257–260. https://doi.org/10.1016/j.maturitas.2011.12.018
15. Li, D., Li, Y., Bai, X., Wang, M., Yan, J., & Cao, Y. (2022). The Effects of Aromatherapy on Anxiety and Depression in People With Cancer: A Systematic Review and Meta-Analysis. *Frontiers in Public Health*, *10*(May), 1–13. https://doi.org/10.3389/fpubh.2022.853056
16. Lin, P. W., Chan, W., Ng, B. F., & Lam, L. C. (2007). Efficacy of aromatherapy (Lavandula angustifolia) as an intervention for agitated behaviours in Chinese older persons with dementia: a cross-over randomized trial. International Journal of Geriatric Psychiatry, 22(5), 405–410. https://doi.org/10.1002/gps.1688
17. Luan, J., Yang, M., Zhao, Y., Zang, Y., Zhang, Z., & Chen, H. (2023). Aromatherapy with inhalation effectively alleviates the test anxiety of college students: A meta-analysis. *Frontiers in Psychology*, *13*(January), pp. 1–11. https://doi.org/10.3389/fpsyg.2022.1042553
18. National Association for Holistic Aromatherapy [NAHA] (2024). *General safety guidelines.* https://naha.org/explore-aromatherapy/safety/general-safety-guidelines//

19. Page, C. E., Epperson, C. N., Novick, A. M., Duffy, K. A., & Thompson, S. M. (2024). Beyond the serotonin deficit hypothesis: communicating a neuroplasticity framework of major depressive disorder. *Molecular Psychiatry*. https://doi.org/10.1038/s41380-024-02625-2

20. Pereira, M., Moreira, C. S., Izdebski, P., Dias, A. C. P., Nogueira-Silva, C., & Pereira, M. G. (2022). How Does Hedonic Aroma Impact Long-Term Anxiety, Depression, and Quality of Life in Women with Breast Cancer? A Cross-Lagged Panel Model Analysis. *International Journal of Environmental Research and Public Health*, *19*(15), 1–15. https://doi.org/10.3390/ijerph19159260

21. Pilipović, K., Jurišić Grubešić, R., Dolenec, P., Kučić, N., Juretić, L., & Mršić-Pelčić, J. (2023). Plant-Based Antioxidants for Prevention and Treatment of Neurodegenerative Diseases: Phytotherapeutic Potential of Laurus nobilis, Aronia melanocarpa, and Celastrol. *Antioxidants*, *12*(3). https://doi.org/10.3390/antiox12030746

22. Sánchez-Vidaña, D. I., Po, K. K. T., Fung, T. K. H., Chow, J. K. W., Lau, W. K. W., So, P. K., Lau, B. W. M., & Tsang, H. W. H. (2019). Lavender essential oil ameliorates depression-like behavior and increases neurogenesis and dendritic complexity in rats. *Neuroscience Letters*, *701*(February), 180–192. https://doi.org/10.1016/j.neulet.2019.02.042

23. Sattayakhom, A., Wichit, S., & Koomhin, P. (2023). The Effects of Essential Oils on the Nervous System: A Scoping Review. *Molecules*, *28*(9), 1–23. https://doi.org/10.3390/molecules28093771

24. Shutes, J. & Galper, A. (2022). *Aromatherapy Companion: A Portable Guide to Blending Essential Oils and Crafting Remedies for Body, Mind, and Spirit*. Fair Winds Press

25. Singewald, N., Schmuckermair, C., Whittle, N., Holmes, A., & Ressler, K. J. (2015). Pharmacology of cognitive enhancers for exposure-based therapy of fear, anxiety, and trauma-related disorders. *Pharmacology and Therapeutics*, *149*, 150–190. https://doi.org/10.1016/j.pharmthera.2014.12.004

26. Terzi, A., Gebressalesie, H. T., & Yildirim, Y. (2020). Aromatherapy interventions in older adult: A systematic review. *Journal of*

Traditional Medical Complementary Therapies, *3*(3), 293–303. https://doi.org/10.5336/jtracom.2019-72546

27. Yousefsani, B. S., Jamshidi, A., & Dadmehr, M. (2021). Herbal medicine in cardiovascular medicine: a discussion of herbal medications described by the Persian physician, Avicenna. *European Heart Journal*, *42*(32), 3037–3039. https://doi.org/10.1093/eurheartj/ehab053

Therapeutic Neuroplasticity: The Neuroscience of Psychotherapy

Selin Sert Yurdakul

1. Introduction to the Relationship Between Psychotherapy and Neuroplasticity

The human brain is capable of remarkable transformations throughout life. Neuroplasticity allows experiences, learning, development and recovery to reshape neural networks that control cognition and behavior. However, maladaptive changes in brain structure and function resulting from trauma can lead to mental disorders. Psychotherapy helps to guide these changes in a positive direction. It's a valuable tool for mental health experts. Just as trauma impacts the nervous system, psychotherapy can also change the brain, therefore, we can assume that it's a neurobilogical treatment. (Bijedić & Ercegović, 2024).

Researchers have used various methods to study brain changes from psychotherapy. They examined psychiatric disorders, therapy's effects, and specific interventions. Neuroimaging has been crucial, but it can't prove direct cause-and-effect. An innovative approach looks for correlations between therapy techniques and their effects on brain correlates in different regions. By studying the interaction between therapy and brain wiring, researchers gain insights into neuroplastic changes. This approach sheds light on how psychotherapy fosters neuroplasticity, revealing its transformative mechanisms (Tatti et al., 2022).

Psychotherapy improves cognitive, emotional, and sensory processes during treatment. It enhances the functioning of critical regions in the brain's complex networks. Researchers use advanced imaging to understand these networks. However, studying network changes has limits. It doesn't always show observable changes in brain structure. Still, this research is key. It helps us understand the mind and improves therapy (Price & Duman, 2020).

Researchers have proposed theoretical frameworks linking certain psychological processes to neurobiological brain changes leading to functional outcomes. These models suggest a sequence: first, a cognitive process changes, then brain structure does. They also predict that steps in between physiological events and predicted alterations in brain neuroanatomy should be consistent with these models (Faustino, 2022).

2. Theoretical Frameworks Linking Psychotherapy and Neuroplasticity

Many theories provide a conceptual framework for the interplay between psychotherapy and neuroplasticity. Understanding these concepts helps us developing a more universal theory.

At larger scales, neuroscience guides the study the relationship between psychotherapy and neuroplasticity. Feedback loops connect the nervous system, body, and environment. These loops operate across various complexity levels. They range from subcellular, cellular, microcircuits, microcircuits to macro networks. As obeying such organizational rules mental states activate corresponding specific neurocircuits. By this way conditionally, modulates (enhancing or attenuating) their functional efficiency and plasticity. (Solomon & Siegel, 2017).

The relationship between psychotherapy and neuroplasticity can be understood through multiple theoretical frameworks that offer different perspectives on how the mind, brain, and behavior interact. At the micro level of abstraction, five paradigms can provide a richer understanding:

1. Representationalist Paradigm: This concept sees the brain as a computational system that represents symbolic information. It fits into classical cognitive science, where mental processes follow rules to manipulate symbols according to formal rules. The theory suggests psychotherapy can change brain flexibility by altering these symbols. So, in therapy, interventions change "symbolic codes," affecting thoughts and feelings (Pylyshyn, 1984).
2. Enactive Paradigm: This theory suggests cognition arises from environmental interactions, not just internal symbols (Varela et al., 1991). It views neuroplasticity as a change in response to the interaction with the environment (therapist), rather than just a shift in internal representation of symbols. Thus, therapy can promote neuroplastic changes by encouraging new ways of interactions with the world. This process

reshapes thoughts and feelings through embodied experiences in therapy sessions.

3. Symbolic and Subsymbolic Computational Approaches: Classical AI uses symbolic methods, seeing mental processes as symbols and rules. In contrast, subsymbolic methods, like connectionist models, distribute processes across simple network units, much like brain neurons (Rumelhart et al., 1986). Connectionist and complex dynamical systems theories focus on activation patterns in neural networks and avoid explicit symbolic representations. For example, in psychodynamic therapy, they suggest treatment changes neural activation patterns (subsymbolic) rather than higher-level symbolic structures.
4. Mind-Brain Correspondence Interpretation: This theory has two parts. First, "equivalence postulation" says mental and brain states are the same. Second, "analogical postulation" indicates mental processes correspond to brain processes in a way that is not strictly same but follows an analogical structure. Moreover, constructivist views suggest our minds actively create knowledge and understanding (Glasersfeld, 1995). These concepts imply that psychotherapy could spark neuroplasticity by changing how mental and brain states relate. This leads to more adaptive cognitive and emotional responses.
5. Free Energy Principle: Karl Friston developed the Free Energy Principle (FEP). It explains how biological systems, like the brain, aim to reduce "free energy." This term refers to prediction error or the gap between expected and actual sensory inputs. The brain acts as a predictive machine, constantly updating its models to avoid surprises and stay balanced (Friston, 2010). Reducing free energy links closely to neuroplasticity. This process requires the brain to adjust its connections and circuits based on new experiences. Recent studies have connected the FEP to psychotherapy's effects on the brain. Psychotherapy targets unhealthy beliefs, emotions, and behaviors. It aims to update the brain's models, lowering prediction errors and psychological distress. This process fosters neuroplasticity, leading to healthier thinking and feeling patterns.

These paradigms offer different views on psychotherapy and neuroplasticity. The representationalist and enactive paradigms explain cognitive change in distinct ways. This affects how we study psychotherapy. Meanwhile, the symbolic and subsymbolic approaches show how neural processing changes at the neural level. The **mind-brain correspondence interpretation** opens pathways for linking these changes to subjective experiences. Future research can use these frameworks to form hypotheses. It can employ methods like neuroimaging, computational modeling, and qualitative analysis to explore psychotherapy's effects.

3. Introduction to Different Psychotherapy Modalities: Neurobiological Foundations of One-Person and Two-Person Therapies

Psychotherapy is often seen as "just talking," yet it effectively reduces mental health symptoms, especially compared to no treatment (Leichsenring, 2022). The exact workings of psychotherapy are not fully understood, but it's clear that it causes various neurobiological changes. Advances in neuroimaging now allow us to study these effects in terms of brain chemistry (like neurotransmitter changes) and structure (such as synaptic growth and new neuron formation). This knowledge can potentially help improve psychotherapy and make it more widely available (Robinson et al., 2020).

Psychotherapy includes various techniques. We'll briefly explain some to enhance understanding. Therapy modalities can be broadly divided into two categories: "one-person" and "two-person." Each has unique effects and neuroscientific principles (Ozakkas, 2018 a). One-person therapies include Cognitive Behavioral Therapy (CBT), Mindfulness-Based Cognitive Therapy (MBCT), Acceptance and Commitment Therapy (ACT), Humanistic Therapy, and Schema Therapy. These focus on individual's internal processes like cognitive restructuring, emotion regulation, and self-reflection. They aim to foster internal change, enhancing self-awareness, emotional resilience, and growth, often without relying on the therapist-client relationship. In CBT, the therapist helps the patient identify and change harmful thoughts and behaviors. This method uses neuroplasticity to form new neural pathways and reduce maladaptive patterns (Beck, 2011). It activates the prefrontal cortex to improve control and downregulate the amygdala activity to alleviate anxiety and depressive symptoms (Siegle et al., 2007).

Therapies such as Interpersonal Psychotherapy (IPT), Emotion-Focused Therapy (EFT), Gestalt, and Existential therapy focus on the relationship between therapist and client. They see this relationship as a central mechanism to managing emotions and bringing about change. By leveraging relational dynamics, these therapies aid in healing (Perls et al., 1951; Yalom 1980).

Psychodynamic psychotherapies are often called two-person therapies. This is because they focus on the dynamic relationship between therapist and client as a central mechanism for therapeutic change. These therapies stem from Freudian psychoanalysis. They aim to uncover unconscious conflicts, emotions, and past experiences that influence current behaviors and relationships (Gabbard, 2017). Previously, people thought analysts should be neutral and passive, like a blank page. Therapy was often seen as a solo journey, typically one-on-one. However, recent developments in psychoanalytic literature have shifted this view (Ozakkas, 2018 b). Modern therapy emphasizes the therapist and client interaction, using techniques such as free association, transference, and countertransference to explore and address unconscious thoughts and feelings (Leichsenring 2022). Transference involves the client projecting feelings and attitudes from past relationships onto the therapist. At the same time, countertransference refers to the therapist's emotional responses to the client, which can offer valuable insights into the client's unconscious processes (Lemma, 2016). By analyzing these dynamics within the therapeutic relationship, psychodynamic therapy aims to promote self-awareness, emotional insight, and psychological growth (Shedler, 2010). Psychodynamic psychotherapies affect neural networks involved in self-referential thinking, memory, and emotional regulation, including the default mode network (DMN) and limbic system structures (Solms, 2013).

IPT focuses on improving interpersonal skills, addressing role transitions, resolving conflicts, and dealing with grief. It modifies neural circuits that manage emotion regulation and social cognition, particularly in the limbic system and mirror neuron network (Stuart, 2006). Meanwhile, EFT centers on emotional experience and expression in the context of attachment relationships. It seeks to change maladaptive emotions by promoting deeper emotional processing, engaging the amygdala, anterior cingulate cortex, and insula (Greenberg, 2015). Both IPT and EFT activate neural circuits related to social cognition, empathy, and emotion regulation, such as the mirror neuron system, the anterior insula, and the anterior cingulate cortex. These areas are activated during interpersonal engagement and emotional processing.

The differences between one-person and two-person therapies lie in their different therapeutic focuses and mechanisms of action. One-person therapies focus on internal change, aiming to shift cognitive and emotional processes. They often use cognitive restructuring or insight. In contrast, two-person therapies aim for healing through relationships, using the therapeutic bond to foster change. This involves affective attunement, empathy, and emotional connection. Such therapies differ in how they affect neural networks related to emotion, cognition, and social interaction. Both methods tap into different neural pathways for mental well-being, showing psychotherapy's impact on brain health and diverse paths to wellness (Cozolino, 2016).

Body-oriented psychotherapies, like Somatic Experiencing and Sensorimotor Psychotherapy, Integral Somatic Psychology, and Bodynamic and Bioenergetic Analysis, fall under two-person therapies. Yet, they also include aspects of one-person approaches. These therapies integrate body and mind, seeing psychological issues are often deeply connected to bodily experiences. Healing involves working through both mental and somatic levels.

In body-oriented psychotherapies, the relationship between therapist and client is crucial. The therapist tunes into the client's bodily sensations, movements, and non-verbal cues. This information guides the therapist in helping the client become more aware of, regulate, and express their bodily experiences. Such relationships are key to change (Ogden et al., 2006). Moreover, the therapist creates a safe environment for clients to explore and express emotions and traumas stored in their bodies with coregulation. This process helps regulate the client's autonomic nervous system responses (Levine, 2010).

Body-oriented psychotherapies share traits with one-person therapies. They focus on a person's bodily awareness and self-regulation. Techniques like mindfulness of bodily sensations, grounding exercises, and somatic tracking help clients manage distress and build resilience. This is similar to how Cognitive Behavioral Therapy (CBT) or psychoanalysis fosters self-reflection and insight (Van der Kolk, 2014).

These therapies activate brain areas linked to self-awareness and social connection. They engage the insula, which connects bodily sensations with emotions, and the vagus nerve, crucial for regulating the autonomic nervous system. They also stimulate the right hemisphere, which processes non-verbal cues and emotions, and the prefrontal cortex, which modulates stress responses (Schore, 2012). By focusing on bodily experiences and incorporating relational dynamics, body-oriented therapies bridge the gap between one-person and two-person therapy models.

4. Neurobiological Changes in the Brain Through Psychotherapy

Psychotherapy combines top-down mental tasks (conscious, deliberate cognitive activities) with indirect influences on concurrent psychophysiological processes. These processes are proposed to impact neuroplasticity by inducing changes in specific structural and functional neural characteristics. This process is thought to operate at multiple levels within the brain's architecture, from localized changes in particular neural circuits to broader, more integrated modifications across large-scale neural networks.

4.1. Top-Down Mental Tasks and Neuroplasticity

Psychotherapy often uses top-down methods. These include cognitive restructuring in CBT, emotional regulation in EFT, and self-reflection in psychoanalysis. Such methods work on higher-level cognitive functions. They use conscious efforts to influence lower-level neural processes. Engaging specific brain circuits repeatedly strengthens or rewires connections. This can lead to neuroplastic changes (Siegel, 2012). For example, CBT reduces amygdala activity. It also boosts connectivity between the prefrontal cortex and limbic structures, improving emotion regulation through repetitive cognitive reappraisal tasks (Davidson & McEwen, 2012).

4.2. Indirect Psychophysiological Influences

Psychotherapies can indirectly boost neuroplasticity by altering the autonomic, endocrine, and immune systems. For instance, body-oriented therapies like Somatic Experiencing and dance movement therapies use physical activity. These activities can affect physiological states like heart rate variability and cortisol levels. This, in turn, affects brain plasticity by modulating stress responses (Levine, 2010). These changes lead to local, selective neuroplastic adaptations as specific neural networks respond to psychophysiological changes triggered by the therapeutic intervention (Porges, 2011).

4.3. Local and Selective Neuroplastic Changes

The model suggests psychotherapy causes neuroplastic changes that are specific to certain brain areas and targeted toward particular pathways. These changes can be lasting or temporary, depending on the therapy's type and intensity. For example, Interpersonal Therapy (IPT) aims to modify social behaviors and attachment. Initially, it may lead to changes in brain areas linked to social understanding, like the medial prefrontal cortex (Fonagy, 2010). Over time, these changes can promote more complex brain interactions. The local adaptations will integrate into larger neural networks. These networks control global brain functions.

4.4. Integration Across Mesoscopic Neural Networks

Psychotherapy boosts interactions within extensive mesoscopic networks. These networks control multiple brain regions and are globally responsive to overt, regulated mental activities. It enhances thinking and feeling by improving connectivity and harmony in the brain. According to complex systems theory, the brain adapts to repeated mental tasks in therapy. This leads more resilient and adaptive network configurations (Bassett & Sporns, 2017)

4.5. Macroscopic Dynamics and Circuitry Constraints

Psychotherapy can bring significant local and global changes, but these changes aren't isolated. They're affected by the brain's large-scale activity, which occurs over seconds to minutes. This global activity is limited by circuitry-design constraints, meaning that changes in specific areas are modulated by the broader network's structure and function by the brain's circuitry. (Deco et al., 2013). Thus, while psychotherapy can produce targeted neuroplastic changes, it must also operate within the limits imposed by the brain's inherent connectivity and dynamic properties.

Understanding psychotherapy's impact on neuroplasticity at multiple levels—from local neural circuits to large-scale brain networks—can provide new insights into how different therapeutic modalities contribute to mental health improvements. Future research can explore these mechanisms using advanced neuroimaging techniques, computational modeling, and longitudinal studies to assess the dynamic interplay between local and global neuroplastic changes.

5. Neurochemical Changes Induced by Psychotherapy

Psychotherapy is a key psychological intervention. It also significantly influences the brain's neurochemistry and neuroplasticity. This section examines how psychotherapy impacts neurochemical systems. These systems include neurotransmitters, neuropeptides, and hormones. The aim is to boost brain adaptability and improve mental health.

5.1. Serotonin: Modulation of Mood, Anxiety, and Synaptic Plasticity

Serotonin (5-HT) is key for regulating mood, anxiety, and emotion processing. It's synthesized in the raphe nuclei and affects the prefrontal cortex, hippocampus, and amygdala, which regulate emotions and cognitive functions. Higher serotonin levels are linked to positive emotions and social behavior. Studies show those with low moods have lower serotonin function. In contrast, a better mood links to higher serotonin. The editorial suggests that improving mood might also raise serotonin, indicating a two-way relationship (Young, 2007). Researchers hypothesize that cognitive behavioral therapy (CBT) boosts serotonin through better emotion control and reduced stress. In a study, bulimia patients showed increased serotonin transporter density after CBT. This suggests CBT enhances serotonin function in these patients, possibly improving their mental health (Brambilla et al., 2010).

Some researchers suggest psychotherapy can lead to changes in gene expression related to serotonin through learning. It alters synaptic connections and changes neuron structure. Serotonin boosts synaptic plasticity by increasing growth and density of dendritic spines. It also controls the release of brain-derived neurotrophic factor (BDNF). Thus, psychotherapy might raise serotonin levels. This increase could enhance neurogenesis in the hippocampus contributing to the recovery from mood and anxiety disorders (Hirvone et al., 2011).

A study on obsessive-compulsive disorder (OCD) found that successful treatment with cognitive behavioral therapy (CBT) or SSRIs increased brain-wide serotonin synthesis capacity. This suggests CBT might enhance overall serotonergic tone. Additionally, in the study, the initial serotonin levels in the raphe nuclei were correlated positively to clinical improvement, regardless of whether the treatment was CBT or medication (Lissemore et al., 2018)

In a Finnish study showed that MDD patients who had psychodynamic psychotherapy increased their serotonin 5-HT1A receptor density in PET scans. However, those on fluoxetine did not. This suggests psychotherapy and medication affect serotonin receptors differently. This might explain why psychotherapy has lower relapse rates than antidepressants. (Hirvone et al., 2011).

5.2. Dopamine: Reward Processing, Motivation, and Cortical Remodeling

Dopamine (DA) is a neurotransmitter made in the brain's ventral tegmental area and substantia nigra. It is key for reward, motivation, and decision-making. Psychotherapy, like CBT and psychodynamic therapy, can boost dopamine pathways. These therapies do this by involving patients in rewarding meaningful activities and improving emotional control. Neuroimaging studies show they increase dopamine release and receptor activity in the mesocorticolimbic pathway. This pathway includes the prefrontal cortex, striatum, and nucleus accumbens (Goldapple et al., 2004).

Cervenka et al. (2012) found that cognitive behavioral therapy (CBT) significantly reduced social anxiety symptoms. This reduction was linked to changes in dopamine D2 receptor binding in the medial prefrontal cortex and hippocampus. Patients who responded well to CBT showed increased receptor binding in these areas. This suggests a direct link between symptom improvement and changes in brain neurotransmission. The findings highlight dopamine's role in social behavior and anxiety. They also support that psychotherapy can enhance neuroplasticity by modifying neurotransmission systems vital for emotional and social processing (Cervenka et al., 2012).

Dopamine is vital for learning and motivation. Dopamine plays a central role in **prediction error** signaling when there's a mismatch between expectations and reality. This is key in reward learning and fear extinction learning, where dopamine signals help update and correct dysfunctional expectations. In psychotherapy, especially exposure therapy, breaking false beliefs (like fearing social situations) helps patients learn more adaptive responses. Psychotherapy-induced increases in dopamine levels can help reinforce positive behavioral changes and cognitive restructuring by promoting long-term potentiation and enhancing the brain's capacity to adapt to new learning experiences (Papalini, 2020).

5.3. Noradrenaline and Cortisol: Arousal, Attention, Synaptic Modulation and Stress Regulation

Noradrenaline, mainly synthesized in the locus coeruleus, controls arousal, attention, and stress response. It's vital for long-term potentiation (LTP) in the hippocampus. Here, it acts through β-receptors, especially in the dentate gyrus. This boosts synaptic strength, contributing to memory consolidation and learning. It also improves the encoding of emotionally important memories by enhancing signal-to-noise ratios in neuronal circuits (Sara, 2009).

Hyperactivity in the norepinephrine system often leads to increased anxiety and alertness in PTSD and generalized anxiety disorder (GAD). Prolonged stress can disrupt the noradrenergic system, causing anxiety disorders, PTSD, and depression. Moreover, stress increases noradrenaline, which enhances memory retention of emotionally charged events, contributing to PTSD (Sanacora et al., 2004).

Psychotherapy can lower noradrenaline hyperactivity, enhancing cognitive flexibility and reorganizing neural circuits. This aids in breaking maladaptive patterns of thought and behavior by improving mood and cognition. It's possible that psychotherapy can indirectly lower noradrenaline levels, but direct evidence is lacking. More studies are needed to explore noradrenaline levels before and after psychotherapy without medication (Ambade et al., 2009).

Cortisol is a stress hormone released by the adrenal glands. It's regulated by the HPA axis. This axis is crucial for the body's stress response. When it malfunctions, it can lead to disorders like depression, anxiety, and PTSD. Psychotherapies such as CBT, MBSR, and ACT can normalize HPA axis function. They do this by reducing perceived stress and enhancing emotional control (Gaab et al., 2005).

Chronic high cortisol from HPA axis issues can harm brain adaptability. It reduces new neuron growth and dendrite development, especially in the hippocampus. Psychotherapy, however, can lower cortisol levels. This action then shields the brain from stress, encouraging synapse growth and new neuron development. Such protection is vital for keeping mental sharpness and emotional strength in those with stress-related disorders (Lupien et al., 2009).

5.4. Glutamate and GABA: Balancing Excitation and Inhibition, Understanding Anxiety and Depression

Glutamate is the brain's main excitatory neurotransmitter. It's vital for learning, memory, and synaptic plasticity. Studies show patients with major depressive disorder (MDD)) exhibit region-specific alterations in glutamate levels. Specifically, glutamate is lower in the prefrontal and anterior cingulate cortex. However, it's higher in the occipital cortex. This indicates a possible imbalance in neurotransmission, crucial for mood regulation and cognitive function. Thus, it's suggested that MDD involves a disruption in this excitatory neurotransmission, especially in circuits that control mood and cognition (Abdallah et al., 2014).

Abdallah et al. (2014) found that effective cognitive behavioral therapy (CBT) for major depressive disorder (MDD) significantly lowered occipital glutamate levels. This decrease correlated with less severe depression. However, GABA levels showed no significant change. The study suggests CBT may trigger neurobiological changes in the brain, particularly in glutamate regulation. This could serve as a biomarker for treatment response to MDD.

Glutamate is key for synaptic plasticity, aiding in long-term potentiation (LTP). LTP strengthens synapses, supporting learning and memory. By normalizing glutamate, psychotherapy reduces excitotoxicity, increase cognitive flexibility, and aids neural reorganization. This process alleviates symptoms and improves cognition (Plitman et al., 2014).

Gamma-aminobutyric acid (GABA) is the brain's main inhibitory neurotransmitter. It lowers neural excitability and maintains synaptic stability. GABA activity is key in managing anxiety, stress responses, and sleep. Reduced GABA activity is related to anxiety and mood disorders, including depression and epilepsy. Drugs that boost GABA activity are often used to improve mood and reduce anxiety. This shows GABA's important role in regulating anxiety (Streeter et al., 2010)

Imaging studies show low GABA levels in the occipital cortex are key in melancholic depression. GABA helps calm hyperactive neural circuits. This might explain why MDD patients are more sensitive to stress and negative emotions. In a 2010 study by Streeter et al., a link was found between higher thalamic GABA levels and lower anxiety scores, as per the STAI-State anxiety scale. This provides strong evidence that improvements in anxiety control is tied to enhanced GABA activity in the brain. Therapies like mindfulness-based cognitive therapy (MBCT) and relaxation techniques can increase GABA levels, promote calmness, and reduce anxiety (Siegel, 2012).

GABA is vital for neuroplasticity. It balances excitation and inhibition in the brain. This balance is key for synaptic stability and preventing over-excitation. Psychotherapy boosts GABA transmission. This, in turn, promotes synaptic balance, reduces neural overactivity, and supports cognitive and emotional flexibility, enhancing mental health (Goddard et al., 2019).

5.5. Oxytocin: Social Bonding, Attachment, and Neuroplasticity

Oxytocin is a neuropeptide made in the hypothalamus and released by the pituitary gland. It's key for social bonding, attachment, trust, and empathy. It affects brain areas like the amygdala, prefrontal cortex, and nucleus accumbens, which are crucial for emotional regulation and social cognition. Therapies that focus on the relationship between therapist and patient, such as Emotion-Focused Therapy (EFT) and psychodynamic therapies, are believed to boost oxytocin levels. This increase enhances feelings of trust and safety (Gonzalez-Liencres et al., 2013).

Oxytocin is vital for brain adaptability. It affects synaptic transmission and neuronal excitability, enhancing the brain's ability to process and adapt to social stimuli. It lowers the activation threshold for LTP in areas like the hippocampus and olfactory bulb. Additionally, it enables quick synaptic changes through disinhibition and strengthen neural circuits for social behaviors, such as maternal care. Furthermore, oxytocin works with serotonin to induce long-term synaptic modifications, highlighting its potential in psychotherapy and treating mental disorders (Froemke, 2021)

Oxytocin boosts neuroplasticity by enhancing synaptic connections and fostering new dendritic spines in areas tied to social and emotional processing. It also dampens amygdala activity, reducing anxiety and improving emotional control. This creates a more stable environment for neuroplastic changes. Higher oxytocin levels during therapy can strengthen the therapeutic alliance, aid emotional healing, and promote cognitive and behavioral change (Carter, 2014).

A strong therapeutic alliance is key to successful psychotherapy. It often matters more than the type of therapy. Oxytocin release is crucial for this bond. Higher oxytocin levels lead to better alliances and outcomes.

Conversely, lower levels are linked to worse responses and less improvement in depression symptoms (Guina, 2024).

5.6. Brain-Derived Neurotrophic Factor (BDNF): Neurogenesis and Cognitive Function

Brain-derived neurotrophic factor (BDNF) is vital for neuron growth and survival and differentiation. Therapies like cognitive behavioral therapy (CBT), which engage the mind and process emotions, boost BDNF levels. This increase reflects better cognitive function and neuroplasticity (Sen et al., 2008). Moreover, psychotherapy that enhances cognition and emotional regulation can increase BDNF production. This, in turn, improves neurogenesis and synaptic plasticity, key for recovering from mood disorders and cognitive impairments. (Duman & Monteggia, 2006).

A review by Claudino et al. (2020) found that individual therapy often boosts BDNF levels in patients with PTSD, bulimia, insomnia, and borderline personality disorder. This increase is correlated to symptom reduction. However, therapy alone didn't consistently raise BDNF levels in those with major depressive disorder. This suggests that combining therapy with medication might be better for these patients. Notably, EMDR and Exposure Therapy, especially with added physical activity, significantly raised BDNF in PTSD patients. BDNF can be a potential biomarker for treatment response but more research is needed to confirm these findings across different mental disorders (Claudino et al., 2020).

Research indicates that BDNF gene methylation could also be an essential biomarker for psychotherapy outcomes, especially in patients with trauma-related or personality disorders like borderline personality disorder. Moreover, a reduction in methylation after therapy suggests that psychotherapy may not only alter brain function but also reverse epigenetic changes linked to mental health disorders (Piotrkowicz et al., 2021).

Psychotherapy significantly impacts various neurochemical systems, promoting neuroplasticity and, aiding mental health recovery. By adjusting neurotransmitters, neuropeptides, and hormones, boosts synaptic connections and encourages neural reorganization. Moreover, it reduces the harmful effects of chronic stress. Understanding these processes reveals how therapy works and helps in creating foundation for developing more targeted and effective treatments.

6. The Impact of Psychotherapy on Long-Term Potentiation (Ltp)

Long-term potentiation (LTP) boosts learning and memory by strengthening synapses with repeated stimulation. Psychotherapy, especially Cognitive Behavioral Therapy (CBT) and psychodynamic therapies, is thought to enhance LTP. These therapies promote cognitive and emotional changes that improve synaptic plasticity.

CBT targets maladaptive thought patterns, encouraging cognitive restructuring. It activates brain areas involved in cognitive control and emotional regulation, like the prefrontal cortex and hippocampus. By repeatedly engaging these circuits, CBT is believed to promote LTP and open new learning pathways, aiding in behavior change and symptom relief (Millan et al., 2021).

Psychodynamic therapies focus on unconscious conflicts and emotional expression within the therapeutic relationship. This can improve emotional regulation and lower stress, aiding synaptic plasticity for long-term potentiation (LTP). The therapeutic alliance and emotional processing activate neural circuits for attachment, self-reflection, and emotional regulation. This may support LTP in areas like the medial prefrontal cortex and hippocampus (Javanbakht, 2019).

Psychotherapy may enhance long-term potentiation (LTP) by altering brain circuits, strengthening synapses, and supporting long-term psychological resilience. It does this by adjusting neurotransmitter levels, improving emotional control, and promoting adaptive learning and cognitive changes. Examining how different therapies affect LTP reveals the neurobiological mechanisms behind their success. This insight helps tailor treatments to individual needs.

7. Generalized Effects of Psychotherapy on Brain Structure and Function

Psychotherapy, including Cognitive Behavioral Therapy (CBT) and psychodynamic therapies, brings significant structural and functional changes to the brain. It enhances emotional regulation, cognitive flexibility, and overall mental health. These benefits span various therapies, showing that all can promote neuroplasticity and adaptive brain changes, despite their different methods (Wenger et al, 2021).

Research highlight therapies like CBT and psychodynamic approaches follows Karl Friston's free energy principle. CBT, for example, helps patients tackle and reshape distorted thoughts. This process leads to new predictions and lessens cognitive biases. Such changes are linked to brain structure improvements in areas vital for decision-making and emotional regulation like the prefrontal cortex and hippocampus (Gunia, 2024). Similarly, psychodynamic therapies aim to uncover unconscious conflicts and unresolved emotions. They help revise internal models. This promotes connectivity between the anterior cingulate cortex and the medial prefrontal cortex. These brain regions are crucial for emotional regulation and self-awareness (Buchheim et al., 2012).

Solms' work in neuropsychoanalysis builds on the Free Energy Principle. He suggests that psychotherapy links different brain regions, like the amygdala, basal ganglia, and prefrontal cortex. This improves emotional regulation and cognitive function. Moreover, psychotherapy aids in emotional regulation by activating the brain's reward and motivational systems. This process may boost neuroplasticity, creating a better balance between emotional and cognitive activities (Solms, 2021).

7.1. Structural Changes in the Brain

Recent research shows psychotherapy can change the brain's structure. For example, it can increase gray matter and thickness. A meta-analysis by Patel et al. (2022) found that therapies like CBT and psychodynamic therapy boost gray matter in the prefrontal cortex and hippocampus. These brain areas are vital for memory, executive function, and emotional regulation. This suggests psychotherapy improves the brain's ability to handle emotions and stress.

For instance, Månsson et al. (2016) found that CBT increased gray matter in the DLPFC and hippocampus of social anxiety patients, reducing their anxiety. Similarly, Kuhn et al. (2018) noted that psychodynamic therapy increased thickness in the mPFC in depressed patients, This was correlated with improvement in their symptoms. These studies suggest psychodynamic therapy enhances neuroplasticity in areas linked to self-reflection and emotion.

Mancke et al. (2018) examined BPD patients after 12 weeks of DBT. They found that DBT boosted gray matter in the ACC and inferior frontal gyrus, areas vital for emotion and cognitive control. This growth matched improvements in emotional regulation and cognitive functions like mentalizing. The study underscores that therapy can alter brain structures involved in emotional regulation, supporting DBT's effectiveness in BPD treatment.

7.2. Functional Changes in Brain Networks

Psychotherapy not only changes the size or activity of certain brain regions but also alters their connectivity, improving symptoms. It also changes brain networks involved in emotional regulation, self-reflection, and cognitive control. Studies using fMRI have shown that psychotherapy affects the default mode network, salience network, and fronto-limbic circuits. These are crucial for emotional balance and stress management (Marwood et al., 2018).

Trauma-focused therapies, such as EMDR, lower connectivity between the visual cortex and left temporal lobe. This reduction may lessen flashbacks. They also boost connectivity between the right temporal lobe and superior frontal gyrus, aiding in the control of intrusive thoughts. Meanwhile, Dialectical Behavior Therapy (DBT) improves connectivity in patients with borderline personality disorder. It especially affects areas tied to executive functioning and attention, like the dorsal anterior cingulate cortex (Guina, 2024).

Barsaglini et al. (2014) reviewed how psychotherapy affects brain changes in various mental disorders. For instance, in OCD, CBT lowers activity in the orbitofrontal cortex and caudate nucleus, normalizing brain function similar to pharmacotherapy. In depression, therapy boosts fronto-limbic circuitry, exerting a "top-down" regulatory effect on emotional processing, distinct from the "bottom-up" changes seen with medication. Panic disorder patients benefit from reduced limbic hyperactivity in the amygdala and insula, improving emotional regulation. Exposure-based therapies for PTSD and specific phobias also calm fear-related brain activity like the amygdala. In schizophrenia, cognitive remediation therapy improves fronto-temporal connectivity and cognition. These studies show psychotherapy's role in altering brain function and reducing symptoms in various mental health issues.

Psychodynamic therapy affects subcortical structures, like the amygdala and basal ganglia. It also impacts their connections to the prefrontal cortex.

This improves the integration between emotion and cognition. (Solms, 2013). This enhanced brain integration improves emotional regulation and decision-making, vital for healthy functioning. For example, psychodynamic therapy boosts connections in the brain related to self-awareness and emotional regulation. Buchheim et al. (2012) used fMRI to study this. They found that patients with major depressive disorder who went through psychodynamic therapy showed increased connectivity between the anterior cingulate cortex (ACC), the insula, and the dorsolateral prefrontal cortex (DLPFC). This improved connectivity linked to fewer depressive symptoms. Thus, it supports psychodynamic therapy's role in enhancing adaptive neural network function.

Psychodynamic therapy aims to make unconscious thoughts conscious. This process alters the brain's reward and motivation systems, especially in the ventral striatum and dopamine pathways. As a result, patients gain better control over their emotional responses and behaviors, leading to lasting psychological change. (Solms, 2021).

Psychotherapy, whether CBT or psychodynamic therapy, alters brain structure and function, boosting its therapeutic effects. Psychotherapy also impacts neuroplasticity, neurogenesis, and neurotoxicity through changes in the HPA axis and serotonergic systems. Mindfulness-based cognitive therapy (MBCT) triggers genetic changes, particularly in single nucleotide polymorphisms (SNPs), that lower inflammation and oxidative stress, factors linked to mental illness. With psychotherapy, emotional-cognitive reintegration and consolidation is achieved. Cellular and neural plasticity changes mediate this process. By enhancing neuroplasticity, psychotherapy increases the brain's flexibility and resilience against mental disorders (Guina, 2024).

8. Takeaways

1. **Neuroplasticity as a Foundation for Psychotherapy:** Psychotherapy capitalizes on neuroplasticity to restructure maladaptive neural circuits, facilitating recovery from psychological disorders.
2. **The Role of Neuroimaging and Theoretical Frameworks:** Advances in neuroimaging have enabled direct observation of brain changes during psychotherapy, with diverse theoretical models explaining the mechanisms of neuroplasticity.

3. **Distinct Modalities of Psychotherapy**: One-person therapies focus on individual cognitive and emotional processes, while two-person therapies emphasize relational dynamics as core to psychological transformation.
4. **Neurobiological Mechanisms Underlying Different Therapies**: One-person therapies (e.g., CBT, MBCT, ACT) enhance cognitive control by modifying prefrontal and amygdala pathways, whereas two-person therapies (e.g., Psychodynamic Psychotherapies, IPT, EFT) modulate neural circuits involved in empathy and interpersonal connection.
5. **Body-oriented Psychotherapies as Integrative Approaches**: Body-oriented therapies integrate somatic awareness and relational dynamics, engaging neural systems responsible for self-regulation and social bonding.
6. **Neurochemical and Neuroendocrine Changes Induced by Psychotherapy**: Psychotherapy modulates multiple neurochemical systems, including serotonin, dopamine, norepinephrine, acetylcholine, glutamate, GABA, cortisol, and oxytocin, enhancing synaptic plasticity and mitigating the neurotoxic effects of chronic stress.
7. **Impact on Long-Term Potentiation (LTP)**: Cognitive behavioral and psychodynamic therapies strengthen synaptic connectivity via long-term potentiation, promoting emotional regulation and psychological resilience.
8. **Role of the Therapeutic Alliance**: The therapeutic relationship influences neuroplasticity by enhancing emotional regulation, reducing stress, and fostering neurochemical changes (like increased BDNF and oxytocin levels), supporting new neural connections and promoting psychological well-being.
9. **Structural Changes in the Brain**: Psychotherapy has been shown to produce measurable structural changes in the brain, such as increased gray matter volume and cortical thickness in regions like the prefrontal cortex and hippocampus leading to improvements in executive functioning, memory, and emotional regulation.
10. **Functional Changes in Brain Networks**: Psychotherapy enhances functional connectivity in critical brain networks such as the default mode network (DMN), salience network, and frontolimbic circuits, promoting adaptive emotional regulation and cognitive flexibility. It also improves the integration of emotional and mental functions by influencing subcortical and cortical connections, supporting better decision-making, emotion regulation, and psychological resilience.

Abbreviations

ACC: Anterior cingulate cortex
ACT: Acceptance and Commitment Therapy
AI: Artificial intelligence
BDNF: Brain-derived neurotrophic factor
BPD: Borderline personality disorder
CBT: Cognitive behavioral therapy
DBT: Dialectical behavioral therapy
DLPFC: Dorsolateral prefrontal cortex
DMN: Default mode network
EFT: Emotion-focused therapy
EMDR: Eye movement desensitization and reprocessing
FEP: Free energy principle
fMRI: Functional magnetic resonance imaging
GABA: Gama aminobutyric acid
GAD: Generalized anxiety disorder
HPA: Hypothalamus pituitary axis
IPT: Interpersonal psychotherapy
LPT: Long-term potentiation
MBCT: Mindfulness-based cognitive therapy
MBSR: Mindfulness-based stress reduction
MDD: Major depressive disease
mPFC: Medial prefrontal cortex
OCD: Obsessive-Compulsive disorder
PET: Positron emission tomography
PTSD: Posttraumatic stress disorder
SSRI: Selective serotonin reuptake inhibitor

Bibliography

1. Abdallah, C. G., Niciu, M. J., Fenton, L. R., Fasula, M. K., … , & Sanacora, G. (2014). Decreased occipital cortical glutamate levels in response to successful CBT and pharmacotherapy for major depressive disorder. *Psychotherapy and Psychosomatics, 83*(5), 298–307. https://doi.org/10.1159/000361078
2. Ambade V, Arora MM, Singh P, Somani BL, Basannar D. (2009) *Adrenaline, Noradrenaline and Dopamine Level Estimation in Depression:*

Does it Help? Med J Armed Forces India. Jul;65(3):216–20. https://doi.org/10.1016/S0377-1237(09)80006-3

3. Barsaglini, A., Sartori, G., Benetti, S., Pettersson-Yeo, W., & Mechelli, A. (2014*). The effects of psychotherapy on brain function: A systematic and critical review.* Progress in Neurobiology, 114, 1–14. https://doi.org/10.1016/j.pneurobio.2013.10.006
4. Bassett, D. S., & Sporns, O. (2017). Network neuroscience. *Nature Neuroscience, 20*(3), 353–364. https://doi.org/10.1038/nn.4502
5. Beck, J. S. (2011). *Cognitive behavior therapy: Basics and beyond* (2nd ed.). Guilford Press. www.guilford.com/p/beck4
6. Bijedić, M. & Ercegović, Z. (2024). Interdisciplinary Approach To Psychotherapy Through Neuroscience. *Research in Education and Rehabilitation*, 7(1), 146–158. https://rer.ba/index.php/rer/article/view/195
7. Brambilla, F., Dalle Grave, R., Calugi, S., Marchesini, G., Baroni, S., & Marazziti, D. (2010). Effects of cognitive-behavioral therapy on Eating Disorders: Neurotransmitter secretory response to treatment. *Psychoneuroendocrinology, 35*(5), 729–737. https://doi.org/10.1016/j.psyneuen.2009.10.014
8. Buchheim, A., Viviani, R., Kessler, H., Kächele, H., Cierpka, M., Roth, G., Georges, D., Brunner, R., & Taubner, S. (2012). Changes in prefrontal-limbic function in major depression after 15 months of long-term psychotherapy. *PLoS ONE,* 7(3). https://doi.org/10.1371/journal.pone.0033745
9. Carter, C. S. (2014). Oxytocin pathways and the evolution of human behavior. *Annual Review of Psychology, pp. 65*, 17–39. https://doi.org/10.1146/annurev-psych-010213-115110
10. Cervenka, S., Hedman, E., Ikoma, Y., Djurfeldt, D. R., Rück, C., Halldin, C., & Lindefors, N. (2012). *Changes in dopamine D2-receptor binding are associated with symptom reduction after psychotherapy in social anxiety disorder.* Translational Psychiatry, 2(5), e120. https://doi.org/10.1038/tp.2012.40
11. Claudino, F. C. A., Gonçalves, L., Schuch, F. B., Martins, H. R. S., & Rocha, N. S. (2020). *The effects of individual psychotherapy in BDNF levels of patients with mental disorders: A systematic review.* Frontiers in Psychiatry, *11*, 445. https://doi.org/10.3389/fpsyt.2020.00445
12. Cozolino, L. (2016). The neuroscience of psychotherapy: Healing the social brain (3rd ed.). W.W. Norton & Company. ISBN : 9780393712643

13. Davidson, R. J., & McEwen, B. S. (2012). Social influences on neuroplasticity: Stress and interventions to promote well-being. *Nature Neuroscience, 15*(5), 689–695. https://doi.org/10.1038/nn.3093

14. Deco, G., Jirsa, V., & McIntosh, A. R. (2013). Resting brains never rest: Computational insights into potential cognitive architectures. *Trends in Neurosciences, 36*(5), 268–274. https://doi.org/10.1016/j.tins.2017.12.007

15. Duman, R. S., & Monteggia, L. M. (2006). A neurotrophic model for stress-related mood disorders. *Biological Psychiatry, 59*(12), 1116–1127. https://doi.org/10.1016/j.biopsych.2006.02.013

16. Faustino, B. (2022). Minding my brain: Fourteen neuroscience-based principles to enhance psychotherapy responsiveness. *Clinical psychology & psychotherapy*, Jul;29(4):1254–1275. https://doi.org/10.1002/cpp.2719

17. Fonagy, P. (2010). The changing shape of the psychotherapy network: Implications for science and practice. *Psychological Medicine 40* (8), 1289–1300. https://doi.org/10.1080/02668731003590139

18. Friston, K. (2010). The free-energy principle: A unified brain theory? *Nature Reviews Neuroscience, 11*(2), 127–138. https://doi.org/10.1038/nrn2787

19. Froemke, R. C., & Young, L. J. (2021). *Oxytocin, Neural Plasticity, and Social Behavior.* Annual Review of Neuroscience, *44*, 359–381. https://doi.org/10.1146/annurev-neuro-102320-102847

20. Gaab, J., Rohleder, N., Nater, U. M., & Ehlert, U. (2005). Psychological determinants of the cortisol stress response: The role of anticipatory cognitive appraisal. *Psychoneuroendocrinology, 30*(6), 599–610. https://doi.org/10.1016/j.psyneuen.2005.02.001

21. Gabbard, G. O. (2017). *Psychodynamic Psychiatry in Clinical Practice* (5th ed.). American Psychiatric Publishing. SBN 978-1-58562-443-0

22. Glasersfeld, E. V. (1995). *Radical Constructivism: A Way of Knowing and Learning*. Routledge.

23. Goldapple K, Segal Z, Garson C, Lau M, Bieling P, Kennedy S, Mayberg H. Modulation of cortical-limbic pathways in major depression: treatment-specific effects of cognitive behavior therapy.

Arch Gen Psychiatry. 2004 Jan;61(1):34–41. https://doi.org/ 10.1001/archpsyc.61.1.34

24. Gonzalez-Liencres, C., Shamay-Tsoory, S. G., & Brüne, M. (2013). Oxytocin, empathy, and the human brain. *Neuroscience & Biobehavioral Reviews,* Sep;37(8):1537–48. https://doi.org/ 10.1016/j.neubiorev.2013.05.001
25. Greenberg, L. S. (2015). *Emotion-focused therapy: Coaching clients to work through their feelings*. American Psychological Association.
26. Guina, J., Kay, J., & Yatim, F. (2024). Neurobiological Foundations of Psychotherapies. In *Tasman's Psychiatry* (pp. 897–917). Springer International Publishing. https://doi.org/10.1007/978-3-030-51366-5_168
27. Hirvonen J, Hietala J, Kajander J, et al. (2011) Effects of antidepressant drug treatment and psychotherapy on striatal and thalamic dopamine D2/3 receptors in major depressive disorder studied with [11C] raclopride PET. J Psychopharmacol. 2011 Sep 9 https://doi.org/10.1177/0269881110376691
28. Javanbakht, A. (2019). A theory of everything: Overlapping neurobiological mechanisms of psychotherapies of fear and anxiety related disorders. Frontiers in Behavioral Neuroscience, *12*. https://doi.org/10.3389/fnbeh.2018.00328
29. Leichsenring, F., Steinert, C., Rabung, S., & Ioannidis, J. P. (2022). *The efficacy of psychotherapies and pharmacotherapies for mental disorders in adults: an umbrella review and meta-analytic evaluation of recent meta-analyses*. World Psychiatry, 21(1), 133–145. https://doi.org/ 10.1002/wps.20941
30. Lemma, A. (2016). *Introduction to the Practice of Psychoanalytic Psychotherapy* (2nd ed.). Wiley-Blackwell.
31. Levine, P. A. (2010). *In an Unspoken Voice: How the Body Releases Trauma and Restores Goodness*. North Atlantic Books.
32. Lissemore JI, Sookman D, Gravel P, Berney A, Barsoum A, Diksic M, …, Benkelfat C. (2018) *Brain serotonin synthesis capacity in obsessive-compulsive disorder: effects of cognitive behavioral therapy and sertraline*. Transl Psychiatry. 2018 Apr 18;8(1):82. https://doi.org/10.1038/s41398-018-0128-4

33. Lupien, S. J., McEwen, B. S., Gunnar, M. R., & Heim, C. (2009). Effects of stress throughout the lifespan on the brain, behavior, and cognition. *Nature Reviews Neuroscience, 10*(6), 434–445. https://doi.org/10.1038/nrn2639

34. Mancke, F., Schmitt, R., Winter, D., Niedtfeld, I., Herpertz, S. C., & Schmahl, C. (2018). *Assessing the marks of change: how psychotherapy alters the brain structure in women with borderline personality disorder.* Journal of Psychiatry & Neuroscience, *43*(3), 171–179. https://doi.org/10.1503/jpn.170132

35. Millan, M. J., Agid, Y., Brüne, M., Bullmore, E. T., Carter, C. S., … & Young, L. J. (2021). Cognitive dysfunction in psychiatric disorders: Characteristics, causes, and the quest for improved therapy. *Nature Reviews Drug Discovery, 20*(9), 646–664. https://doi.org/10.1038/nrd3628

36. Ogden, P., Minton, K., & Pain, C. (2006). *Trauma and the Body: A Sensorimotor Approach to Psychotherapy*. W.W. Norton & Company. https://doi.org/10.1080/03069880903166509

37. Özakkaş, T. (2018 a). *Alan Kuramının Diğer Psikanalitik Kuramlarla İlişkisi*: A Review Türkiye Bütüncül Psikoterapi Dergisi. 2(1), 1–28.

38. Özakkaş, T. (2018 b). *Psikoterapi Tarihi ve Bütüncül Psikoterapi:* Türkiye Bütüncül Psikoterapi Dergisi/ 1(1), 1–24.

39. Papalini, S., Beckers, T., & Vervliet, B. (2020). Dopamine: from prediction error to psychotherapy. In *Translational Psychiatry* (Vol. 10, Issue 1). Springer Nature. https://doi.org/ 10.1038/s41398-020-0814-x

40. Perls, F., Hefferline, R. F., & Goodman, P. (1951). *Gestalt Therapy: Excitement and Growth in the Human Personality*. Julian Press.

41. Piotrkowicz, M., Janoska-Jazdzik, M., Koweszko, T., & Szulc, A. (2021). The influence of psychotherapy on peripheral brain-derived neurotrophic factor concentration levels and gene methylation status: A systematic review. In *Journal of Clinical Medicine* (Vol. 10, Issue 19). https://doi.org/10.3390/jcm10194424

42. Plitman, E., Nakajima, S., de la Fuente-Sandoval, C., Gerretsen, P., Chakravarty, M. M., Kobylianskii, J., … & Mizrahi, R. (2014). Glutamate-mediated excitotoxicity in schizophrenia: A review. *European Neuropsychopharmacology, 24*(10), 1591–1605. https://doi.org/10.1016/j.euroneuro.2014.07.015

43. Porges, S. W. (2011). *The Polyvagal Theory: Neurophysiological Foundations of Emotions, Attachment, Communication, and Self-Regulation*. W.W. Norton & Company.

44. Price, R. B. & Duman, R. (2020). Neuroplasticity in cognitive and psychological mechanisms of depression: an integrative model. Molecular psychiatry, Mar;25(3):530–543. https://doi.org/ 10.1038/s41380-019-0615-x

45. Pylyshyn, Z. W. (1984). *Computation and Cognition: Toward a Foundation for Cognitive Science*. MIT Press. SBN: 9780262660587

46. Robinson, L., Delgadillo, J., & Kellett, S. (2020). The dose-response effect in routinely delivered psychological therapies: A systematic review. Psychotherapy Research. https://doi.org/10.1080/10503307.2019.1566676

47. Rumelhart, D. E., McClelland, J. L., & the PDP Research Group (1986). *Parallel Distributed Processing: Explorations in the Microstructure of Cognition*. MIT Press.

48. Sanacora, G., Gueorguieva, R., Epperson, C. N., Wu, Y. T., Appel, M., Rothman, D. L., & Mason, G. F. (2004). *Subtype-specific alterations of gamma-aminobutyric acid and glutamate in patients with major depression.* Archives of General Psychiatry, *61*(7), 705–713. https://doi.org/10.1001/archpsyc.61.7.705

49. Sara, S. J. (2009). The locus coeruleus and noradrenergic modulation of cognition. *Nature Reviews Neuroscience, 10*(3), 211–223. https://doi.org/10.1038/nrn2573

50. Schore, A. N. (2012). *The Science of the Art of Psychotherapy*. W.W. Norton & Company. https://doi.org/ 10.1080/07421656.2015.1060564

51. Sen, S., Duman, R., & Sanacora, G. (2008). Serum brain-derived neurotrophic factor, depression, and antidepressant medications: Meta-analyses and implications. *Biological Psychiatry, 64*(6), 527–532. https://doi.org/10.1016/j.biopsych.2008.05.005

52. Shedler, J. (2010). The efficacy of psychodynamic psychotherapy. *American Psychologist, 65*(2), 98–109. https://doi.org/10.1037/a0018378

53. Siegel, D. J. (2012). *The Developing Mind: How Relationships and the Brain Interact to Shape Who We Are*. Guilford Press.

54. Siegle, G. J., Thompson, W., Carter, C. S., Steinhauer, S. R., & Thase, M. E. (2007). Increased amygdala and decreased dorsolateral prefrontal BOLD responses in unipolar depression: Related and independent features. *Biological Psychiatry, 61*(2), 198–209. https://doi.org/10.1016/j.biopsych.2006.05.048
55. Solms, M. (2013). The conscious id. *Neuropsychoanalysis, 15*(1), 5–19.
56. Solms, M. (2021). *The Hidden Spring: A Journey to the Source of Consciousness*. W. W. Norton & Company. https://doi.org/10.53765/20512201.28.11.153
57. Streeter, C. C., Whitfield, T. H., Owen, L., Rein, T., Karri, … , & Jensen, J. E. (2010). *Effects of yoga versus walking on mood, anxiety, and brain GABA levels: A randomized controlled MRS study.* The Journal of Alternative and Complementary Medicine, *16*(11), 1145–1152. https://doi.org/10.1089/acm.2010.0007
58. Stuart, S. (2006). Interpersonal psychotherapy: A guide to theory and practice. *American Journal of Psychotherapy, 60*(1), 107–124. https://doi.org/10.1521/pdps.2024.52.3.370
59. Tatti, E., Phillips, A. L., Paciorek, R., Romanella, S. M., Dettore, D., Di Lorenzo, G., … & Santarnecchi, E. (2022). Boosting psychological change: Combining non-invasive brain stimulation with psychotherapy. Neuroscience & Biobehavioral Reviews, 142, 104867. https://doi.org/10.1016/j.neubiorev.2022.104867
60. Varela, F. J., Thompson, E., & Rosch, E. (1991). *The Embodied Mind: Cognitive Science and Human Experience*. MIT Press. https://doi.org/10.29173/cmplct8718
61. Wenger, E., & Kühn, S. (2021). Neuroplasticity. Cognitive training: An overview of features and applications, 69–83.
62. Yalom, I. D. (1980). *Existential Psychotherapy*. Basic Books.
63. Young, S. N. (2007). How to increase serotonin in the human brain without drugs. *Journal of Psychiatry and Neuroscience*, 32(6), 394–399. PMID: 18043762

Neuroplastic Mechanisms in Addiction: From Development to Recovery

Nahide Ayça Özgül

1. Introduction

Addiction results from repeated substance use or behaviors involving mental and physical aspects. It leads to a strong urge for a substance or behavior, difficulty in control, and prioritizing these over other activities. This often causes compulsion, increased tolerance, and withdrawal symptoms. Recent examples include internet and gambling addictions (Michael B. First, 2022; Popescu et al., 2021).

Stigma and outdated views on addiction hinder treatment. These views wrongly link addiction to personality traits. Modern treatment now sees addiction as a chronic disease linked to brain changes (Semaan Abdo & Khan Meşal K., 2023). Factors like neuroplasticity and genetics are crucial in treatment. Understanding how drugs and behaviors affect the brain is critical to advancements.

Differences in addiction tendencies come from the balance between reward sensitivity and deficiency. Mental health issues like depression and anxiety, along with factors such as genetics and drug exposure, play a significant role in addiction. This section will explore these topics further (Valentino et al., 2024).

2. Neurobiology of Addiction

Addiction develops through three stages: 1. Excessive use, 2. Withdrawal, and 3. Anticipation. The key brain areas involved are the basal ganglia, extended amygdala, and prefrontal cortex. Important nuclei include the ventral tegmental area, nucleus accumbens (NAc), and amygdala. These areas significantly affect behavior. The prefrontal cortex influences thought and behavior and is affected by changes in other nuclei. Regular drug use leads to lasting changes in brain networks, including the prefrontal cortex (Luciana & Collins, 2022; Nall et al., 2021).

Research on addiction has mainly focused on the VTA's (Ventral tegmental area) dopaminergic projections. These projections drive addictive behavior in the NAc and its circuits (Koob & Volkow, 2016; Volkow et al., 2017). Recent studies, however, have revealed the VTA's complex structure and the different roles of its cell populations in aversion and reward (Morales & Margolis, 2017; Mathis & Kenny, 2019).

3. Mechanism of Addiction Development

It has been demonstrated that various addictive substances increase extracellular levels of striatal dopamine. There is a relationship between this dopamine-related reward mechanism and the misuse of substances. In particular, projections from dopamine neurons in VTA to the nucleus accumbens (NAc) form the mesolimbic dopamine pathway, linked to natural reward mechanisms and rewards associated with substances and addictive drugs. Dopamine releases from VTA to the PFC and from the substantia nigra (SN) to the dorsal striatum are involved in reward-related decision-making and habit formation (Koob & Volkow, 2016; Volkow et al., 2017). In the early stages of addiction, these striatal dopamine projections related to reward facilitate the conditioning of the stimulus-stimulus, stimulus-response, and response-reinforcement relationships (Nall et al., 2021). As addiction develops, the reward and intoxication associated with mesoaccumbens dopamine shift toward habitual drug consumption linked to nigral-dorsal striatal dopamine. In response, homeostatic mechanisms that reduce the drug's dopamine response are activated, leading to a compensatory increase in drug intake (Koob & Volkow, 2016). Following this stage, withdrawal symptoms, such as intense irritability, dysphoria, anhedonia, lethargy, and some physical symptoms (e.g., flushing, tremors), make behavioral control more difficult (Luciana & Collins, 2022). A prolonged state of withdrawal following treatment and reduced striatal dopamine sensitivity may contribute to relapse in addiction (Trifilieff et al., 2017).

During acute stress, the release of the peptide corticotropin-releasing factor (CRF) is activated. In withdrawal, CRF acts within the extended amygdala; substances such as alcohol, cocaine, cannabinoids, opioids, and nicotine, as well as CRF antagonists, alleviate anxiety and similar conditions resulting from withdrawal. This negative reinforcement motivates substance intake (Koob & Volkow, 2016).

In addition to dopamine, other neurotransmitters such as serotonin, acetylcholine, glutamate, GABA, endogenous opioids, and cannabinoids also play significant roles in the development of addiction (Nall et al., 2021; Luciana & Collins, 2022).

4. The Role of the Nucleus Accumbens (Nac) in Addiction

NAc is critical in the brain's reward system, crucial for neuroplasticity and addiction. This makes it a prime target for addiction research and treatment. The brain reward system (BRS) is complex, involving various parts and neurotransmitters. Key components include the NAc, VTA, amygdala, and striatum, as well as the hippocampus and entorhinal cortex.

The NAc is essential for several reasons: it connects to many brain areas, adapts to regulate synaptic activity, supports different neurotransmitter systems, and relies on neuroglia. Therefore, studying the NAc's structure alongside molecular, genetic, and behavioral data is vital to understanding its role (Bayassi-Jakowicka et al., 2021).

5. The Importance of Pfc in Addiction

The prefrontal cortex (PFC) is critical for flexible behavior and is aided by neuroplasticity (Luciana & Collins, 2022).

It contains synapses, presynaptic and postsynaptic neurons, interneurons, and glial cells.

The PFC has six layers and is near the paralimbic area, which includes the anterior cingulate cortex (ACC). Its dorsolateral part connects to the mediodorsal thalamus and the dorsal striatum. It also links to limbic and paralimbic areas like the ventromedial and orbitofrontal regions, ACC, amygdala, ventral striatum, and insula.

Large pyramidal neurons in layers III and V use glutamate and connect widely. Glutamate, along with dopamine, is crucial in addiction.

The PFC's functional connections are vital for executive control, detecting essential stimuli, and maintaining the brain's default mode (Menon & D'Esposito, 2022; Menon, 2011).

6. The Role of the Habenula in Addiction

The habenula regulates neurotransmitters like dopamine and serotonin, linked to anxiety and avoidance (Antunes et al., 2023; Boulos et al., 2017). The Habenula is shown in Figure 1.

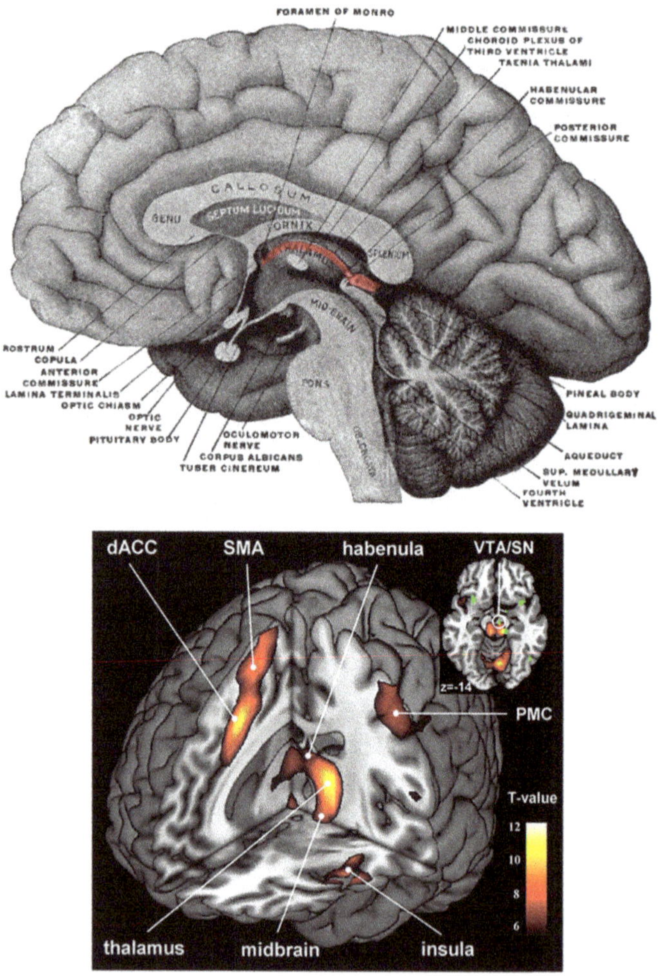

Figure 1: Habenula
(<https://upload.wikimedia.org/wikipedia/commons/thumb/8/83/Brainstem_and_thalamus_ja_ja_135.png/250px-Brainstem_and_thalamus_ja_ja_135.png>)

Its lateral and medial parts target different areas. They are critical in the anti-reward circuit, promoting avoidance (Mathis & Kenny, 2019). The habenula may regulate drug receptors by communicating with other systems and responding to internal signals.

Brain non-neuronal cells include glial, ependymal, endothelial, myeloid, pericytes, and tanycytes. Once seen as merely supportive, these cells are vital for brain balance, synapse formation, and substance transport. Oligodendrocytes and astrocytes show significant transcriptional changes in NAc of rats and mice (Reiner et al., 2022; Valentino et al., 2024).

7. The Role and Importance of the Amygdala in Addiction

The shell of the nucleus accumbens (NAshell), the central amygdala, and the sublenticular extended amygdala are key in addiction. Substances like cocaine, amphetamine, nicotine, alcohol, and tetrahydrocannabinol disrupt the brain's reward system. This leads to "tolerance," where more substance is needed for the same effect. After stopping use, the body shows "withdrawal" signs, needing more for rewards. In the extended amygdala, dopamine and serotonin levels drop while CRF levels rise. These changes contribute to addiction (Koob, 2021; Koob & Volkow, 2016). The Amydala is shown in Figure 2.

In a study, stimulating the right amygdala caused fear and sadness. Meanwhile, stimulating the left amygdala led to both happiness and negative emotions like fear, anxiety, and sadness. This suggests that the left amygdala is crucial for the brain's reward system (Lanteaume et al., 2007) (Paul & Elizabeth, 2009).

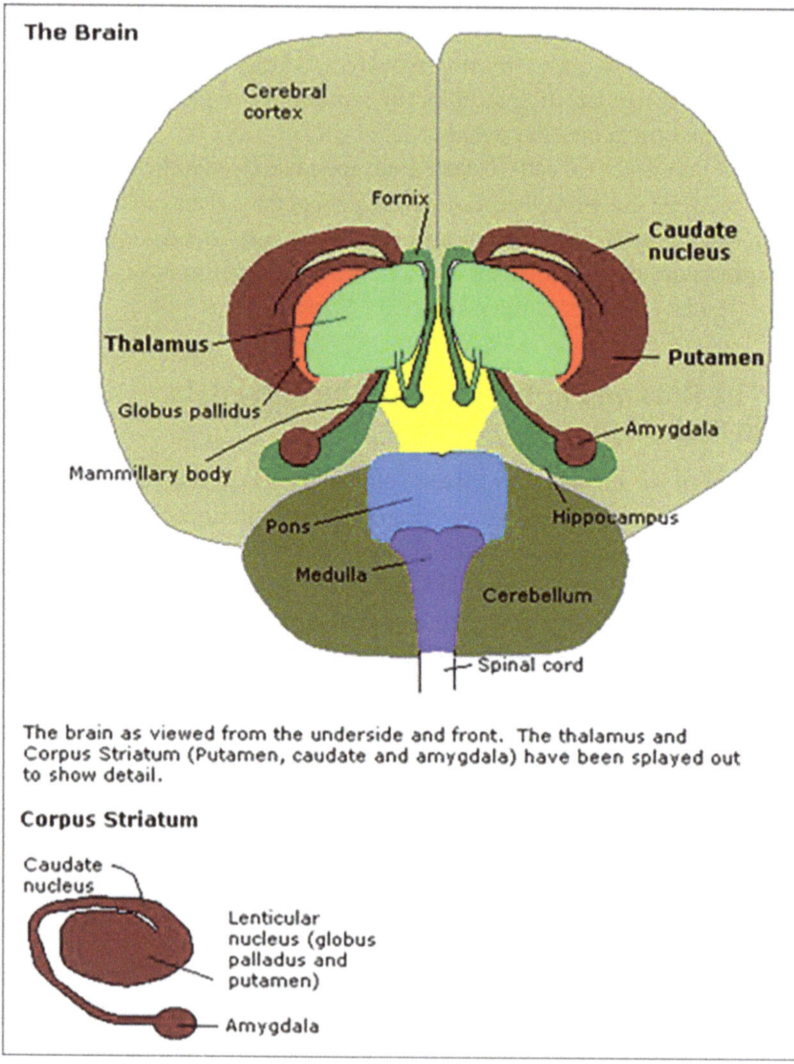

Figure 2: Amygdala
(<https://en.wikipedia.org/wiki/Amygdala>)

8. Mechanism of Addiction Development at the Synaptic Level

The Hebbian theory states that neurons firing together strengthen their connections. In 1973, Bliss and Lomo discovered long-term potentiation (LTP) and depression (LTD). LTP boosts synaptic sensitivity after active neuron pairs. Conversely, LTD occurs without enough neuron activity (Bliss and Lomo, 1973). Metaplasticity, influenced by past activities, affects synapse adaptability (Wickliffe c. Abraham & Mark F. Bear, 1996). It plays a role in learning, memory, and addiction.

Glutamate binds to AMPA (Alpha-amino-3-hydroxy-5-methyl-4-isoxazole propionic acid) and NMDA (N-methyl-d-aspartate) receptors. NMDA receptor activation can block LTP (Robert C. Malenka, 1992). It also mediates stress effects on LTP and LTD (Kim et al., 1996). Changes in glutamatergic synapses, crucial for memory, also affect addiction-related brain areas (Kader Julie a. & Malenka Robert C, 2007; Kourosh-Arami et al., 2022; Lüscher & Malenka, 2011).

In NAc, metaplasticity relates to drug memory. It integrates signals from dopamine-producing and glutamate-producing neurons (Chiamulera et al., 2021).

9. Functioning of Dopamine Receptors in Addiction

Addictive substances are rewarding. They change NAc, a vital part of the brain's reward system. This increases dopamine there. PFC has D1 and D2 dopamine receptors. D2 receptors, with higher affinity, activate at lower dopamine levels. Usually, the cortex gets a steady, low dopamine flow from VTA. VTA neurons fire rapidly in response to unexpected rewards or aversive events, raising dopamine levels temporarily. This surge activates D1 receptors, crucial for the full reward effect. Drugs, especially psychostimulants, mimic this surge. They activate both D1 and D2 receptors. D1 receptors trigger reward through memory and conditioning, involving the amygdala, medial orbitofrontal cortex (OFC), and hippocampus, with pathways affecting the striatum and cortex (Koob & Volkow, 2016).

The four main dopaminergic pathways are mesolimbic, mesocortical, nigrostriatal, and tuberoinfundibular. The mesolimbic and mesocortical pathways form the mesocorticolimbic system. Additionally, the hypothalamo-spinal and incertohypothalamic pathways are also critical. The dopaminergic pathways are shown in Figure 3.

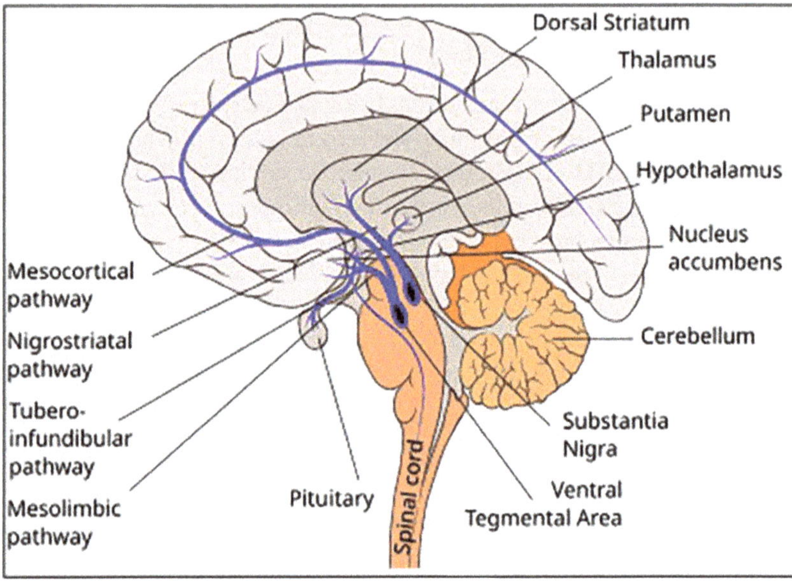

Figure 3: Dopaminergic Pathways
(<https://en.wikipedia.org/wiki/Dopaminergic_pathways>)

10. The Impact of Nicotine in Addiction

The use of tobacco and tobacco products is one of the most significant health problems we face today. It is the cause of numerous diseases, including chronic lung disease, circulatory disorders, and heart attacks. Smoking behavior often begins due to curiosity or the desire to fit in socially, and the addiction that follows develops through changes in the brain linked to negative neuroplasticity. Despite knowing the harmful effects of smoking, and although many smokers wish to quit, the number of those who successfully do so is relatively low. Therefore, developing methods that are easy to implement and lead to positive results in smoking cessation is crucial (Sezik et al., 2016).

Nicotine exerts its addictive effect through nicotinic cholinergic receptors in the mesocorticolimbic dopaminergic pathways. When nicotine stimulates cholinergic receptors, the increased activity in the dopaminergic pathways, which extend from the ventral tegmental area to the limbic system and cortex, activates the reward mechanism. Prolonged exposure to nicotine leads to neuroadaptations, resulting in an increase in receptor numbers and the development of tolerance (Leone & Evers-Casey, 2022).

It is thought that individual differences in nicotine metabolism rate play a crucial role in determining the level of nicotine addiction. Variations in the CYP2A6 gene, which encodes a liver enzyme responsible for metabolizing and inactivating nicotine, result in significant differences in nicotine metabolism and clearance among individuals. People who metabolize nicotine rapidly need more nicotine to maintain its levels, whereas those who metabolize it slowly have been observed to quit smoking more efficiently and successfully (Tran et al., 2019; López-Flores et al., 2017; Clark et al., 2023).

Adolescents who were exposed to nicotine prenatally have been shown to have higher rates of drug use and, in particular, a smaller volume in the orbitofrontal cortex (OFC) compared to the overall prefrontal cortex volume (Ceceli, 2022).

11. The Importance of Serotonin (5-Ht) in Addiction

The majority of the nuclei of serotonergic neurons are located in the brainstem, particularly in the raphe nuclei. Here, 5-HT (serotonin) cells are multipolar and have projections to almost all parts of the central nervous system, including the limbic structures, basal ganglia, brainstem, cerebral cortex, and gray matter of the spinal cord. This structure consists of heterogeneous serotonergic cells, morphologically forming nine core groups from B1 to B9. For example, the dorsal raphe nucleus (DRN) comprises B6 and B7.

Although they are located in different neuronal regions, the same 5-HT receptor subtypes act synergistically. Thus, the raphe cell bodies and postsynaptic 5-HT1A receptors in the limbic region regulate impulse through the postsynaptic excitability of 5-HT raphe neurons (Pierucci et al., 2014).

Although the mesocorticolimbic dopaminergic system is significant in psychostimulant addiction due to its rewarding effects, research suggests that the brain's serotonin (5-HT) system is also essential in the development of addiction. Psychostimulants act by inhibiting dopamine transporters, and they also inhibit 5-HT reuptake (Miszkiel et al., 2011).

The most effective 5-HT receptor in nicotine addiction is 5HT2C. Therefore, selective 5-HT2C agonists can be considered an effective adjunct treatment in nicotine addiction (Pierucci et al., 2014).

After exposure to alcohol, the serotonergic system is activated; however, this effect reverses with chronic alcohol consumption. These adaptations in the serotonergic system resulting from chronic alcohol use also contribute to the development of addiction. Hence, research on serotonin receptors may be considered for alcohol addiction treatment (Müller & Homberg, 2015).

12. Therapeutic Effect of Melatonin in Addiction

Melatonin, a circadian hormone, is also effective in regulating the biological clock and the microbiota. Additionally, it has anti-inflammatory, antidepressant, analgesic, and neuroprotective effects due to its epigenetic properties. Disruption of the circadian rhythm plays a significant role in the development of addiction. For instance, regulating sleep can be a helpful factor in quitting smoking, as the stimulation of 5-HT neurons is also related to sleep and wakefulness.

Furthermore, genes such as Clock and Per 2, which can cause circadian rhythm disorders, have been found to be associated with a predisposition to addiction.

Recent studies suggest that addictive substances interact with the gut microbiota, which, therefore, could be effective in addiction treatment (Jia et al., 2022).

13. Imaging Studies

Neuroimaging studies have shown that deficits in cognitive and emotional control in substance use disorders (SUDs) are due to negative neuroplastic changes that develop in the frontostriatal, frontolimbic, and frontocerebellar networks (Mahoney et al., 2020; Sullivan et al., 2019; Zilverstand et al., 2017). In addition, fundamental differences in the structure and function of the PFC may create vulnerability to addiction (Henderson J. Scott, 2018). This makes it challenging to confirm preclinical and post-addiction models (Koob, 2021;

Zhang & Volkow, 2019; Luciana & Collins, 2022). Furthermore, deficits in working memory, cognitive flexibility, and decision-making are prominent in SUDs. Observations vary according to the different substances of abuse (Luciana & Collins, 2022).

13.1. Positron Emission Tomography (PET) Studies

PET studies have demonstrated the development of neuroadaptation toward reduced dopamine sensitivity in chronic alcohol use disorders (AUD) (Ravan et al., 2014). Compared to healthy individuals, recently detoxified individuals with AUD have shown reduced availability of dopamine transporters (DAT) and D2/D3 receptors in the striatum, as well as in subcortical and cortical regions, including PFC (Luciana & Collins, 2022). Reduced striatal dopamine sensitivity has been observed for many substances of abuse, such as cocaine, methamphetamine, opioids, and nicotine (Trifilieff et al., 2017). Methylphenidate administration revealed lower striatal dopamine responses in SUD groups compared to healthy controls, alongside reduced metabolic activity across OFC, dorsolateral prefrontal cortex (DLPFC), and anterior cingulate cortex (ACC) (Trifilieff et al., 2017; Volkow et al., 2017). These regions are components of networks that are prominent in addiction models. For instance, the ACC is involved in salience processing, the OFC is in the reward system, and the DLPFC is in executive control (Nall et al., 2021).

Preclinical research has shown that neuroplasticity within and between these networks due to addiction is associated with increased drug craving, impulsive risk-taking, and the compulsivity of addictive behavior (Zorrilla & Koob, 2019).

Neuroimaging studies of chronic users of psychostimulants, such as cocaine and methamphetamine, which increase dopamine release, have demonstrated lower dopamine sensitivity patterns. This reduction is linked to decreased prefrontal brain activation and executive function (EF) task normalization following stimulant-induced changes (Bosker et al., 2017).

13.2. Magnetic Resonance Imaging (MRI) Studies

MRI studies have also shown differences in PFC between substance users and non-users, which have been linked to other addiction-related networks (Luciana, 2020; Lees et al., 2020).

Three separate longitudinal studies, spanning childhood or adolescence to middle age, have examined alcohol use. Consistent with alcohol use, smaller gray matter volumes (GMV) have been observed in the right insula and the DLPFC and dorsomedial prefrontal cortex (DMPFC).

The Enhancing Neuro-Imaging Genetics through Meta-Analysis (ENIGMA) project, which conducts large-scale sMRI studies, published results in 2019 (Mackey et al., 2019). This study involved 3,240 individuals, 2,140 of whom were addicted to alcohol, nicotine, cocaine, methamphetamine, or cannabis, across early to mid-adulthood. Findings indicated smaller volumes in the amygdala, hippocampus, and nucleus accumbens (NAc), alongside thinner cortical regions in parts of every lobe, including the OFC and DLPFC. Some effects were substance specific. Alcohol addiction, which commonly accompanies other substance addictions, showed the most widespread and robust effects. These significant and multifactorial effects were also observed in nicotine addiction. The results for these substances were markedly different from those of the control groups. Another ENIGMA report conducted a meta-analysis of case-control MRI studies focused on alcohol and cannabis addiction (Navarri et al., 2022). Consistent with previous findings, lower subcortical volumes and cortical thickness were observed in various brain regions linked to alcohol addiction. This effect was similar to, and even more pronounced than, those seen in emotional disorders and schizophrenia, suggesting the possibility of different stages or transitions of neuroplasticity between these conditions. In cannabis addiction, the effects were more minor in scale and confined mainly to subcortical regions (Navarri et al., 2022; Luciana & Collins, 2022).

Additionally, due to a predisposition to inflammation in white matter, substance use in addicted individuals has been associated with lower fractional anisotropy (FA) values, as shown by diffusion-weighted MRI. This decrease in FA, observed across multiple white matter tracts, including the corpus callosum, is secondary to inflammation. Fractional anisotropy (FA) indicates the consistency of water molecule orientation within a particular direction in white matter tracts. A low FA value suggests demyelination and reduced axonal integrity, meaning a decrease in the coherence of diffusion in the main direction.

For example, alcohol use has been linked to low FA in widespread fronto-cingular pathways, while tobacco use has shown low FA in the anterior corpus callosum, anterior cingulum white matter, and in connections extending from the nucleus accumbens, habenula, and motor cortex to the PFC (Ceceli, 2022).

14. Genetics and Addiction

Genome-wide association studies (GWAS) have significantly identified genetic variant associations. GWAS compares the DNA (Deoxyribonucleic acid) of individuals with different phenotypes. Through studies with patient and control groups, GWAS has identified single nucleotide polymorphisms (SNPs) and DNA variants related to the disease in question. While SNPs explain part of the variability in substance addiction, they are still not sufficient to fully explain addiction. More research is needed in this area. The first GWAS on addiction focused on nicotine dependence. (Bierut et al., 2007; Hancock et al., 2018; Popescu et al., 2021). Looking at addiction and genetics from a broader perspective, the relationship between genetic and environmental factors brings mental factors into focus. It has been shown that environmental factors can modulate genetic factors. For example, a study conducted on adolescent twin smokers found that parental control reduces the risk of nicotine addiction. (Trifu et al., 2020; Popescu et al., 2021).

Additionally, childhood experiences, urban living, stressful life events, and low education levels have been shown to affect genes involved in alcohol metabolism, as well as dopaminergic and serotonin transporter genes (Popescu et al., 2021).

15. The Role of Epigenetic Changes in Addiction

Epigenetics studies modifications that result in changes in the function and regulation of DNA, RNA (Ribonucleic acid), and proteins without altering the DNA sequence. DNA methylation and histone modifications are the most researched mechanisms that provide epigenetic diversity. Additionally, epigenetic enzymes are involved in demethylation, which plays essential roles in learning, memory, neurodevelopment, and some psychiatric phenomena (Gräff & Mansuy, 2009; Popescu et al., 2021).

Moreover, in the context of addiction, changes in epigenetics have been demonstrated in studies, particularly regarding cocaine and opioid use. It has been shown that opioids trigger high levels of permissive histone acetylation, low levels of repressive histone methylation, and changes in DNA methylation patterns and non-coding RNA regions within the brain's reward circuits (Browne et al., 2019).

16. The Role of Inflammation in Addiction and Recent Approaches to Addiction

In recent years, it has been understood that beyond dopamine and stress hormones, other neurotransmission systems, such as endocannabinoids and oxytocin, play a crucial role in understanding the path to addiction in specific individuals. The glutamatergic system, which enhances neuronal excitability and regulates neuroplasticity, is also significant in addiction.

The immune system has been identified as another important system in the development and maintenance of addiction. It has been shown that the central nervous system (CNS) receives and processes signals from the immune system. One of the most significant examples of this bidirectional communication is manifested in the form of "sickness behavior." (Pfau et al., 2024; Hofford et al., 2019; Ferrer-Pérez et al., 2024).

Sickness behavior presents as social withdrawal, apathy, and anhedonia. The actual cause of these behaviors is the body's energy redirected toward fever and antibody production. To conserve energy, these responses are inevitable (Dantzer, 2019; Ferrer-Pérez et al., 2024).

Proinflammatory cytokines are protein substances secreted by immune cells in response to pathogens and play a key role in sickness behavior. These cytokines manage the defense response by attracting immune cells to the infection site and supporting inflammation. Sometimes, these cytokines can cross the blood-brain barrier, affecting the CNS. When microglia—the brain's resident immune cells—encounter these cytokines, they also secrete cytokines, triggering an inflammatory response and initiating inflammation in the brain. The release and increase of these cytokines in the brain (neuroinflammation) cause sickness behavior and the underlying condition (Pascual et al., 2012; Ferrer-Pérez et al., 2024).

Many drugs have high inflammatory effects and impact the immune system. Alcohol is an example of such a substance. The inflammatory effects of alcohol on the liver are well-known, and they arise because alcohol is a xenobiotic foreign to the body. It also activates microglia (Montesinos et al., 2015).

In cases of chronic alcohol consumption, chronic inflammation occurs in the brain. This inflammation leads to atrophy, neuronal degeneration, and death. As the immune system is activated to clear the resulting dead tissues, inflammation further intensifies. This mechanism operates similarly for many drugs, yielding similar outcomes. These neuroinflammatory processes explain how chronic drug use causes cognitive impairment.

The inflammation theory also explains how stress increases susceptibility to drug use. Stress increases inflammation in both the body and the brain. This stress-induced inflammation, through the hormonal axis, enhances the dopaminergic reward response to drugs and triggers addiction (Ferrer-Pérez et al., 2018; Ferrer-Pérez et al., 2024).

17. Behavioral Effects of Addiction

The impairment and damage resulting from addiction lead to negative changes in learning, cognitive control, and reward-oriented behavior. The primary sources of inhibitory control dysfunction are the DLPFC (dorsolateral prefrontal cortex), IFG (inferior frontal gyrus), and ACC (anterior cingulate cortex). A preclinical study examining the effects of cocaine on decision-making observed a disruption in cognitive flexibility, affecting the ability to make decisions (Ceceli, 2022).

18. Addiction Treatment

18.1. Recent Developments In Addiction Treatment

Addiction is a complex psychological disorder influenced by genetics, environmental factors, and the effects of the drug or behavior involved (Solinas et al., 2021).

The reduction in GMV in PFC observed in addiction is a common feature resulting from the use of many addictive substances, including alcohol, heroin, methamphetamine, and cocaine. Therefore, pharmacological studies aimed at restoring the structure and function of the PFC are thought to be effective in treating various substance addictions. As a result, treatments that enhance activity in the corticolimbic circuits are also considered potentially beneficial for addiction.

Another emerging therapeutic approach involves the transformation and repair of atrophic cortical neurons. Drugs designed for this purpose, referred to as "psychoplastogens," are intended to promote brain plasticity. Ketamine, for example, is a psychoplastogen and an NMDA receptor antagonist (Olson, 2018).

Ketamine works rapidly in the PFC and induces structural flexibility. It is also expected to be effective in the treatment of addiction, along with other atrophy-related disorders such as PTSD (Post-traumatic stress disorder) and depression.

Most classical psychedelics, although classified as controlled substances, are not addictive and have shown efficacy in addiction treatment. Their psychoplastogenic effects are manifested through the growth of dendritic extensions and the formation of new synapses (dos Santos et al., 2016).

Some researchers believe that the hallucinogenic effects of psychedelics may also play a role in their therapeutic effects. However, the solid psychoplastogenic effects of psychedelics suggest that their therapeutic potential should not be attributed solely to psychological or neurobiological mechanisms (Yaden & Griffiths, 2021).

The controlled use of hallucinogens, along with their costly therapeutic process, has led to the development of new plastogens.

In recent years, a psychoplastogen called tabernanthalog (TBG) has been designed. TBG is structurally similar to ibogaine and 5-MeO-DMT but is non-hallucinogenic (Cameron et al., 2021).

Although ibogaine and 5-MeO-DMT are hallucinogenic and cardiotoxic, they are effective in addiction treatment.

TBG, on the other hand, has been developed with an improved safety profile compared to ibogaine and 5-MeO-DMT. TBG has proven effective in the treatment of alcohol and heroin addiction. It is also likely to be effective for other addictive substances such as nicotine, methamphetamine, and cocaine. TBG is promising for the simultaneous treatment of multiple addictions (Peters & Olson, 2021).

Genetic variations can also be used to assess the risk of developing addiction and predict the potential response to treatment. However, since genetic factors cannot be practically altered, there is a need for interventions within the genetic framework of addiction (Solinas et al., 2021).

DAT (dopamine transporter) activators cause a decrease in extracellular dopamine levels, while D3R (dopamine D3 receptor) antagonists reduce the neuron's sensitivity to dopamine. However, both increase withdrawal symptoms later. The D3R partial agonist SK608, discovered based on these findings, works by desensitizing D3R and has shown promise for addressing withdrawal attacks.

Additionally, drugs targeting serotonergic or GABAergic neurons, as well as anti-inflammatory drugs, may be beneficial in addiction treatment. Drugs that support synaptic plasticity are also being considered for addiction therapy (Yuan et al., 2024).

18.2. Neuroplastic Effects of Addiction Treatment

In alcohol-dependent individuals seeking treatment, significant increases in brain volumes have been observed in the frontal, parietal, and occipital gray and white matter, total cortical gray matter, and total lobar gray matter, as well as in the cerebellum and thalamus, following treatment and abstinence periods of 1 week, 1 month, and 7.5 months. After 7.5 months of abstinence, a reduction in ventricular volume was also noted, indicating recovery-related neuroplastic changes (Timothy C. Durazzo, 2015).

19. Epigenetics, Genetics, and Current Treatments in Opioid Addiction

19.1. Epigenetic and Transcriptional Mechanisms in Opioid Addiction

Epigenetic studies related to opioid addiction focus on the manipulation of enzymes responsible for opioid dependency. Through this approach, epigenetic modifications and regulation can be achieved, revealing the link between epigenetics and opioid addiction. Epigenetic enzymes function in three primary ways: 1) writers, 2) readers, and 3) erasers. Among these epigenetic regulators, non-specific pharmacological inhibitors of histone deacetylases (HDAC), which play a significant role in opioid addiction behavior, remove acetyl groups from histone tails. However, treatment studies using HDAC inhibitors have shown varied results across different opioids (Browne et al., 2020).

19.2. CRISPR Method in the Treatment of Opioid Addiction

CRISPR (Clustered Regularly Interspaced Short Palindromic Repeats) has recently become a gene-editing method. Many studies focusing on addiction utilize the CRISPR method, with some proving successful in animal trials. Research into opioid addiction using the CRISPR method is still ongoing. These studies focus on the GABA-A (gamma-aminobutyric acid type A) receptors, which include the $\alpha 2$ subunit (Gabra2) present in most brain regions. These receptors are linked to GABA, an inhibitory neurotransmitter that plays a significant role in addiction. Gabra2 has become a particular focus due to its association with psychiatric disorders, epilepsy, alcoholism, and drug use when altered. Mulligan and colleagues successfully edited the Gabra2 region in mice using CRISPR, suggesting that gene therapy for drug use is possible through Gabra2 (Mulligan et al., 2019; Wang et al., 2019).

20. The Role of Behavior and Self-Awareness in Addiction Treatment

Positive neuroplasticity, supported by behavior in line with the principle of "use it or lose it," plays a decisive role in brain reorganization. In this context, self-awareness is crucial for the conscious guidance of behavior and brain reorganization. Cognitive Behavioral Therapy (CBT), Existential Therapy, Person-Centered Therapy, Emotion-Focused Therapy (EFT), Mind-Body Therapy (MBT), Music Therapy, Art Therapy, Transactional Analysis, Solution-Focused Therapy, Eye Movement Desensitization and Reprocessing (EMDR), Mindfulness, and Hypnosis are all methods that support self-awareness and have shown efficacy in addiction treatment. The Awareness Integration Theory (AIT), which integrates these methods and emphasizes awareness, creates a new model by combining these techniques. AIT helps elevate unconscious self-awareness, allowing individuals to observe themselves objectively through reflection and introspection and make conscious, positive behavioral changes (Zeine et al., 2024).

The continuous formation, adaptation, and reorganization of new synapses through constant learning, adaptation, and reorganization, triggered by these methods, leads to positive neuroplasticity in the brain. Positive synaptic plasticity is harmonious with lifestyle changes, stress management, and the resulting emotional, environmental, and physiological stimuli.

The brain is an organic muscle. Therefore, just like other muscles, the brain requires constant and regular exercise to maintain its condition and foster its development. Brain exercise not only triggers positive synaptic plasticity but also prevents synaptic pruning (homeostatic plasticity), which leads to cortical and overall brain atrophy if these brain exercises are neglected. Thus, exercise and consistency are crucial in the continuity of neuroplasticity, helping to preserve and even enhance the brain's structure and function (Sakai, 2020; Zeine et al., 2024).

Recently discovered gene-editing techniques like CRISPR produce revolutionary effects in behavior and spirituality. With CRISPR, a programmable RNA can target and alter specific genome regions. This method holds promise for significant advances in addiction treatment (Zeine et al., 2024).

21. Conclusion

The relationship between addiction and neuroplasticity cannot be overlooked. Neuroadaptation mechanisms are at the forefront of addiction. When delving into the specifics of neuroplasticity in addiction, it is clear that synaptic metaplasticity and epigenetic mechanisms contribute significantly to the deepening of neuroplastic addiction levels. Additionally, the role of individual and structural differences in the development of addiction is evident, with genetic and social diversity being some of the most apparent reasons. Further elucidation of brain structures, such as the ventral tegmental area, NAc, amygdala, PFC, and habenula, and their communication pathways, will likely lead to significant advancements in addiction treatment in the coming years. While addiction treatment is considered a long and challenging process, it has become evident that positive neuroplasticity can lead to addiction being a chronic illness that can be almost or entirely cured. Moreover, the diversification of treatments, considering genetic and individual variations, is crucial for adapting addiction treatments to society.

22. Takeaways

1. Mental health issues like depression and anxiety, along with factors such as genetics and drug exposure, play a significant role in addiction.
2. The key brain areas involved are the basal ganglia, extended amygdala, and prefrontal cortex. Important nuclei include the ventral tegmental area, nucleus accumbens (NAc), and amygdala.
3. Recent studies, have revealed the VTA's complex structure and the different roles of its cell populations in aversion and reward.
4. Dopamine releases from VTA to the PFC and from the substantia nigra (SN) to the dorsal striatum are involved in reward-related decision-making and habit formation.
5. In withdrawal, CRF acts within the extended amygdala; substances such as alcohol, cocaine, cannabinoids, opioids, and nicotine, as well as CRF antagonists, alleviate anxiety and similar conditions resulting from withdrawal.
6. The NAc is essential for several reasons: it connects to many brain areas, adapts to regulate synaptic activity, supports different neurotransmitter systems, and relies on neuroglia.
7. Large pyramidal neurons in layers III and V of PFC use glutamate and connect widely. Glutamate, along with dopamine, is crucial in addiction.
8. The left amygdala is crucial for the brain's reward system.
9. Metaplasticity, influenced by past activities, affects synapse adaptability. It plays a role in learning, memory, and addiction.
10. D1 receptors trigger reward through memory and conditioning, involving the amygdala, medial orbitofrontal cortex (OFC), and hippocampus, with pathways affecting the striatum and cortex.
11. Nicotine stimulates cholinergic receptors, the increased activity in the dopaminergic pathways, which extend from the ventral tegmental area to the limbic system and cortex, activates the reward mechanism.
12. Variations in the CYP2A6 gene, which encodes a liver enzyme responsible for metabolizing and inactivating nicotine, result in significant differences in nicotine metabolism and clearance among individuals.
13. The most effective 5-HT receptor in nicotine addiction is 5HT2C. Therefore, selective 5-HT2C agonists can be considered an effective adjunct treatment in nicotine addiction.

14. Research on serotonin receptors may be considered for alcohol addiction treatment.
15. Melatonin, a circadian hormone, is also effective in regulating the biological clock and the microbiota. Additionally, it has anti-inflammatory, antidepressant, analgesic, and neuroprotective effects due to its epigenetic properties.
16. Recent studies suggest that addictive substances interact with the gut microbiota, which, therefore, could be effective in addiction treatment.
17. Neuroimaging studies have shown that deficits in cognitive and emotional control in substance use disorders (SUDs) are due to negative neuroplastic changes that develop in the frontostriatal, frontolimbic, and frontocerebellar networks.
18. PET studies have demonstrated the development of neuroadaptation toward reduced dopamine sensitivity in chronic alcohol use disorders (AUD).
19. Lower subcortical volumes and cortical thickness were observed in various brain regions linked to alcohol addiction.
20. Environmental factors can modulate genetic factors.
21. Epigenetic enzymes are involved in demethylation, which plays essential roles in learning, memory, neurodevelopment, and some psychiatric phenomena.
22. The immune system has been identified as another important system in the development and maintenance of addiction.
23. Many drugs have high inflammatory effects and impact the immune system.
24. In cases of chronic alcohol consumption, chronic inflammation occurs in the brain. This inflammation leads to atrophy, neuronal degeneration, and death.
25. "Psychoplastogens" are intended to promote brain plasticity.
26. Many studies focusing on addiction utilize the CRISPR method, with some proving successful in animal trials.
27. Positive neuroplasticity, supported by behavior in line with the principle of "use it or lose it," plays a decisive role in brain reorganization.
28. With CRISPR, a programmable RNA can target and alter specific genome regions. This method holds promise for significant advances in addiction treatment.

Abbreviations

ACC: Anterior cingulate cortex
AIT: Awareness integration theory
AMPA: Alpha-amino-3-hydroxy-5-methyl-4-isoxazole propionic acid
AUD: Chronic alcohol use disorders
BRS: Brain reward system
CBT: Cognitive behavioral therapy
CNS: Central nervous system
CRF: Corticotropin-releasing factor
CRISPR: Clustered regularly interspaced short palindromic repeats
DAT: Dopamine transporter
DLPFC: Dorsolateral prefrontal cortex
DMPFC: Dorsomedial prefrontal cortex
DNA: Deoxyribonucleic acid
DRN: Dorsal raphe nucleus
D3R: D3 dopamine receptor
EF: Executive function
EFT: Emotion-focused therapy
EMDR: Eye movement desensitization and reprocessing
ENIGMA: Enhancing NeuroImaging Genetics through Meta-Analysis Project
GABA: Gamma-aminobutyric acid
GABA-A: Gamma-aminobutyric acid type A
GMV: Gray matter volume
GWAS: Genome-wide association studies
HDAC: Non-specific pharmacological inhibitors of histone deacetylases
LTD: Long-term depression
LTP: Long-term potentiation
MBT: Mind-Body therapy
MGluR: Metabotropic glutamate receptor
NAc: Nucleus accumbens
NAshell: Nucleus accumbens shell
NMDA: N-methyl-d-aspartate
OFC: Orbitofrontal cortex
PFC: Prefrontal cortex
PTSD: Post-traumatic stress disorder
RNA: Ribonucleic acid

TBG: Tabernanthalog
SN: Substantia nigra
SNP: Single nucleotide polymorphism
SUD: Substance use disorder
VTA: Ventral tegmental area

Bibliography

1. Antunes, G. F., Campos, A. C. P., Martins, D. de O., Gouveia, F. V., Rangel Junior, M. J., Pagano, R. L., & Martinez, R. C. R. (2023). Unraveling the Role of Habenula Subnuclei on Avoidance Response: Focus on Activation and Neuroinflammation. *International Journal of Molecular Sciences*, *24*(13). https://doi.org/10.3390/ijms241310693
2. Atsız Sezik, H., Can, H., Kılınç, E. A., Çulha, G., et al. (2016). Fighting Tobacco and Tobacco Products. Clinical Medicine Family Medicine "Tütün ve Tütün Ürünleriyle Mücadele. Klinik Tıp Aile Hekimliği", 8(6), 42–47. http://www.kliniktipdergisi.com/
3. Bayassi-Jakowicka, M., Lietzau, G., Czuba, E., Steliga, A., Waśkow, M., & Kowiański, P. (2021). Neuroplasticity and a multilevel system of connections determine the integrative role of the nucleus accumbens in the brain reward system. In *International Journal of Molecular Sciences* (Vol. 22, Issue 18). MDPI. https://doi.org/10.3390/ijms22189806
4. Bierut, L. J., Madden, P. A. F., Breslau, N., Johnson, E. O., Hatsukami, D., Pomerleau, O. F., Swan, G. E., Rutter, J., Bertelsen, S., Fox, L., Fugman, D., Goate, A. M., Hinrichs, A. L., Konvicka, K., Martin, N. G., Montgomery, G. W., Saccone, N. L., Saccone, S. F., Wang, J. C., … Ballinger, D. G. (2007). Novel genes identified in a high-density genome-wide association study for nicotine dependence. *Human Molecular Genetics*, *16*(1), 24–35. https://doi.org/10.1093/hmg/ddl441
5. Bliss T.V.P. and Lomo T. (1973), Long-lasting potentiation of synaptic transmission in the dentate area of the anaesthetized rappit following stimulation of the perforant path, J. Physiol. (1973), 232, pp. 331–356, ttps://static1.squarespace.com/static/589f6511db29d6b160f930bc/t/5cf97c6be2066e000142688f/1559854188349/Bliss+and+Lomo%2C+1973.pdf

6. Bosker, W. M., Neuner, I., & Shah, N. J. (2017). The role of impulsivity in psychostimulant- and stress-induced dopamine release: Review of human imaging studies. In *Neuroscience and Biobehavioral Reviews* (Vol. 78, pp. 82–90). Elsevier Ltd. https://doi.org/10.1016/j.neubiorev.2017.04.008
7. Boulos, L. J., Darcq, E., & Kieffer, B. L. (2017). Translating the Habenula—From Rodents to Humans. In *Biological Psychiatry* (Vol. 81, Issue 4, pp. 296–305). Elsevier USA. https://doi.org/10.1016/j.biopsych.2016.06.003
8. Browne, C. J., Abela, A. R., Chu, D., Li, Z., Ji, X., Lambe, E. K., & Fletcher, P. J. (2019). Dorsal raphe serotonin neurons inhibit operants that respond to a reward via inputs to the ventral tegmental area. Still, no nucleus accumbens evidence from studies combining optogenetic stimulation and serotonin reuptake inhibition exists. *Neuropsychopharmacology*, *44*(4), 793–804. https://doi.org/10.1038/s41386–018-0271-x
9. Browne, C. J., Godino, A., Salery, M., & Nestler, E. J. (2020). Epigenetic Mechanisms of Opioid Addiction. In *Biological Psychiatry* (Vol. 87, Issue 1, pp. 22–33). Elsevier USA. https://doi.org/10.1016/j.biopsych.2019.06.027
10. Cameron, L. P., Tombari, R. J., Lu, J., Pell, A. J., Hurley, Z. Q., Ehinger, Y., … & Olson, D. E. (2021). A non-hallucinogenic psychedelic analogue with therapeutic potential. *Nature*, *589*(7842), 474–479. https://doi.org/10.1038/s41586-020-3008-z
11. Ceceli, A. O., Bradberry, C. W., & Goldstein, R. Z. (2022). The neurobiology of drug addiction: cross-species insights into the dysfunction and recovery of the prefrontal cortex. In Neuropsychopharmacology (Vol. 47, Issue 1, pp. 276–291). Springer Nature. https://doi.org/10.1038/s41386-021-01153-9
12. Chiamulera, C., Piva, A., & Abraham, W. C. (2021). Glutamate receptors and metaplasticity in addiction. In *Current Opinion in Pharmacology* (Vol. 56, pp. 39–45). Elsevier Ltd. https://doi.org/10.1016/j.coph.2020.09.005
13. Clark, A., Tate, B., Urban, B., Schroeder, R., Gennuso, S., Ahmadzadeh, S., McGregor, D., Girma, B., Shekoohi, S., & Kaye, A. D. (2023). Bupropion Mediated Effects on Depression, Attention

Deficit Hyperactivity Disorder, and Smoking Cessation. *Health Psychology Research*, *11*. https://doi.org/10.52965/001c.81043

14. Dantzer, R. (2019). From Stress Sensitization to Microglial Priming and Vice Versa: A New Era of Research in Biological Psychiatry. In *Biological Psychiatry* (Vol. 85, Issue 8, pp. 619–620). Elsevier USA. https://doi.org/10.1016/j.biopsych.2019.02.002
15. dos Santos, R. G., Osório, F. L., Crippa, J. A. S., Riba, J., Zuardi, A. W., & Hallak, J. E. C. (2016). Antidepressive, anxiolytic, and antiaddictive effects of ayahuasca, psilocybin and lysergic acid diethylamide (LSD): a systematic review of clinical trials published in the last 25 years. In *Therapeutic Advances in Psychopharmacology* (Vol. 6, Issue 3, pp. 193–213). SAGE Publications Ltd. https://doi.org/10.1177/2045125316638008
16. Ferrer-Pérez, C., Escrivá Martinez, T., Montagud-Romero, S., Ballestín, R., Reguilón, M. D., Miñarro, J., & Rodríguez-Arias, M. (2018). Indomethacin blocks the increased conditioned rewarding effects of cocaine-induced by repeated social defeat. *PLoS ONE*, *13*(12). https://doi.org/10.1371/journal.pone.0209291
17. Ferrer-Pérez, C., Montagud-Romero, S., & Blanco-Gandía, M. C. (2024). Neurobiological Theories of Addiction: A Comprehensive Review. *Psychoactives*, *3*(1), 35–47. https://doi.org/10.3390/psychoactives3010003
18. Gräff, J., & Mansuy, I. M. (2009). Epigenetic dysregulation in cognitive disorders. In *European Journal of Neuroscience* (Vol. 30, Issue 1, pp. 1–8). https://doi.org/10.1111/j.1460-9568.2009.06787.x
19. Hancock, D. B., Markunas, C. A., Bierut, L. J., & Johnson, E. O. (2018). Human Genetics of Addiction: New Insights and Future Directions. *Current Psychiatry Reports*, *20*(2). https://doi.org/10.1007/s11920-018-0873-3
20. Henderson J. Scott. (2018). Modulating Neural Circuits with Transcranial Magnetic Stimulation: Implications for Addiction Treatment Development. *Aspet*, *70*(3), 661–683. https://doi.org/10.1124/pr.116.013649
21. Hofford, R. S., Russo, S. J., & Kiraly, D. D. (2019). Neuroimmune mechanisms of psychostimulant and opioid use disorders.

In *European Journal of Neuroscience* (Vol. 50, Issue 3, pp. 2562–2573). Blackwell Publishing Ltd. https://doi.org/10.1111/ejn.14143

22. Jia, S., Guo, X., Chen, Z., Li, S., & Liu, X. an. (2022). The roles of the circadian hormone melatonin in drug addiction. In *Pharmacological Research* (Vol. 183). Academic Press. https://doi.org/10.1016/j.phrs.2022.106371
23. Kader Julie a. & Malenka Robert C. (2007). Synaptic plasticity and addiction. *Nature*, *8*, 844–858. https://doi.org/10.1038/nrn2234
24. Kim, J. J., Foy4, M. R., & Thompson, R. F. (1996). Behavioral stress modifies hippocampal plasticity through N-methyl-D-aspartate receptor activation (learning/memory/long-term potentiation/long-term depression/learned helplessness). In *Neurobiology* (Vol. 93). https://www.pnas.org
25. Koob, G. F. (2021). Drug addiction: Hyperkatifeia/negative reinforcement as a framework for medication development. *Pharmacological Reviews*, *73*(1), 163–201. https://doi.org/10.1124/PHARMREV.120.000083
26. Koob, G. F., & Volkow, N. D. (2016). Neurobiology of addiction: a neurocircuitry analysis. In *The Lancet Psychiatry* (Vol. 3, Issue 8, pp. 760–773). Elsevier Ltd. https://doi.org/10.1016/S2215-0366(16)00104-8
27. Kourosh-Arami, M., Komaki, A., & Gholami, M. (2022). Addiction-induced plasticity in underlying neural circuits. In *Neurological Sciences* (Vol. 43, Issue 3, pp. 1605–1615). Springer-Verlag Italia s.r.l. https://doi.org/10.1007/s10072-021-05778-y
28. Lanteaume, L., Khalfa, S., Régis, J., Marquis, P., Chauvel, P., & Bartolomei, F. (2007). Emotion induction after direct intracerebral stimulations of the human amygdala. *Cerebral Cortex*, *17*(6), 1307–1313. https://doi.org/10.1093/cercor/bhl041
29. Lees, B., Meredith, L. R., Kirkland, A. E., Bryant, B. E., & Squeglia, L. M. (2020). Effect of alcohol use on the adolescent brain and behavior. In *Pharmacology Biochemistry and Behavior* (Vol. 192). Elsevier Inc. https://doi.org/10.1016/j.pbb.2020.172906
30. Leone, F. T., & Evers-Casey, S. (2022). Tobacco Use Disorder. In *Medical Clinics of North America* (Vol. 106, Issue 1, pp. 99–112). W.B. Saunders. https://doi.org/10.1016/j.mcna.2021.08.011

31. López-Flores, L. A., Pérez-Rubio, G., Ramírez-Venegas, A., Ambrocio-Ortiz, E., Sansores, R. H., & Falfán-Valencia, R. (2017). Data on polymorphisms in CYP2A6 associated with risk and predispose to smoking-related variables. *Data in Brief, 15*. https://doi.org/10.1016/j.dib.2017.09.013
32. Luciana, M. (2020) Risks versus consequences of adolescent and young adult substance use: a focus on executive control. Current addiction reports, 7, 453–463.https://doi.org/10.1007/s40429-020-00341-9/ Published
33. Luciana, M., & Collins, P. F. (2022). Annual Review of Clinical Psychology Neuroplasticity, the Prefrontal Cortex, and Psychopathology-Related Deviations in Cognitive Control. *The Annual Review of Clinical Psychology Is Online At 18*, 443–469. https://doi.org/10.1146/annurev-clinpsy-081219
34. Lüscher, C., & Malenka, R. C. (2011). Drug-Evoked Synaptic Plasticity in Addiction: From Molecular Changes to Circuit Remodeling. In *Neuron* (Vol. 69, Issue 4, pp. 650–663). https://doi.org/10.1016/j.neuron.2011.01.017
35. Mackey, S., Allgaier, N., Chaarani, B., Spechler, P., Orr, C., Bunn, J., Allen, N. B., Alia-Klein, N., Batalla, A., Blaine, S., Brooks, S., Caparelli, E., Chye, Y. Y., Cousijn, J., Dagher, A., Desrivieres, S., Feldstein-Ewing, S., Foxe, J. J., Goldstein, R. Z., … Garavan, H. (2019). Mega-analysis of gray matter volume in substance dependence: General and substance-specific regional effects. *American Journal of Psychiatry, 176*(2), 119–128. https://doi.org/10.1176/appi.ajp.2018.17040415
36. Mahoney, J. J., Hanlon, C. A., Marshalek, P. J., Rezai, A. R., & Krinke, L. (2020). Transcranial magnetic stimulation, deep brain stimulation, and other forms of neuromodulation for substance use disorders: Review of modalities and implications for treatment. In *Journal of the Neurological Sciences* (Vol. 418). https://doi.org/10.1016/j.jns.2020.117149
37. Mathis, V., & Kenny, P. J. (2019). From controlled to compulsive drug-taking: The role of the habenula in addiction. In *Neuroscience and Biobehavioral Reviews* (Vol. 106, pp. 102–111). Elsevier Ltd. https://doi.org/10.1016/j.neubiorev.2018.06.018

38. Menon, V. (2011). Large-scale brain networks and psychopathology: A unifying triple network model. In *Trends in Cognitive Sciences* (Vol. 15, Issue 10, pp. 483–506). https://doi.org/10.1016/j.tics.2011.08.003

39. Menon, V., & D'Esposito, M. (2022). The role of PFC networks in cognitive control and executive function. In *Neuropsychopharmacology* (Vol. 47, Issue 1, pp. 90–103). Springer Nature. https://doi.org/10.1038/s41386-021-01152-w

40. Michael B. First. (2022). Diagnostic and Statistical Manual of Mental Disorders, 5th edition. In *Merican Psychiatric Association: Diagnostic and Statistical Manual of Mental Disorders, 5th edition. American Psychiatric Association, Washington, DC.*

41. Miszkiel, J., Filip, M., & Przegaliński, E. (2011). Role of serotonin (5-HT)1B receptors in psychostimulant addiction. In *Pharmacological Reports* (Vol. 63, Issue 6, pp. 1310–1315). Elsevier B.V. https://doi.org/10.1016/s1734-1140(11)70695-8

42. Montesinos, J., Pascual, M., Pla, A., Maldonado, C., Rodríguez-Arias, M., Miñarro, J., & Guerri, C. (2015). TLR4 elimination prevents synaptic and myelin alterations and long-term cognitive dysfunctions in adolescent mice with intermittent ethanol treatment. *Brain, Behavior, and Immunity, 45*, 233–244. https://doi.org/10.1016/j.bbi.2014.11.015

43. Morales, M., & Margolis, E. B. (2017). Ventral tegmental area: Cellular heterogeneity, connectivity, and behavior. In *Nature Reviews Neuroscience* (Vol. 18, Issue 2, pp. 73–85). Nature Publishing Group. https://doi.org/10.1038/nrn.2016.165

44. Müller, C. P., & Homberg, J. R. (2015). The role of serotonin in drug use and addiction. In *Behavioural Brain Research* (Vol. 277, pp. 146–192). Elsevier. https://doi.org/10.1016/j.bbr.2014.04.007

45. Mulligan, M. K., Abreo, T., Neuner, S. M., Parks, C., Watkins, C. E., Houseal, M. T., Shapaker, T. M., Hook, M., Tan, H., Wang, X., Ingels, J., Peng, J., Lu, L., Kaczorowski, C. C., Bryant, C. D., Homanics, G. E., & Williams, R. W. (2019). Identification of a functional non-coding variant in the GABAA receptor α2 subunit of the C57BL/6J mouse reference genome: Major implications for neuroscience research. *Frontiers in Genetics, 10*(MAR). https://doi.org/10.3389/fgene.2019.00188

46. Nall, R. W., Heinsbroek, J. A., Nentwig, T. B., Kalivas, P. W., & Bobadilla, A. C. (2021). Circuit selectivity in drug versus natural reward seeking behaviors. In *Journal of Neurochemistry* (Vol. 157, Issue 5, pp. 1450–1472). John Wiley and Sons Inc. https://doi.org/10.1111/jnc.15297

47. Navarri, X., Afzali, M. H., Lavoie, J., Sinha, R., Stein, D. J., Momenan, R., Veltman, D. J., Korucuoglu, O., Sjoerds, Z., van Holst, R. J., Hester, R., Orr, C., Cousijn, J., Yucel, M., Lorenzetti, V., Wiers, R., Jahanshad, N., Glahn, D. C., Thompson, P. M., … Conrod, P. J. (2022). How do substance use disorders compare to other psychiatric conditions on structural brain abnormalities? A cross-disorder meta-analytic comparison using the ENIGMA consortium findings. *Human Brain Mapping*, *43*(1), 399–413. https://doi.org/10.1002/hbm.25114

48. Olson, D. E. (2018). Psychoplastogens: A Promising Class of Plasticity-Promoting Neurotherapeutics. In *Journal of Experimental Neuroscience* (Vol. 12). SAGE Publications Ltd. https://doi.org/10.1177/1179069518800508

49. Pascual, O., Achour, S. Ben, Rostaing, P., Triller, A., & Bessis, A. (2012). Microglia activation triggers astrocyte-mediated modulation of excitatory neurotransmission. *Proceedings of the National Academy of Sciences of the United States of America*, *109*(4). https://doi.org/10.1073/pnas.1111098109

50. Paul J. Whalens & Elizabeth a. Phelps. (2009). *The Human Amygdala*.

51. Peters, J., & Olson, D. E. (2021). Engineering Safer Psychedelics for Treating Addiction. *Neuroscience Insights*, *16*. https://doi.org/10.1177/26331055211033847

52. Pfau, M. L., Ménard, C., & Russo, S. J. (2024). Inflammatory Mediators in Mood Disorders: Therapeutic Opportunities. *Annual Review of Pharmacology and Toxicology Annu. Rev. Pharmacol. Toxicol*, *22*(8), 411–439. https://doi.org/10.1146/annurev-pharmtox

53. Pierucci, M., Chambers, S., Partridge, L., De Deurwaerdère, P., & Di Giovanni, G. (2014). Role of central serotonin receptors in nicotine addiction. In *Nicotinic Receptors* (pp. 279–305). Springer New York. https://doi.org/10.1007/978-1-4939-1167-7_14

54. Popescu, A., Marian, M., Drăgoi, A., & Costea, R.-V. (2021). Understanding the genetics and neurobiological pathways behind

addiction (Review). *Experimental and Therapeutic Medicine*, *21*(5). https://doi.org/10.3892/etm.2021.9976

55. Ravan S., Martinez D., Slifstein M., Abi-Dargham A., 2014, Molecular imaging in alcohol dependence, Handbook of Clinical Neurology Volume 125, 2014, Pages 293–311. https://doi.org/10.1016/B978-0-444-62619-6.00018-5
56. Reiner, B. C., Zhang, Y., Stein, L. M., Perea, E. D., Arauco-Shapiro, G., Ben Nathan, J., Ragnini, K., Hayes, M. R., Ferraro, T. N., Berrettini, W. H., Schmidt, H. D., & Crist, R. C. (2022). Single nucleus transcriptomic analysis of rat nucleus accumbens reveals cell type-specific patterns of gene expression associated with volitional morphine intake. *Translational Psychiatry*, *12*(1). https://doi.org/10.1038/s41398-022-02135-1
57. Robert C. Malenka. (1992). The Influence of Prior Synaptic Activity on the Induction of Long-Term Potentiation. *Science*, *255*(5045), 730–733. https://doi.org/ 0.1126/science.1346729
58. Sakai, J. (2020). How synaptic pruning shapes neural wiring during development and, possibly, in disease. *Proceedings of the National Academy of Sciences of the United States of America*, *117*(28), 16096–16099. https://doi.org/10.1073/pnas.2010281117
59. Semaan A. (2023). Neurobiology of Addiction, Semaan A., Khan M. https://europepmc.org/article/NBK/nbk597351
60. Solinas, M., Chauvet, C., Lafay-Chebassier, C., Jaafari, N., & Thiriet, N. (2021). Environmental enrichment-inspired pharmacological tools for the treatment of addiction. In *Current Opinion in Pharmacology* (Vol. 56, pp. 22–28). Elsevier Ltd. https://doi.org/10.1016/j.coph.2020.09.001
61. Sullivan, E. V., Zahr, N. M., Saranathan, M., Pohl, K. M., & Pfefferbaum, A. (2019). Convergence of three parcellation approaches demonstrating cerebellar lobule volume deficits in Alcohol Use Disorder. *NeuroImage: Clinical*, *24*. https://doi.org/10.1016/j.nicl.2019.101974
62. Timothy C. Durazzo, Anderson Mon, Stefan Gazdzinski, Ping-Hong Yeh, Dieter J. Meyerhoff, 2015, Serial longitudinal magnetic resonance imaging data indicate non-linear regional gray matter volume recovery in abstinent alcohol-dependent individuals https://doi.org/10.1111/adb.12180

63. Tran, A. X., Ho, T. T., & Varghese Gupta, S. (2019). Role of CYP2B6 pharmacogenomics in bupropion-mediated smoking cessation. In *Journal of Clinical Pharmacy and Therapeutics* (Vol. 44, Issue 2). https://doi.org/10.1111/jcpt.12783

64. Trifilieff, P., Ducrocq, F., van der Veldt, S., & Martinez, D. (2017). Blunted Dopamine Transmission in Addiction: Potential Mechanisms and Implications for Behavior. In *Seminars in Nuclear Medicine* (Vol. 47, Issue 1, pp. 64–74). W.B. Saunders. https://doi.org/10.1053/j.semnuclmed.2016.09.003

65. Trifu, S., Tudor, A., & Radulescu, I. (2020). Aggressive behavior in psychiatric patients in relation to hormonal imbalance (Review). *Experimental and Therapeutic Medicine*. https://doi.org/10.3892/etm.2020.8974

66. Valentino, R. J., Nair, S. G., & Volkow, N. D. (2024). Neuroscience in addiction research. *Journal of Neural Transmission*, *131*(5), 453–459. https://doi.org/10.1007/s00702-023-02713-7

67. Volkow, N. D., Wise, R. A., & Baler, R. (2017). The dopamine motive system: Implications for drug and food addiction. In *Nature Reviews Neuroscience* (Vol. 18, Issue 12, pp. 741–752). Nature Publishing Group. https://doi.org/10.1038/nrn.2017.130

68. Wang, S. C., Chen, Y. C., Lee, C. H., & Cheng, C. M. (2019). Opioid addiction, genetic susceptibility, and medical treatments: A review. In *International Journal of Molecular Sciences* (Vol. 20, Issue 17). MDPI AG. https://doi.org/10.3390/ijms20174294

69. Wickliffe c. Abraham & Mark F. Bear. (1996). Metaplasticity: the plasticity of synaptic plasticity. *Trends in Neuroscience*, *19*(4), 126–130. https://doi.org/ 10.1016/S0166-2236(96)80018-X

70. Yaden, D. B., & Griffiths, R. R. (2021). The Subjective Effects of Psychedelics Are Necessary for Their Enduring Therapeutic Effects. In *ACS Pharmacology and Translational Science* (Vol. 4, Issue 2, pp. 568–572). American Chemical Society. https://doi.org/10.1021/acsptsci.0c00194

71. Yuan, S., Jiang, S. C., Zhang, Z. W., Li, Z. L., & Hu, J. (2024). Substance Addiction Rehabilitation Drugs. In *Pharmaceuticals* (Vol. 17, Issue 5). Multidisciplinary Digital Publishing Institute (MDPI). https://doi.org/10.3390/ph17050615

72. Zeine, F., Jafari, N., Nami, M., & Blum, K. (2024). Awareness integration theory A Psychological and genetic path to self-directed Neuroplasticity. *Health Sciences Review*, *11*, 100169. https://doi.org/10.1016/j.hsr.2024.100169

73. Zhang, R., & Volkow, N. D. (2019). Brain default-mode network dysfunction in addiction. In *NeuroImage* (Vol. 200, pp. 313–331). Academic Press Inc. https://doi.org/10.1016/j.neuroimage.2019.06.036

74. Zilverstand, A., Parvaz, M. A., & Goldstein, R. Z. (2017). Neuroimaging cognitive reappraisal in clinical populations to define neural targets for enhancing emotion regulation. A systematic review. *NeuroImage*, *151*, 105–116. https://doi.org/10.1016/j.neuroimage.2016.06.009

75. Zorrilla, E. P., & Koob, G. F. (2019). Impulsivity derived from the dark side: Neurocircuits that contribute to negative urgency. In *Frontiers in Behavioral Neuroscience* (Vol. 13). Frontiers Media S.A. https://doi.org/10.3389/fnbeh.2019.00136

Notes on Contributors

Selen Gür Özmen
*Bahçeşehir University, Faculty of Health Sciences, Physiotherapy and Rehabilitation Department, Istanbul, Turkiye
**Bahçeşehir University, Graduate School of Educational Sciences, Lecturer in The Programs of Neuroscience, Istanbul, Turkiye
selen.gurozmen@bau.edu.tr

Asena Himmetoğlu Padgett
*Bahçeşehir University, Graduate School, Neuroscience Doctoral Program, Istanbul, Turkiye
asenahim@gmail.com

Zeynep Gül
*Bahçeşehir University, Vocational School of Health Services, Istanbul, Turkiye
** Bahçeşehir University, Graduate School, Neuroscience Doctoral Program, Istanbul, Turkiye
zeynep.gul@bau.edu.tr

Irmak Özcan
*Bahçeşehir University, Graduate School, Speech and Language Therapy Master Program, Istanbul, Turkiye
irmak.ozcan@bahcesehir.edu.tr

SEDA EYİLİKEDER TEKİN
*Bahçeşehir University, Faculty of Health Sciences, Department of Speech and Language Therapy, Istanbul, Turkiye
seda.eyilikedertekin@bau.edu.tr

YAHYA DAHHAN
* Bahçeşehir University, Graduate School, Neuroscience Master Program, Istanbul, Turkiye dahhanyahya@yahoo.com

SELİN YILMAZ
* Bahçeşehir University, Graduate School, Neuroscience Master Program, Istanbul, Turkiye
**Marmara University, Social Sciences Institute, Marketing Doctoral Program, Istanbul, Turkiye
selin.ylz@outlook.com

HACER DOĞAN
* Bahçeşehir University, Graduate School, Neuroscience Doctoral Program, Istanbul, Turkiye
**Erenköy Physical Therapy and Rehabilitation Hospital, Istanbul, Turkiye
hacer.dogan@bahcesehir.edu.tr

MERVE BİLGİÇ KOÇ
* Bahçeşehir University, Graduate School, Physiotherapy and Rehabilitation Doctoral Program, Istanbul, Turkiye
**Department of Public Health, General Directorate of Public Health, Ministry of Health, Republic of Turkiye
mrv.blgc17@gmail.com

SELİN SERT YURDAKUL
* Bahçeşehir University, Graduate School, Neuroscience Doctoral Program, Istanbul, Turkiye
selin.yurdakul@bahcesehir.edu.tr

Tuğba Duygu Özmet
* Istanbul Medipol University, Faculty of Health Sciences, Department of Nursing, Istanbul, Turkiye
tugba.ozmet@medipol.edu.tr

Nahide Ayça Özgül
*Beşiktaş Sait Çiftçi State Hospital, Istanbul, Turkiye
**Bahçeşehir University, Graduate School, Neuroscience Doctoral Program, Istanbul, Turkiye
aycaozgul51@gmail.com

www.ingramcontent.com/pod-product-compliance
Ingram Content Group UK Ltd.
Pitfield, Milton Keynes, MK11 3LW, UK
UKHW021828100926
8556IPUK00004B/230